KV-017-627

CONTROL TECHNOLOGY AND PERSONAL COMPUTERS

System Design and Implementation

ML

Books are to be returned on or before
the last date below.

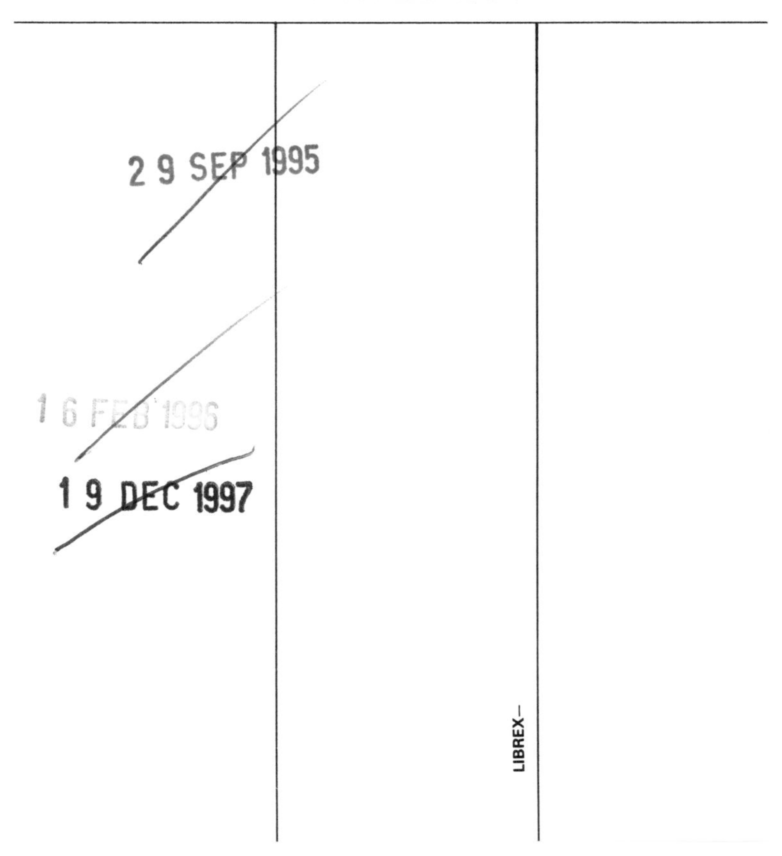

#24909136

CONTROL TECHNOLOGY AND PERSONAL COMPUTERS

System Design and Implementation

Michael F. Hordeski

VNR
VAN NOSTRAND REINHOLD
New York

Copyright © 1992 by Van Nostrand Reinhold

Library of Congress Catalog Card Number 91-40161
ISBN 0-442-00568-7

All rights reserved. No part of this work covered by the copyright hereon may be reproduced or used in any form or by any means—graphic, electronic, or mechanical, including photocopying, recording, taping, or information storage and retrieval systems—without written permission of the publisher.

Manufactured in the United States of America

Published by Van Nostrand Reinhold
115 Fifth Avenue
New York, New York 10003

Chapman and Hall
2-6 Boundary Row
London, SE1 8HN, England

Thomas Nelson Australia
102 Dodds Street
South Melbourne 3205
Victoria, Australia

Nelson Canada
1120 Birchmount Road
Scarborough, Ontario M1K 5G4, Canada

D
629.895
HOR

16 15 14 13 12 11 10 9 8 7 6 5 4 3 2 1

Library of Congress Cataloging-in-Publication Data

Hordeski, Michael F.
Control technology and personal computers: system design and implementation / Michael F. Hordeski.
p. cm.
Includes bibliographical references.
ISBN 0-442-00568-7
1. Automatic control—Data processing. 2. Microcomputers.
I. Title.
TJ223.M53H653 1992
629.8'95—dc20 91-40161
CIP

CONTENTS

Preface

The use of personal computers (PCs) in control and automation has been growing at an astounding rate. Initial applications for the PC and its associated software packages involved basic office and business automation. These applications consisted mainly of word processing, spreadsheet analysis, and data base management. Today, there are many application packages for various types of engineering computation and analysis, production and quality control, and computer-aided design.

Among the fastest-growing nonoffice applications is data acquisition measurement and control. This involves collecting data and monitoring the status of processes and equipment.

The availability of powerful software and PC versions has made it possible to apply PCs directly in control or indirectly through programmable or process controllers. Because of the digital interface of PC buses, analog data acquisition requires the use of an analog-to-digital converter to transform measurement data into a PC bus-compatible format. Then the information can be easily transformed, analyzed, stored, or displayed using low-cost, readily available hardware and software.

A major application of PCs in industrial applications is low-cost programmable control systems. Another large application is data acquisition used in conjunction with programmable logic controllers (PLCs) or process controllers. Direct connection to a process can be made through I/O expansion boards.

Older PCs lacked real-time capabilities. Since they operated at too slow a speed, they were used primarily to monitor and control simple processes. With the advent of the 386 and 486 processors, they can be expected to play a major role in the monitoring and control of devices or processes.

PCs are available in several configurations for control and automation use.

These include the common table or desktop versions, portable, battery powered units, rack-mounted units, and National Electrical Manufacturers Association (NEMA)-enclosed industrial versions. A common configuration for control devices in U.S. plants is rack mounting. NEMA enclosures are generally required in plants with corrosive environments.

Most applications software has been directed at the IBM-PC family because of its greater market share and availability in versions designed for industrial environments. Some of these computers are provided with special cooling devices, such as removable filters and special cases.

Industrial versions of the IBM-PC family are available from suppliers in several different forms. IBM's Model 7552 Industrial Computer, for example, uses battery backup for protection against transients.

IBM compatibility should continue to be a major consideration for PC selection in the future. Other companies that provide industrial PC versions include General Electric, Hewlett Packard, and AT&T.

An important development in PC control and automation growth was Intel's 32-bit microprocessor, the 80386. This was followed by the next model of the Intel family, the 80486, also known as the i486.

Some i486-based PCs and workstations are faster than some of the mainframes and minicomputers in networking and multiuser Unix systems. The 80486 microprocessor is a 32-bit chip containing 1,180,235 transistors on a single silicon chip. It is capable of running more than one processor simultaneously, so that one chip can run a network while another one is free to multitask the applications used on the network. The instruction set for the 486 is basically the same as for the 386, but with a built-in math coprocessor, 8K cache, and cache-controller circuitry.

The Extended Industry Standard Architecture (EISA) bus will accept old 8-bit and 16-bit cards, as well as new expansion boards designed specifically for the EISA 32-bit channel. The EISA bus is faster than the Micro Channel circuitry (MCA) introduced by IBM in 1987.

The bus interface unit (BIU) includes a burst bus control unit, which can access code and data from a RAM cache at a rate of up to 106 Mb/second through a data path 128 bits wide—nearly twice as fast as the 33-MHz 80386 microprocessor, which has a data path width of 32 bits.

The i486 uses a construction that reduces the number of clock cycles needed to execute each computing instruction, making it two to three times faster than clock speed and data path alone would provide. With the Complex Reduced Instruction Set Processing (CRISP), the i486 can perform a complete instruction in less than two clock cycles. In chips like the 80286 and 80386, each computing instruction takes four to six clock cycles.

Drive arrays are another improvement in the i486; these double transfer rates. In a drive array, the controller splits data and writes the two halves to two drives simultaneously. It can also write the same data to two or more hard drives at once, so if a drive fails, the data will not be lost.

Chapter 1 introduces the PC workstation and discusses its place in factory applications. The characteristics of the typical AT-compatible 286 PC are considered in detail, including ROM, RAM, and disk limitations, crystals, monitors, backup, and power supplies. The uses of 386 systems are discussed, as well as system-board considerations such as memory expanders for memory upgrades. Compatibility is explained in terms of performance terminology like disk interleaves, defragmentation, and caches. Compact disks for the PC are also explained, as well as WORM drives and rewritable optical media. The use of math coprocessors is discussed.

Chapter 2 introduces measurement and control applications using industrial PC configurations. The 80386 and 80486 microprocessors are considered for these applications. Methods of PC data acquisition are detailed. PCs are also useful as development tools. Direct control and environmental concerns are addressed, as well as standardization and training needs. The role of the PC in control system integration is considered, and inhouse software development is addressed. Guidelines are given for developing PC automation projects. Statistical control is explained, and the use of PLCs and PCs is considered.

Chapter 3 explains the use of operational amplifiers and details their characteristics. The use of isolation amplifiers is discussed for high-noise applications. Ground loops, shields, floating inputs, and wiring and cabling considerations are also described. Amplifier selection and frequent system problems are considered, and general troubleshooting advice is given.

Chapter 4 explains digital coding, D/A conversion, voltage-current conversion, A/D conversion, and multiplexing. Error sources are discussed, as well as the use of multiplexers in high-noise environments. Sample holds are explained, including sample hold errors and performance specifications. Configuring the PC interface is explained, and a data-acquisition design example is given.

The various types of user input devices are explained in Chapter 5, including touch panels, trackballs, light pens, joysticks, and mice. Optical data input is considered, including video digitizers, laser scanners, and digital cameras. Verbal input is discussed, along with speech-recognition problems and voice systems interfacing. OCR technology is also considered.

The PC/AT bus is explained in Chapter 6, along with the PS/2 Micro Channel bus and the Micro Channel option setup scheme. Parallel interfacing is discussed, as well as the RS-232 and RS-449 standards. Current loop techniques are explained, along with the RS-422, 423 and 449 standards. Asynchronous and synchronous communication methods are compared, and IEEE-488 is discussed. The characteristics of IEEE-488 interface boards are detailed, as well as those of RS-232/422 boards. Programming examples are given for these boards. Software and noise problems are discussed, and troubleshooting tools are detailed.

Chapter 7 explains the common techniques used for pressure measurement. It shows how to interface these transducers to the PC using counters and programmable timers. It illustrates how to use counter/timer I/O boards, as well as A/D converter/timer counter boards. Interrupts and DMA are also explained, and the use of multiplexer boards is discussed.

Chapter 8 details the most common types of temperature measurement, including resistance temperature detectors, semiconductors, and thermocouples. The use of low-speed analog and digital I/O interface boards for interfacing these transducers is explained. Resistance sensor input boards and expansion multiplexers are also discussed.

Chapter 9 is concerned with the interfacing of resistance, capacitive, inductive, and digital displacement sensors. The use of the 8255 programmable peripheral interface is explained, as well as its operation in digital I/O board applications. The use of high-speed parallel digital interface boards is explained for digital encoder applications.

Chapter 10 includes an introduction to differential pressure flow transducers and the use of orifice plates, Venturi tubes, and flow nozzles. Turbine flowmeters, variable area meters, fluid characteristic sensors, and electromagnetic flowmeters are considered and interfacing techniques are demonstrated.

Chapter 11 explains PC network configurations and the general network organization for distributed process control. It shows how to perform a network needs evaluation. Private branch exchange, baseband, and broadband network technologies are compared, and network implementation tips are given.

Chapter 12 is concerned with network protocols. IEEE-802 and the Manufacturing Automation Protocol (MAP) standard are explained. A typical MAP application is given. Operations such as collision detection are explained.

Thanks are given to the many people in the control and instrumentation field who provided inspiration and guidance on this project. Special thanks go to Dr. Berry Phillips of Keithley-MetraByte, Inc., for providing specifications and program examples of their IBM PC expansion boards and to Dee at Jablon Computer for her assistance in completing this book.

Author's note: Additional information on products discussed in this book are available from Jablon Computer, 2250 Monterey Road, Atascadero, California 93422, 805 466-3209. There are also tutorial disks for IBM PCs available from the same source.

CONTROL TECHNOLOGY AND PERSONAL COMPUTERS

System Design and Implementation

1

Introduction to Personal Computers

This chapter explains the various components of the personal computer (PC) that are important for its use as a workstation in factory applications. The important characteristics of the typical IBM-compatible computer include read-only memory (ROM), random access memory (RAM), disk and disk drive limitations, crystal speed, type of monitor required, backup, and power supplies. Expansion board considerations are also important, since the expansion slots allow you to expand memory and make other upgrades. Compatibility is an important characteristic; it is explained in terms of performance capabilities like disk interleaves, defragmentation, and caches. Compact disks for the PC are also explained, as well as write-once/read many (WORM) drives and rewritable optical media. In addition, the use of math coprocessors is discussed.

Many different monitors can be found in IBM PCs. These range from monochrome monitors to (red-green-blue) RGB dual frequency types, enhanced color graphics (EGA, VGA, super VGA,)* multisync, and multimode types, as well as special nonstandardized monitors. All of these are discussed and compared in this chapter. Dual display techniques, which are often neglected, can be very useful in many applications.

The power supply is critical to the proper operation of the computer system. It becomes more important as expansion boards are added to the system.

Most applications are memory dependent. It is necessary to know the differences between expanded and extended memory, how to use memory expanders, how memory upgrades are accomplished, and what task switching involves.

The floppy drive is the most common technique for getting programs and other information into the system. In order to make use of the various types of floppy disks that are available, you will need to know what types are available and what the specifications of the disks and disk drives are.

*EGA - Enhanced Graphics Adapter
VGA - Video or Visual Graphics Adapter

Disk caches are becoming more important in the use of PCs. These are discussed, as well as the use of compact disks for the PC. Write-once/read-many (WORM) drives and rewritable optical disks are also explained. Tape drives are a useful backup and mass storage technique; the techniques and hardware used for tape backup are described. Coprocessors are used to speed up math processing; they can be installed quite easily. We will examine their use and installation.

In just a few years, the PC has grown from a simple curiosity into a formidable tool for the engineer and manager. What started as a breadboard curiosity tediously programmed with an array of lights and switches has become a neatly packaged, powerful tool which can be used with thousands of commercially available software packages to satisfy some of the most demanding requirements of today's control and automation applications. As PCs come to pack more power into less expensive packages, they will be used for more applications.

PCs have many advantages, such as low cost and small size. What many people do not realize is that PCs are very powerful, especially the newer 386 and 486 systems. In addition, PCs do not require special air conditioning or line power.

In addition to these physical advantages, there are others. The availability of software from many vendors is high on this list. It is possible that a package already exists for your application.

From the systems standpoint, a company can purchase the required number of systems for current needs, rather than trying to configure a minicomputer with the proper disk storage and number of ports. As needs grow, microcomputer systems can be expanded on a per workstation basis to prevent the acquisition of unused hardware and to provide a low incremental cost per workstation.

As PCs continue to expand in power and functionality, their use will become even more widespread. Reducing the size and power consumption of the PC is an active area of development. Many companies now offer powerful laptop systems that can be carried like a briefcase. The user does not need to record data and bring them back to the computer system for processing; the data can be captured in the field and processed instantly.

The efficiency of a process can be monitored on a real-time basis by a central supervisory computer receiving data from PCs located throughout the factory. Data from additional PC-based workstations can be used to perform incoming testing and inspection; final test results can also be fed into the central computer. The central computer, in turn, can send messages or control signals back to the local workstations, thus modifying ongoing processes for improved efficiency.

The rapid development of many types of local area networks (LANs) is easing the communication between personal workstations. These networks span the technological and monetary ranges, offering high-speed (10 megabit (Mbits/second), high-cost LANs for distributed file systems and databases and lower-speed (1 Mbit/second), lower-cost LANs for file transfer and remote electronic messaging.

THE PERSONAL WORKSTATION

A personal workstation consists of a PC, entry and display devices, and the software and hardware necessary to provide complete functionality for the performance of a task or tasks. Software can take the form of turnkey packages such as electronic spreadsheets, word processing, data base management, or computer-aided design, or it can consist of specialized programming languages such as those designed for data acquisition and control. Additional hardware can include accessories such as mass storage devices (tape and hard disk drives) and communications equipment, as well as special-purpose hardware such as analog-to-digital (A/D) and digital to analog (D/A) converters for inputting and outputting voltage signals directly to and from the computer.

The personal workstation harnesses the power of a PC to provide the manager, engineer, or technician with the needed tools. A workstation can be configured as a computer-aided design center, a statistical analysis system for quality control data, an automated test station, or almost any other application required.

Separate workstations can be connected together using one of the commercially available networking schemes, making the resources of each station available to any other station on the network. A user can tap into the central computer with a personal workstation to receive the information needed for decision making. The user can examine the information directly or feed it into a program such as an electronic spreadsheet for analysis.

Direct access over a network can be made available to any workstation in the network for a closer look at a particular point in the process. The user can gain a fuller comprehension of the production process in real time. This enables the workstation user to fine-tune the operation for maximum efficiency.

EVOLUTION OF FACTORY APPLICATIONS

A typical factory with already operating automatic machinery may not need major changes. The available temperature, pressure, flow, level, speed, load, voltage, and current-sensing capabilities represent inputs that can be utilized for monitoring and control. After processing by an industrial computer, mechanical linkages may or may not be required for operation. The control outputs in an oven, for example, are heaters, solenoids, and motors.

In most control applications, it is possible to conceive different levels of integration and automation, resulting in a hierarchical implementation of computer-controlled systems. While there is great promise for future total automation concepts, it will take time before total implementation of many of these schemes is possible. In most cases, implementation will be predicated on the availability of the required hardware and software systems.

The first level of implementation in discrete manufacturing is the automation of tasks at individual work centers such as incoming inspection, production testing, and field service testing. Work center automation generally means that the test systems or work centers within each function (such as incoming inspection or production tests) are networked to facilitate test data collection and analysis. Automating the work center eliminates tedious and inefficient manual data collection and processing. The result is a real-time information and analysis system.

Having the ability to collect and analyze large amounts of data contributes to process control. In the incoming inspection work center, for example, process control begins with vendor control. It is necessary to qualify vendors in order to establish a source of quality components. This qualification process may involve collecting detailed data on parts to determine which vendors have high-quality products.

THE USE OF SENSORS

Sensors are a key link in process control. The system may feel pressure, level, flow, force, outlines, or shapes; see color, light/dark, shapes, heal vibration, noises; sense smell; and detect oxygen or other gases, including pollutants or hazardous materials. With the right training modified by a knowledge base, the system becomes sensory intelligent. Data processing computers provide information processing and factory control; automation computers require a significant level of sensor interfaces for physical variables processing. The fundamental purpose of a control engineering system is to impart to the computer some of what is called *human* or *engineering judgment*.

An industrial microcomputer with a BASIC interpreter and direct analog and digital input/output can support the typical programming needed for factory automation, testing, and measurement. The computer can have several expansion slots for independent, high-speed processing which are coordinated by the central processor.

A typical measurement and control application measures specific parameters which determine the degree of control over the process. The monitoring process attempts to ensure conformance to specific criteria. Exception conditions programmed into the computer can issue an alert if variations occur. Such conditions as low yields, consecutive identical failures, and excessive numbers of repeat failures can trigger a warning that alerts the user to the problems. It is also possible to generate summary and trend reports, either automatically or on demand, to assist in locating problem areas.

THE TYPICAL AT-COMPATIBLE 286 PERSONAL COMPUTER

The AT class represents the minimum computing environment needed for a typical industrial workstation. A 286 system is defined as an expandable, upgradable microcomputer which uses the same 16-bit microprocessor as the IBM AT. A typical 286 comes with the following standards:

1. High-capacity, 1.2-megabyte floppy disk drive.
2. Combination hard disk drive and floppy disk drive controller card.
3. A 101-key adjustable keyboard.
4. SETUP utility in ROM.
5. Socket for the 80287 math coprocessor.
6. Six 16-bit expansion slots and two 8-bit expansion slots.
7. Runs at 8- to 20-MHz clock speed.
8. A 190- to 200-watt power supply, switchable from 110 to 220 volts.

The complete 286 system includes a display adapter, monitor, hard disk drive, and disk operating system (DOS).

ROM, RAM, AND DISKS

Read-only memory (ROM) and random-access memory (ROM) are the PC's basic electronic memory. ROM and RAM have some features in common. Both are contained on computer chips. Both are measured in terms of kilobytes (K) or megabytes (Mb). ROM is called *read-only* because you can read its contents but not change them. It is permanently burned into the chip. ROM contains basic instructions that help the PC function, ranging from booting-up (starting) instructions to instructions for hardware.

RAM is also contained on microchips, but it is reusable. RAM provides the temporary storage that DOS needs to process program instructions and data. RAM houses data and program instructions only while you work with them. When power is shut down, RAM loses its contents.

Disk storage capacity is also expressed in terms of kilobytes and megabytes (a 40-MB hard disk drive, for example, or a 360K floppy disk). Items stored on disk cannot be directly accessed by the central processing unit (CPU); a disk controller is needed to connect the disk to the CPU. Disks provide a place for long-term storage.

LIMITS TO RAM

DOS versions up to 4.01 can only recognize addresses up to 1 Mb of memory. Out of that 1 MB, DOS can access addresses from 0 to 640K for RAM. This 640K total RAM is known as *conventional memory*. The remaining 384K is reserved for system functions such as video and ROM instructions. DOS's original 1-MB limit matches that of the 8088 chip, because the 8088 can access 1 Mb of addresses. The 286 systems can access 16 Mb of memory, that is, 16,777,216 storage addresses. The 386 systems can access 4 gigabytes (Gb) or more than 4 billion addresses. None of this extra storage space exists as far as DOS is concerned.

Although a 286 can access 16 Mb, it usually contains only 1 or 2 Mb of RAM. Even this is more memory than DOS can use. Only 640K of RAM is used to run a program. There are two types of memory that go beyond DOS's limits, known as *expanded memory* and *extended memory*. Both of them use RAM beyond DOS's 640K limit.

Expanded memory has no corresponding physical memory addresses. Expanded memory was originally developed to circumvent the physical memory constraints of the 8088-based XT. On an XT, there are no addresses above 1 Mb. Lotus, Intel, and Microsoft developed the standard for an add-in memory board called the *expanded memory specification* (*EMS*) board. Extended memory, in contrast, is standard with many 286 and 386 computers. Extended memory is linear; its addresses start at the 1 Mb limit.

SYSTEM BOARD RAM

The typical 286 system comes with 512K to 1 Mb RAM on the system board. The systems with 512K may be upgraded to 640K or to 1 Mb, depending on the space available on the system board, or downgraded to 256K using 64K chips. In addition to its system board RAM, the 286 unit can hold up to 16 Mb of extended memory on add-in cards, as discussed above.

DOS is able to support up to 640K of user-addressable base memory. If you add 256K RAM chips to the 286, you will get a total of 1024K of memory. The extra 384K will become extended memory.

System board RAM chips are often arranged in rows, which make up logical banks. For example, as you count from the front of the chassis, the first and third rows can comprise bank 1, while the second and fourth rows can comprise bank 0.

TABLE 1-1 Typical System Board RAM Configurations

Total RAM	Chips in Bank 0	Chips in Bank 1
256K	64K	64K
512K	256K	empty
640K	256K	64K
1 Mb	256K	256K

Each bank must be either completely full or completely empty. You can never leave one row or part of a row empty. The 286 system uses 64K or 256K RAM chips. It should use chips with a speed of 150 nanoseconds or more. The RAM chip configuration of the 286 is selected by a dipswitch on the system board.

Table 1-1 shows typical configurations of the system board RAM and the chips in bank 0 and bank 1 that correspond to each.

Suppose that there is a total of 18 chips per bank. You should buy 18 chips if you are adding one bank, or 36 chips if you are pulling out the existing 256K chips and populating the board entirely with 64K chips.

Each chip has a pin 1, which is marked with a notch or a dot. If you look closely at the green PC card, beneath each empty chip socket you will see a printed white outline of a notched chip in the proper orientation (pin 1 is oriented toward the front of the chassis). You must install each chip in the correct orientation, with all its pins fully seated in the sockets.

The pins that connect each chip to its socket are easily bent out of shape. No force is required to install RAM chips; if you have to press hard, you are probably bending a pin.

If you make a mistake and have to remove a chip from a socket, insert the tip of a small flat-blade screwdriver under one end of the chip and twist gently. When the chip begins to loosen, move the screwdriver to the other end and loosen that side. Work gently back and forth until the chip is freed; then straighten the bent pin and start over.

A program included on the DOS diskette (version 3.1 or later) can be used to create a RAM disk in the system's extended memory. A RAM disk behaves like an extra floppy or hard disk drive installed in the system, but because no disk is mechanically attached, access time is extremely fast.

CRYSTALS

Suppose that your 286 system comes with a 20-MHz crystal. Since computers generally run at half their crystal speed, the 286 runs at 10 MHz. If the peripherals and the rest of the circuitry will run at higher speeds, you may be able to change the crystal. Another crystal that the 286 can accept is a 24-MHz crystal. Switching in this crystal speeds up the system to 12 MHz.

The crystal looks like a small tin pellet on two wires. Its underside is tucked down to the system board by an adhesive pad. To change crystals, simply pull the old crystal loose from the adhesive pad and slide it off the holder. Install the new crystal on the holder, with the printed side face up, and stick it down on the adhesive pad.

MEMORY, I/O, AND MULTIFUNCTION CARDS

The 286 system is capable of addressing a maximum of 16 Mb of memory. You can buy add-in memory cards which occupy the expansion slots of the unit to upgrade the memory to this limit. If you want to connect your PC to a printer, a modem, or another external device, you will need a serial or parallel port (depending on the device). These I/O ports also come on add-in cards. Since many users are interested in both of these options, and since the number of expansion slots inside each computer is limited, some manufacturers offer multifunction cards, which combine additional memory with serial and/or parallel ports and offer additional features on a single card. As with all add-in cards, it is important that the product be designed to run at the system's clock speed.

Other peripherals that are available for your PC include an internal or external modem, a mouse, a light pen, and additional floppy disk drives. If you want more peripherals than your system unit can hold, you can buy an expansion chassis to hold the additional boards.

The 286 system uses a 16-bit bus. However, in many units, two of the eight expansion slots are designed to accommodate an 8-bit card. The left rear section of the system chassis houses the expansion slots.

EXPANSION SLOTS

Expansion slots allow you to add extra circuit boards (the terms *board* and *card* are used interchangeably) to upgrade the system. In most 286 systems, the expansion slots are located in the left rear section of the system box (Figure 1-1). Each slot has a rectangular hole in the back of the chassis. These holes are covered by narrow metal plates called *slot covers* when the slot is empty and by the endplate of the expansion card when the slot is occupied.

When a card is installed in an expansion slot, the slot cover is removed and the board inserted into the data bus connector. The slot cover screw is used to secure the endplate of the card to the back of the chassis.

It is possible to install an 8-bit card in a 16-bit slot, since some 8-bit cards are physically compatible with a 16-bit slot. The difference is in the bottom

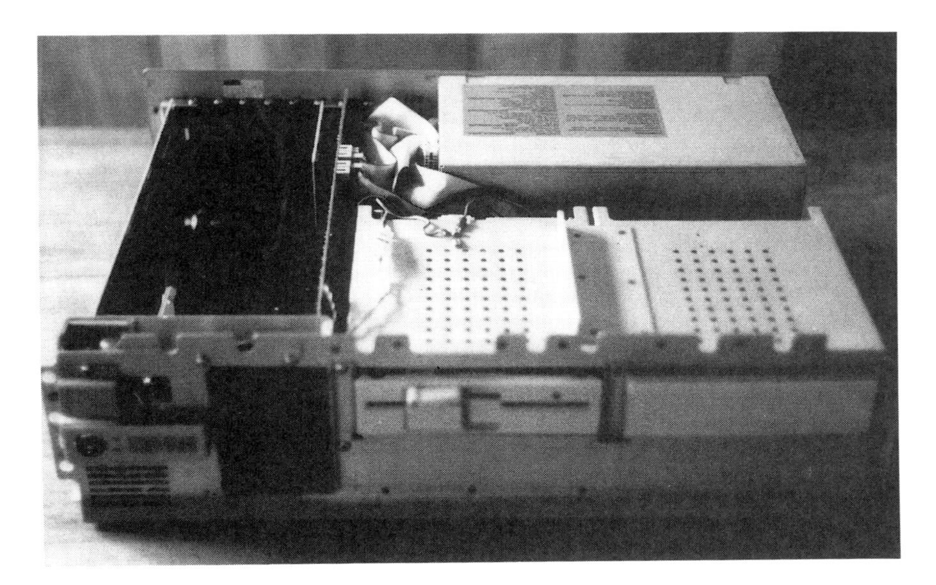

Figure 1-1 286 System box.

edge of the card next to the 8-bit connector pins. If there is a wide cutout on the card next to the 8-bit tab connector or if the edge goes straight out from the tab, it is physically compatible. If the edge of the card dips down again next to the 8-bit tab connector, it is physically incompatible. You cannot install this card in a 16-bit slot because this part will interfere with the second connector on the system board.

The endplate of the expansion card provides space for a rear connector so that a monitor or a printer can be plugged into the card. The endplate bracket also anchors the card to the system chassis and completes the metallic shield to minimize the entry of electromagnetic interference.

Short cards (Figure 1-2) are less than half the length of standard boards. Short cards were invented for the IBM-XT and compatibles, which have one or more expansion slots behind the left-hand drive bank, where only about half of the usual space is available. These short slots work the same way as long slots but require a card that fits into a small space. The 286 system has no short slots; however, it can accommodate short cards. These are all 8-bit cards and require the same selection criteria (speed, configurability, and physical compatibility with 16-bit slots) needed for any 8-bit card.

To install an expansion card with a 16-bit bus (Figure 1-3), you must use a 16-bit slot. If you have a card with an 8-bit bus, install it in one of the 8-bit slots. Some, but not all, 8-bit cards will run in a 16-bit slot.

Figure 1-2 Short 8-bit expansion card.

The primary source of information when installing peripherals is the operations manual that accompanies each product. However, there are a few general rules that apply:

1. Always remember that circuit boards are very sensitive to static electricity. One discharge can permanently damage some electronic chips. You should rid your hands of static electricity by touching the system chassis every time before touching a circuit board.

Figure 1-3 A 16-bit expansion card.

2. Do not eat, drink, or smoke while you are working over an open computer.
3. Remember to screw down the endplate brackets of your cards and keep slot covers in place over every unused slot. This minimizes the possible entry of radio interference.
4. Cards that attach to ribbon cables, such as drive controller cards, should generally be installed to the right of cards without cables. That way the cables are not in the way of other expansion slots.
5. Be careful not to let metal parts such as screws get loose in the computer. If you drop a screw on the system board, you must find it and remove it before you can apply power to the computer, or it may cause a damaging short-circuit.

The actual installation of cards in expansion slots is simple. Choose an empty slot, and remove the screw holding the slot cover to the back of the chassis. Slide the card into place, with its tab(s) meeting the grooves in the expansion slot. The endplate of the card should be in the same place that the slot cover just occupied. When the tab meets the groove, press carefully on the top edge of the card. The tab will snap into place.

DISPLAYS

Pixel is short for *picture element*. Pixels are the individual points of light that make up the patterns on the computer screen. The resolution of a display is defined by the numbers of pixels across and up and down the screen.

A video display adapter is a card that goes in one of the expansion slots and controls the monitor. The display adapter and the monitor must be compatible. A monochrome monitor cannot be used with a display adapter that is designed only for color monitors.

DISPLAY ADAPTER

The display adapter is a card that goes inside your system and connects the monitor to the computer. It acts as the monitor controller. The display adapter type must be compatible with the monitor type. There are display adapters that will run two or more monitors. Some of these offer features such as 132-column extended display, dual displays, and the ability to run color graphics software on a monochrome monitor. Like monitors, display adapters are available in a wide range of quality and price. It is best to purchase the monitor and adapter together to get a matched set that best suits your needs.

The display adapter is installed in one of the expansion slots. There are some 16-bit display adapter cards on the market, but consider buying an 8-bit card that saves a 16-bit slot. Many display adapters are not designed to run reliably at the higher megahertz speeds, so if you expect to run at high speeds, you should ask specifically about this when selecting the display adapter card.

Some 286 units have a slide switch on the system board. This determines whether the system boots up in color or monochrome mode. You must use the correct switch setting. Often a display adapter will have several different display modes, which are set with jumpers or dipswitches. It is important to use the correct switch settings for the system board, since you can damage some monitors by attaching them to a display adapter configured for the wrong mode.

MONOCHROME MONITORS

A monochrome monitor runs at a horizontal frequency of 18.432 kilohertz (kHz) and a vertical frequency of 50 hertz (Hz). It is usually available in amber or green. The color is a personal choice, and one color may seem more pleasant than the other to different users. A monochrome monitor can display a 9 × 16-pixel text character with 25 rows of 80 characters. Most video adapters display 720 × 348-pixel graphics and can emulate color/graphics (CGA) or enhanced graphics (EGA) displays. These CGA and EGA emulation boards display the colors as varying tones of amber or green.

There are often external vertical and horizontal sync controls that a user can adjust. It may be necessary to adjust the sync frequencies to change the monitor's resolution for different video modes. This allows all modes to operate properly on the monitor. In some cases, a temperature change can alter the monitor's sync characteristics. This may cause a monitor to roll vertically. The case may be exposed to direct sunlight, which warms up the monitor and changes its vertical sync frequency. A small adjustment of the vertical sync control can correct this problem. The original IBM monochrome display did not have external sync controls, and even minor adjustments of the sync frequency required a service technician.

The better monochrome monitors use phase lock loop (PLL) circuitry to track drifting of the input sync signal. Many color emulation boards will not work properly unless the monochrome monitor is equipped with PLL. In order to display clean, clear text free of interference from external sources, some monitors filter the input signal to remove noise. Color emulation boards alter the signal characteristics to produce gray shades. The filter circuit interprets this alteration as noise and converts it back to a clean monochrome signal. This can reduce the scaling effect to just a few shades. Monitors that use only a shielded cable for filtering will not have this problem.

RGB MONITORS

An RGB monitor (often called a *CGA monitor*) has an input frequency of 15.75 kHz horizontal and 60 Hz vertical. Separate red, green, blue, and intensity inputs (RGBI format) allow a total of 16 colors to be displayed. The characters are 8 pixels high by 8 pixels deep on an 80 × 25-character display. Graphics can be displayed to a maximum of 640 × 200 pixels.

Multivideo adapters are available which also allow an RGB monitor to display monochrome and EGA graphics. Since the maximum vertical resolution of an RGB monitor is 200 pixels, these adapters use an interlacing technique to produce the 350 lines required for EGA or monochrome graphics. Interlacing involves creating a virtual screen in memory and forming the display with two alternating 200-line segments; the result is a screen with 400 lines. Interlacing does not change the vertical frequency, but it does change the screen refresh rate to 30 Hz and can produce flicker. When interlacing is used, the RGB monitor should have a high-persistence phosphor to reduce the flicker. External sync controls are usually needed if EGA or monochrome emulation boards are to be used with an RGB monitor. Changes in vertical sync characteristics can distort screen images and make the text characters difficult to read.

DUAL FREQUENCY MONITORS

A dual frequency monitor is a display which can operate at either monochrome or color sync frequencies. These monitors usually have a monochrome screen and produce shades when operating as a color display. They are seen in many computers and is also available as a stand-alone unit. These monitors can switch from monochrome sync frequencies (18.432-kHz horizontal, 50 Hz vertical) to color sync frequencies (15.75-kHz horizontal, 60 Hz vertical) by analyzing the input sync frequency. Some monitors check the pin configuration (usually pins 3, 4, and 5) to determine if a color adapter is connected.

When using a dual frequency monitor with a multivideo board, it is best to operate the monitor as a monochrome display. This allows the highest resolution and eliminates the need for interlacing. However, an adapter with the capability to display video on a color monitor will have circuit connections for pins 3, 4, and 5. In order to use monochrome operation, a special cable is needed between the monitor and the video board to disconnect these pins.

EGA MONITORS

An EGA monitor displays an 8 × 14-pixel character on an 80 × 25 display. It can be driven by a CGA adapter (15.75 kHz) or by an EGA adapter (21.8 kHz) to deliver graphic resolutions ranging from 320 × 200 pixels to 640 × 350 pixels. The vertical sync frequency is constant at 60 Hz. A total of 16 colors from among a choice of 64 can be displayed at any one time. Colors are generated through primary and secondary red, green, and blue inputs (RGB format). Special programs allow an EGA monitor to support downloadable character fonts. Most EGA adapters also emulate monochrome text and graphics on the EGA display. Some EGA monitors do not support the 15.75-kHz horizontal scan frequency. This limits the monitor to EGA resolution (640 × 350). These monitors can be recognized by the absence of the 15.75-kHz frequency from the monitor specifications. Some video adapters double-scan CGA graphics to produce better graphics. Most EGA monitors can handle these frequency changes, but some monitors will lose vertical sync.

25-KHZ, 400-LINE MONITORS

The 25-kHz monitor was designed to produce better CGA graphics by double-scanning each of the 200 lines used on a CGA graphic screen. Some adapters are only CGA compatible, while others can emulate EGA and provide their own 640 × 400- and 752 × 410-pixel graphics modes on the 25-kHz monitor. These monitors do not have an established standard. Many manufacturers use their own specifications for video frequencies and color format. Horizontal frequencies can vary from 23 to 27 kHz. Some 25-kHz monitors use a combined horizontal and vertical (composite) sync signal instead of the separate horizontal and vertical signals used in the standard RGB monitor. An important consideration for matching a 25-kHz monitor to a video adapter is the sync polarity. A CGA monitor uses a positive sync for the horizontal frequency and a negative sync for the vertical frequency. An EGA monitor uses a negative sync for both horizontal and vertical frequencies in high-resolution modes and switches to a positive horizontal/negative vertical sync in CGA resolutions. A 25-kHz monitor may use one of these formats, or it may use its own.

VGA MONITORS

VGA stands for Video Graphics Adapter or Visual Graphics Adapter. VGA monitors are the most recent IBM graphics standards. They operate at a horizontal frequency of 31.5 kHz and a vertical frequency of 70 Hz. These monitors provide 640 × 480-pixel resolution. The VGA monitor uses an analog RGB input scheme to achieve the 262,144 different colors available in VGA modes. By switching the vertical and horizontal sync polarities, the VGA adapter signals the monitor to change its vertical sizing characteristics for lower vertical resolution modes.

MULTISYNC AND MULTIMODE MONITORS

Multisync monitors adopt themselves to the sync characteristics of the video board. They can operate with most video boards, including special-purpose and workstation video adapters. Multimode monitors are different from multisync monitors in that they can only match input sync characteristics within one of several frequency ranges. Most multisync and multimode monitors support both digital (monochrome, CGA, and EGA) and analog (VGA and graphics workstations).

Some multisync monitors will not sync on frequencies below 21 kHz. They cannot be used with CGA adapters, but they can be used with video boards that double-scan CGA. With the use of multimode monitors, some multivideo board emulation modes may be within an unsupported frequency range. This can occur with EGA and VGA boards. Some high-resolution modes may not be supported when using these adapters on multimode monitors. Connecting a multisync or multimode monitor to a VGA adapter requires a special cable to mate the video board's PS/1 15-pin connector with the monitor's analog input connector.

A VGA adapter can cause a multisync monitor to experience changes in the vertical display size when the video mode changes. This is because the monitor may not sense the sizing signals sent by the VGA adapter.

SPECIAL MONITORS

There are also high-resolution graphics for computer-assisted design and other applications. These monitors are usually designed to accommodate the monitor manufacturer's proprietary video system. If another video system is used, it is important to match the monitor's sync characteristics with those of the video adapter. The following characteristics must match:

1. Vertical/horizontal sync frequency.
2. Analog/digital signal.
3. Sync polarity.
4. Color format.
5. Composite/separate sync.

DUAL DISPLAY

Dual display is the ability to run two monitors, with different displays on each one, at the same time. This feature is supported by some software packages (such as Lotus 1-2-3 2.0 and Symphony 1.1). With a dual display card running one of these packages, you can view a graph on an enhanced monitor and simultaneously view text on a monochrome monitor. You can type changes on the monochrome monitor, and the graph on the enhanced monitor will modify itself to show the changes. *Dual monitor* support means that the system supports two different types of monitors, while *dual display* refers to the dual, simultaneous, independent display on two separate monitors.

BACKUP

Although hard disks are a reliable form of data storage, they do fail; in addition, valuable information can be erased by an operator error. Since the volume of information contained on a hard disk is large, the failure of a hard disk that is not backed up can be a major loss. There are several possible ways to back up a hard disk. One method is to copy files to floppy disks using the DOS backup and retrieve commands. The advantage of floppy disk backup is that you do not need to add any special equipment, and the use of high-capacity floppy disks reduces the handling and speeds up the time required. However, there are also some disadvantages. Floppy disk backup tends to be time-consuming and tedious, and many users fail to do it often enough. When the hard disk fails, they find themselves without the backup of critical files. Also, the number of floppy disks needed to back up a high-capacity hard disk can quickly become unmanageable. At first, floppy disk backup appears to be inexpensive, since it requires no special hardware or training. However, the cost of the time spent backing up on floppies grow quickly.

The other popular method of hard disk backup involves using a tape drive. There are two basic types of tape drives: floppy tape and streaming tape. Floppy tape runs directly from the floppy disk controller, so it does not take up an extra expansion slot. Streaming tape has the advantages of being very fast, standardized, and reliable. Its drives can be activated automatically at a specified time of day, freeing the operator from backup operations. Since streaming tape is easy to use, people tend to perform backups more regularly.

The industry standard quarter-inch cartridge (QIC) units allow tapes to be interchanged among drives, which is required for archival storage. The disadvantage of streaming tape is cost, which can exceed the cost of the hard disk drive, but this must be weighed against the cost of regenerating lost data after a disk failure. Speed and ease of use are important factors in determining the value of a tape subsystem, and these depend on the quality of the tape software.

Another backup method that is sometimes considered is a second hard disk used for backup. This is an unreliable choice since the problem that causes the main hard first disk to fail, such as an unexpected physical shock to the system chassis, will damage the backup disk at the same time.

An external tape drive requires the installation of a controller card that will use the drive. The question of 16-bit bus compatibility and maximum megahertz operation need to be considered.

If you do purchase an internal tape drive, it should be a half-height drive so that it can be installed under the floppy drive in the right drive bank. The plastic or metal bezel covering the opening is removed. The controller card should be installed in a suitable slot that is close to the drive, since a ribbon cable will be connected from the card to the tape drive. You will also need to connect power cables to power the tape drive.

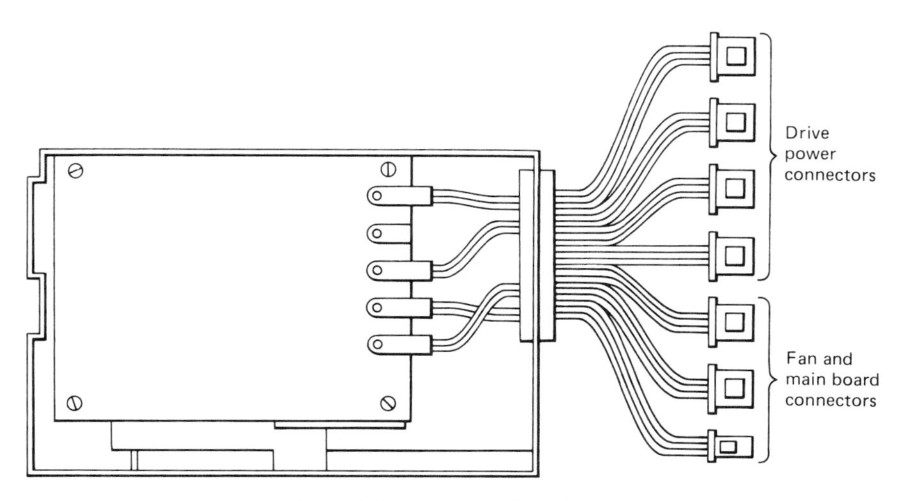

Figure 1-4 A 286 power supply and connectors.

POWER

The power supply (Figure 1-4) is the shielded cube in the right-rear section of the chassis. In the 286 system it is likely to be a 190- to 200-watt power supply, which means that you will probably never need to augment it with an extra power supply. The unit's power-on switch is usually part of the power supply, as are the internal cables designed to be connected to the drives. A switch on many units lets you select 220-volt operation when you use the unit outside the United States. If you need more power cables for additional drives, you can get an inexpensive Y power cable that connects to one of the cables and splits its output in two. You should not use more than one Y cable per system, since it is possible to overload the power supply.

POWER SUPPLY

The power supply inside the system unit provides power for the system board, adapters, diskette drives(s), hard disk drive(s), monitor, and keyboard. The power supply is designed to operate the maximum number of drives that can be added inside the system unit. IBM AT-compatible computers have a total output of 175–200 watts. A 115/230 Vac selectable switch is usually located at the rear of the power supply enclosure.

INPUT CHARACTERISTICS

The power supply is typically designed to operate at a frequency of either 60 ±3 Hz or 50 ±2 Hz, and at 100–130 Vac, 5.0 A or 220/260 Vac, 2.5 A. The

TABLE 1-2 Typical Power Supply Output Characteristics

Output (V)	Load (A)	Tolerance (%)	Ripple (MV)
+5	20	±2	50
+12	7.3	±5	100
−5	0.3	±10	100
−12	0.3	±10	100

voltage is selected by a switch at the rear of the power supply. The input AC voltage requirements then become 100–130/200–260 selectable.

OUTPUT CHARACTERISTICS

The power supply provides +5, −5, +12, and −12 Vdc. Table 1-2 shows the typical load current and regulation tolerance for a 200-watt power supply.

OUTPUT PROTECTION

If any output becomes overloaded, the power supply will switch off within about 20 milliseconds. This will normally prevent an overcurrent condition from damaging the power supply.

If no fixed disk drive is connected to the power supply of many AT-type systems, a dummy load must be connected to simulate the fixed disk drive load. The dummy load is typically a 5-ohm, 50-watt resistor.

Under normal conditions, the output voltage levels track within 300 milliseconds of each other when power is applied to or removed from the power supply, provided that at least the minimum loading is present. When primary power is applied with no load on any output level, the power supply may switch off and a power-on cycle will be required. The power supply requires a minimum load for proper operation.

The power supply typically provides a power-okay signal to indicate proper operation. When the supply is switched off for a minimum of 1 second, then switched on, the power-okay signal is generated, assuming that there are no problems. This signal is a logical AND of the dc output-voltage sense signal and the AC input-voltage sense signal. The power-okay signal is a TTL*-compatible at high levels for normal operations and at low levels for fault conditions. The ac fail signal causes power-okay to go to a low level at least 1 millisecond before any output voltage falls below the regulation limits. The operating point used as a reference for measuring the 1 millisecond is normal operation at minimum line voltage and maximum load.

*Transistor-Transistor-Logic

TABLE 1-3 Typical Power Supply Minimum Sense Levels for Output Voltages

Level (VDC)	Minimum (VDC)
+5	+4.5
−5	−3.75
+12	+10.8
−12	−10.4

The dc output-voltage sense signal holds the power-okay signal at a low level when power is switched on until all output voltages have reached their minimum sense levels. The power-okay signal has a turn-around delay of at least 100 milliseconds but not longer than 500 milliseconds. Table 1-3 shows the minimum sense levels for the output voltages.

386 SYSTEMS

A high CPU speed is typical of 386 microcomputers which are based on the 80386 microprocessor. The 80386 chip speed range is 16 to 33 MHz, and it is important to match the capabilities of the other components to CPU speed. The 386 chip is available in four speeds: 16, 20, 25, and 33 MHz. There are also two different types of 386 chips: the original DX and the less expensive, less powerful SX.

The SX can run any software the 80386DX chip can run, but its external 16-bit data bus and 16-MB memory limit make it act more like the 80286. With fewer pins, a smaller package, and the ability to fit on an AT motherboard whose design has been modified slightly, the SX chip allows manufacturers to produce less expensive 386 computers. The 16-MHz SX and DX chips, as well as the 286 chips, perform about the same when running DOS programs or 286 software such as OS/2 1.0X. It is only when running 386 software, such as OS/2 2.0, or a program using 386 DOS extender technology that the SX falls behind the DX.

SYSTEM BOARD CONSIDERATIONS

Once you have decided on a CPU and a clock speed, you must consider a number of factors in the system board. Most of these factors involve RAM, but a few others are also important. The 386 system board, which holds memory and other system logic in addition to the CPU, is not very different from an AT motherhood. Most 386 systems are built on a modified AT-type I/O bus called *industry standard architecture (ISA)*. A computer bus is the collection of parallel data, address, and control lines over which a computer transacts its information processing business. The system board provides standard 8- and 16-bit slots for the same expansion boards you plug into an AT. Most 386 ISA systems also include one or more nonstandard 32-bit slots on a separate memory bus that provides full-speed, 32-bit access to RAM on proprietary expansion boards.

High-end 386 systems offer the newer bus designs: IBM's Micro Channel Architecture (MCA) or the Compaq-led consortium's competing EISA (enhanced ISA, EISA). IBM's Micro Channel-equipped PS/2 line does not accept ISA boards, but Micro Channel has the advantage of being IBM's current bus standard. EISA, on the other hand, may turn out to be less popular, but it accepts standard expansion cards.

Another system board consideration is the clock speed at which the I/O bus runs. In order to be compatible with expansion cards designed for the slower ATs, 386 systems use 8 MHz for the bus lines that connect to the expansion slots. If you do not intend to use these older expansion cards, you can use a system with a faster bus such as the 16-bit small computer system interface (SCSI, pronounced *scuzzy*). A SCSI hard drive controller on a fast bus can substantially increase the hard drive system's data transfer rate.

MEMORY

Although memory is critical to a PC's productivity, many people are confused about how it works. Part of the confusion stems from the many types of memory used by computers; part of it stems from the additional memory capacities of 286 and 386 computers. The 286 and 386 systems can access much more memory than XTs can. So, there is virtual memory, extended memory, expanded memory, RAM caches, and RAM disks.

Memory is basically storage, so we see the terms *memory* and *storage* used interchangeably. PCs are capable of processing millions of bytes of information. (A byte is the computer's basic unit of storage.) The computer has to be able to find each byte it contains. The number of storage cells (addresses) the CPU can access is determined by the type of microprocessor used as the CPU, 8088; PC, XT series, 80286 (286); AT series, 80386 (386) or 80486 (486).

Some midrange and high-end 386 come with 4 Mb of memory, which is the amount needed to run OS/2 with Presentation Manager. Most programs that will run on a 386 system can use extra memory, either as extended or expanded RAM, to provide more data space or to enhance performance by reducing disk I/O.

There is another advantage of the 80386: built-in memory-paging capability, which allows it to emulate expanded memory (EMS). Emulated EMS accessed at the full CPU clock speed is faster than true EMS accessed on a board at I/O bus speed, usually 8 or 10 MHz. This means that you do not have to install an EMS board in the 386 system.

The memory options on a DX are determined by the number of RAM banks the system supports and by what combinations of 1-Mb and 4-Mb banks that number allows. The DX chip's 32-bit data bus dictates that a bank of RAM must hold 36 memory devices: four 8-bit bytes and an error-checking parity bit for each byte. Usually, 80386DX RAM banks are filled in one of the following ways:

1. 36 256K by 1-bit chips for a 1-Mb bank.
2. 36 1-Mb by 1-bit chips for a 4-Mb bank.
3. Four equivalent single inline memory modules (SIMMs).
4. Single inline packages (SIPs), each of which holds nine devices.

The 80386 SX, with its data bus halved to 16 bits externally, needs half as many chips per bank.

The cost of memory depends on the type of RAM chip the motherboard accepts: dual inline package (DIP), SIP, or SIMM. Another factor is whether 256K or 1-Mb chips are used to fill the banks. The 256K chips cost more per megabyte than the 1-Mb chips.

Memory management hardware such as expanded memory boards adds memory to the system. Memory management software does not add any chips, but it helps to make more efficient use of existing RAM. DOS was originally developed for 8088 and 8086 microprocessors, which cannot work directly with anything over 1 Mb of RAM; 384K of that was set aside for system use, leaving only 640K for programs. Two types of memory, expanded and extended, are available for programs that need more than 640K (Figure 1-5) .

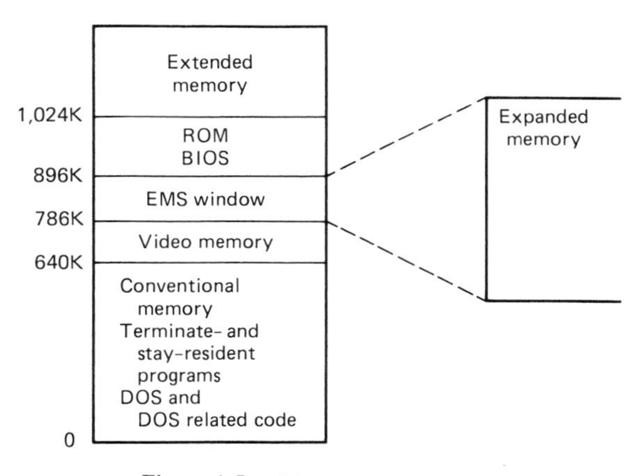

Figure 1-5 Memory partitions.

Expanded memory is RAM on an expansion board. This type of memory is accessible only through a window or page frame in conventional memory. The window holds different pages of memory that are swapped as needed into the processor's address space. To use expanded memory, a memory board and software that follow the Expanded-Memory Specification (EMS)* standard is needed. EMS 4.0 allows the user to run programs and store data in up to 32 Mb of paged RAM. It works on IBM compatibles ranging from XTs to 486 machines.

Extended memory is an extension of RAM. It is not available on 8086- and 8088-based machines. It is possible to add 15 Mb of extended RAM at the end of 1 Mb in an 80286 or 386 SX-based machine. The 386 and 486 machines can handle an additional 4095 Mb.

Some operating systems, such as OS/2 and Unix, can use extended memory directly, but DOS cannot. To run in extended memory, a program must contain a DOS extender that switches the CPU between the real mode, used by DOS, and the protected mode, which can access memory beyond 1 Mb. Three standards are used: Extended Memory Specifications (EMS), Virtual Control Program Interface (VCPI), and DOS Protected-Mode Interface (DPMI).

Extended memory is less cumbersome than the EMS scheme. However, extended memory will not work in XT-class machines and is newer than EMS for 286 and 386 systems, while EMS is established as a popular standard for all three classes. Hardware and software products are available that use either one or a combination of the two memory types.

EXPANDED MEMORY

Expanded memory borrows existing addresses within DOS's 1-Mb address limit that are not being used. Expanded memory swaps additional blocks of memory into those unused addresses by a process called *bank switching*.

The expanded memory manager uses a 64K window called a *page frame* to access the memory located outside RAM. The page frame is located within the 1 Mb that DOS recognizes. The page frame let DOS access up to four blocks or pages of memory at once. Suppose that the program needs data that are located in expanded memory. The expanded-memory manager (EMM) finds the page(s) the program is looking for and puts it into the 64K page frame so that DOS can find it.

You can add expanded memory by adding an EMS board or installing a software emulator; the latter is software that mimics an EMS board. You can add expanded memory to an XT, AT, or 386 machine.

*XMS is also used as an abbreviation.

MEMORY EXPANDERS

Expanded memory provides more memory space for programs that take advantage of EMS. These are called *memory expanders*; the most common examples are the EMS 4.0 memory boards. Several software products emulate expanded memory. These products swap sections of code using the hard drive or extended RAM to simulate EMS's paged RAM. One example is Above Disk. Although it is primarily an EMS emulator, this program also includes utilities for borrowing 96K of EGA or VGA video memory for nongraphics programs and for relocating Novell Netware TSRs (Terminate-and-Stay-Resident programs). It can simulate EMS 4.0's entire 32-Mb limit in extended memory or on a hard disk, and it is one of the easiest memory managers to install. It begins by asking if this is a first-time installation, and which disk and subdirectory to copy files to. Then a configuration screen is used to show the settings for emulating EMS memory in the system. A device driver is added to the CONFIG.SYS file and a memory-resident module to AUTOEXEC.BAT.

Above Disk will not work with some programs, like Microsoft Windows/386. Protected mode works only on 286 or 386 machines with CPUs that are dated August 1987 and that run DOS 3.3 or later models. The package comes with a Lotus 1-2-3 EMS emulator that provides memory management not available in the main program. It can provide three memory management functions:

1. Emulating EMS.
2. Reclaiming unused video RAM.
3. Space for large 1-2-3 worksheets.

TC Power supplies EMS emulation only. The installation procedure requires you to create a new hard disk subdirectory, copy the program files, and accept the default settings before rebooting. The program adds a command line to CONFIG.SYS. It can combine extended, expanded, and hard disk memory in any combination up to the 32B that EMS 4.0 allows. Like Above Disk, TC Power will run on an XT, but since XTs do not support extended RAM, the program uses only the machine's hard disk for EMS emulation. The disk-based emulation can provide the full 32 Mb of EMS 4.0 memory.

Some products make room for the software application by moving TSRs and drivers out of the lower 640K and relocating them to expanded or extended memory, the hard disk, or gaps between 640K and 1 Mb if your system has memory in this range. The latter occurs with some 286 and 386 compatibles with 1 Mb on the motherboard.

Other products stretch DOS RAM to 704 or 736K, depending on the type of video card used. Monochrome and CGA cards do not use the memory addresses immediately above 640K, so a memory manager can assign the empty space to DOS.

There are also some memory management products designed for 386 machines that convert extended memory to expanded memory. These products use the 386 chip's memory-mapping capability to substitute for the circuitry on the EMS boards. They convert extended memory to EMS RAM and use the 386's memory-mapping features to relocate TSRs and drivers.

EXTENDED MEMORY

MS-DOS version 3.X (Micro-Soft Disk Operating System) is able to support only 640K of user-addressable base memory. In the 286 system, memory is linear. That means that it proceeds (theoretically, not physically) in a straight line from bit 1 through the highest bit in memory. Memory up to the 640K point is called *base memory*. DOS can use this memory for any purpose. Memory from 640K to 1 Mb is reserved for ROM. ROM acts as built-in permanent software that the system needs to perform its most basic functions, like starting up, which is called *booting*, and testing the system during this power-up. The user does not have access to the memory in the reserved area between 640K and 1 Mb. Above 1 Mb, extended memory begins. DOS does not recognize extended memory, but many applications programs use it. Extended memory goes from 1 Mb to 16 Mb for a 286 system. If you install 1 Mb of RAM on the system board, the first 640K goes to base memory, and the remaining 384K jumps above the reserved area for ROM to become the first 384K of extended memory.

A common application for extended memory is a RAM disk. DOS versions 3.1 and above usually include a program for creating a RAM disk in extended memory. The RAM disk is electronic memory which acts like a fast disk drive. DOS treats it like an extra floppy or hard disk drive connected to the system. Since no physical disk is involved, the access time to and from the RAM disk depends only on the speed of the memory chips. Even though it acts like storage, a RAM disk is still volatile memory; the contents of the RAM disk must be saved to a true storage device after a computing session if power is to be turned off.

Extended memory is different from expanded memory in that only 286 and 386 machines can use it. Extended memory is linear. Its addresses begin where the DOS limit leaves off, from 1 Mb up to 16 Mb in a 286 system and up to 4996 Mb in a 386 system.

Extended memory does not require bank switching, as expanded memory does; all of it is available simultaneously, instead of being limited to 64K chunks. Extended memory is necessary for multitasking (running several applications at once) and for operating systems that can handle multitasking [Operating System 2 (OS/2) and Unix].

When 286 and 386 machines use extended memory, they are working in protected mode. Some software programs are written specifically for 286-and-above machines, so the programs can use the extended memory. AT and 386 machines can also run in real mode, the conventional 640K RAM limit, using conventional DOS-compatible software. When running in real mode, AT and 386 machines pretend to be XTs.

DOS versions 4.XX and below work within a 1-Mb limit. Expanded memory is memory that lies completely outside that limit but is switched into and out of a 64K window so that DOS can work with it. Extended memory is a linear extension of that 1 Mb, and all of it can be accessed directly by the CPU. Extended memory is available only with 286 and 386 computers.

MEMORY UPGRADES

Some 286 units come with 512K of RAM on the system board. This can be upgraded to 640K or 1 Mb or more by installing more RAM chips on the system board. The RAM chips are located in rows on the system board. Some rows will be filled with chips, while others will have empty sockets. The rows are divided into logical banks.

In a 16-bit system like the 286, each logical bank contains 16 memory chips plus two organizers, while an 8-bit system uses 8 memory chips plus one organizer. That way each bank can send 16 bits, one from each memory chip, to the data bus at one time.

If there are empty rows of sockets on the 286's board, you can upgrade the memory. For example, suppose that you have 512K of memory and you want to upgrade to 640K. This is calculated as follows.

1. Suppose that your board has nine memory chip sockets per row.
2. Since one chip is used for parity testing, eight chips of actual memory are used per row.
3. 64K bits times eight chips equals 64K bytes per row.
4. 64K bytes times two rows equals 128K bytes added to the system.
5. 128K bytes plus the original 512K bytes equals 640K.

If the upgrade goes to 1 Mb of RAM, 18 256K RAM chips are needed. If you want to downgrade the total system memory to 256K, pull two rows of existing 256K chips and fill all four rows with 64K chips. There is no physical difference between a chip used for parity and a chip used for RAM. Any RAM chip can be the organizer if it is installed in the ninth socket of the bank.

RAM chips are available in different speeds, according to how long it takes the chip to send a bit of information to the CPU. Chip speeds are measured in nanoseconds, or one-billionth of a second. The lower the number of nanoseconds (ns), the faster the chip. In the higher-performance machines, the CPU needs chips that can deliver information fast. It cannot be kept waiting for a slow RAM chip. The recommended speed for RAM chips in most 286 machines is 150 nanoseconds.

The important things to remember about memory upgrading are as follows:

1. Capacity: for example, in the discussion above, we used 64K chips for upgrading to 640K and 256K chips for upgrading to 1 Mb.
2. Number: Chips must be added in banks; above, we added 18 chips. You cannot add 9 now and 9 later.
3. Speed: As discussed, 150 nanoseconds or faster is needed for most 286 systems. The fewer the nanoseconds, the faster the chip.

TASK SWITCHING

There are also programs that swap applications or TSRs into and out of the 640K working area. This allows the user to switch between programs without having to exit one program and start another. When tasks are switched, the program saves a "snapshot" of the suspended program while the user works with the new task in the foreground. These products differ from true multitasking environments such as those provided by Desqview or Microsoft Windows, which run several programs simultaneously.

SYSTEM CONSIDERATIONS

A PC or an XT machine can use an EMS expanded memory board that fits into an 8-bit slot. This improves the performance of task-switching programs. A hard disk is a slower but often satisfactory solution with a task switcher or an EMS emulator. In a 286 system, if you use hard disk swapping without adding memory, you may find it too slow. A better solution is to combine these programs with an expanded memory board. In a 386 system it is best to use all the extended RAM, on the motherboard or in a 32-bit memory slot, instead of a slower expansion board slot, if you can. A 386 memory manager will allow EMS access, and a task switcher can be used to make sure that all of the RAM is fully utilized.

FLOPPY DRIVES

There are floppy drives that accommodate single-sided $5\frac{1}{4}$-inch floppy disks; double-sided, double-density $5\frac{1}{4}$-inch floppy disks; double-density $3\frac{1}{2}$-inch floppy disks; and high-capacity floppy disks of both sizes. Most XT systems come with floppy drives that use double-density disks. AT (286), 386, and 486 systems use high-capacity drives and high-capacity disks that hold two to four times the amount of data that can fit on a double-density disk. High-capacity floppy disks offer several advantages. When you back up your hard drive on high-capacity floppy disks, it takes fewer disks and fewer disk swaps during backups than with double-density disks. Also, if you need to transport files between field sites, or from one machine to another, it takes fewer disks if they are of the high-capacity type.

This section will discuss the differences between double-density and high-density controllers, explain how floppy drive controllers work, and review the types available. The floppy drive controller is the circuitry within the machine that controls the physical operation of a floppy drive and directs the head actuator, heads, motor, and disk sensors. It is the interface, or connecting component, between the floppy drive and the rest of the computer. In the first PCs developed in the early 1970s, the floppy drive controller circuitry resided on a card installed in one of the computer's expansion slots. Most of the recent microcomputers have floppy drive controllers built into the computer's system board or combined with other circuitry (usually the hard drive controller) on an expansion card. An AT-class controller card requires a 16-bit expansion slot and an AT-class ROM BIOS that tells the AT's hardware components how to communicate with each other. An XT needs a high-capacity floppy disk controller that is designed for an XT's 8-bit slots.

The XT-class machines use double-density floppy drive controllers designed to work with floppy drives that spin at 300 rpm and transfer data from the disk to the controller at a speed of 250K bits/second. These double-density drives used standard 360K, $5\frac{1}{4}$-inch, double-density disks or 720K, $3\frac{1}{2}$-inch, double-density disks. At the above speed and under ideal conditions, it takes about 1 second to transfer a 32K file from a drive to the computer's memory.

A high-capacity disk holds more data than a double-density disk: 1.44 Mb on a $3\frac{1}{2}$-inch disk and 1.2 Mb on a $5\frac{1}{4}$-inch disk compared with 720K on a $3\frac{1}{2}$-inch disk and 360K on a $5\frac{1}{4}$-inch disk. It also spins faster (360 versus 300 rpm) and transfers data twice as fast: at a rate of 500K bits per second compared to 250K bits per second.

DISKS

A double-density disk will not format correctly if you attempt to format it as a high-capacity disk. Tracks 75–79 will lock out, or a BAD MEDIA error message will appear on the screen. As a general rule, you should not format a double-density $3\frac{1}{2}$-inch disk to 1.44 Mb in a high-capacity $3\frac{1}{2}$-inch drive. This can also cause a BAD MEDIA error message. Even if you succeed in getting a double-density disk to act as a high-density disk, it may not store data reliably.

A high-capacity disk usually has the letters HC (high capacity) or HD (high density) embossed on one corner next to the metal slide. A 720K disk has a write-protect window in one corner that can be opened to prevent writing to the disk. A 1.44 Mb disk has this window, plus another hole directly across from it that tells the floppy drive that the disk is high capacity. Even though a high-capacity drive can read double-density disks, once you format or write to a double-density, $5\frac{1}{4}$-inch disk in a high-capacity $5\frac{1}{4}$-inch drive, it may not be usable in a double-density drive.

High-capacity disks can cost up to three times more than double-density disks, but since you need fewer of them to hold the same amount of data, the cost per bit is about the same.

Suppose that you must decide whether to install a $5\frac{1}{4}$-inch or a $3\frac{1}{2}$-inch drive. Unless you need $5\frac{1}{4}$-inch disks for compatibility, you are probably better off using a $3\frac{1}{2}$-inch drive. The $5\frac{1}{4}$-inch drives will decline in popularity, and once a $5\frac{1}{4}$-inch high-capacity drive formats or writes to a double-density $5\frac{1}{4}$-inch disk, a double-density drive can no longer reliably read the disk. A high-capacity $3\frac{1}{2}$-inch drive can format and write to a double-density $3\frac{1}{2}$-inch disk. In addition to holding more data in a smaller package, a $3\frac{1}{2}$-inch drive gives better compatibility. A $3\frac{1}{2}$-inch disk is also more rugged and fits into your shirt pocket.

DISK COMPATIBILITY

The $5\frac{1}{4}$-inch, 1.2-Mb, high-capacity diskette drive used in most of the 286 machines is capable of reading 160/180-kB, 320/360-Kb, or 1.2-Mb diskettes. The high-density 1.2-Mb diskettes that have been formatted in a 360-Kb double-sided drive can be used interchangeably with 1.2-Mb diskettes in the 286's floppy drive. But because of the way some diskettes are copy protected, the high-capacity drive may not be able to read them. This can occur if copy protection is based on one of the following:

1. Rotation speed—Copy protection uses the time between two events on a diskette.
2. Access speed—The diskette BIOS must set the track access time for the different types of media used on the 286.
3. Diskette change signal—Copy protection may not be able to reset this signal.

HARD DISKS

Hard drives, which are also called *fixed drives*, are popular PC storage devices. While floppy drives accept removable floppy disks, in a hard drive the disks are permanently mounted inside the drive. Hard drives provide more speed and storage than floppy drives. Hard drives can store hundreds of times more information than floppy drives. They can also retrieve the information 10 to 100 times faster.

When you install a program, you transfer the information from the software's floppies to the hard disk. When you want to use the application, you call it up on the hard disk.

In order to understand hard disks, it is necessary to introduce some technical terms. Figure 1-6A shows a typical hard disk. A disk can be made up of several metal platters that are coated on each side with magnetic material. Each side of each platter will have its own magnetic head, which writes data to the disk and reads data from it. When the head reads data from the disk, these data are sent to the disk controller.

To prepare a disk for storing data, it must be formatted twice. The first format is called a *low-level format*. It divides each surface into concentric rings called *tracks*, and then it divides each track into sections called *sectors*. The number of tracks per hard disk and the number of sectors per track vary among hard disk manufacturers. A low-level format can take an hour or more.

If the hard disk is larger than 32 Mb, you can partition it into two or more logical drives of not more than 32 Mb, each using MS-DOS version 3.3. This is done with DOS's FDISK command. Compaq's DOS 3.31 allows a single logical drive of up to 512 Mb. DOS 4.0 and OS/2 allow drives of up to 2 Gb.

After the disk is partitioned, it must be formatted a second time using DOS's FORMAT command. This prepares the disk so that DOS can store and find files and for starting the computer.

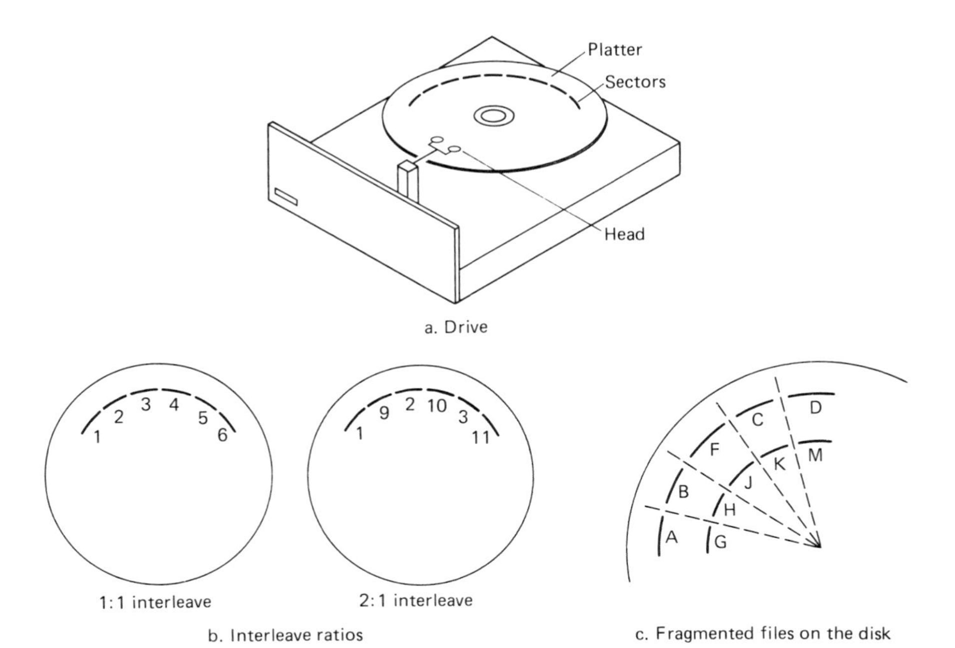

Figure 1-6 Hard disk drives. (a) Drive; (b) interleave ratios; (c) fragmented files on the disk.

PERFORMANCE TECHNOLOGY

Average access time, which is also known as *average seek time*, is the amount of time, on average, that it takes for the magnetic head to move to the track containing the needed data. A typical IBM-PC AT-compatible hard disk has an average access time of about 20 milliseconds. Average transfer rate is the amount of data, on average, that the disk controller can transfer to the computer in 1 second.

Effective access time is the average access time as it appears to the program. It is based on the speed needed for the program to retrieve data from the disk. If two hard disks have the same average access time, but disk 1 has a higher average transfer rate than disk 2, then disk 1 has a faster effective access time. This means that disk 1 retrieves data as if it had a faster average access time.

Interleave refers to the ratio of the number of sectors passed over the magnetic head for every one sector that the head reads (Figure 1-6B). The interleave ratio depends on the hard disk controller and the system bus. Some hard disk controllers cannot transfer the data from one sector to the computer before the head reads the next sector. To compensate for this problem, the controller allows the head to read only every fifth sector or every third sector, providing an interleave ratio of 5 : 1 or 3 : 1. The higher the interleave ratio, the more times the disk must rotate to reach a specific item of information. The most efficient hard disk controllers include Hardcard II's controller and the Western Digital controller used with Swift. Both of these operate with an interleave ratio of 1 : 1.

HARD DISK OPERATION

The order of procedures to perform before you operate your hard disk is as follows:

1. SETUP drive type identification.
2. Physical format.
3. Partitioning.
4. Logical format with system transfer.

These procedures must be done in this order.

Virtually all hard disks have bad tracks. Therefore, the disk is specified at slightly less than its actual capacity. Even when all the bad tracks are counted, the capacity of the disk should be slightly greater than specified. With every hard disk drive, the manufacturer provides a list of the locations of the bad tracks. This list is a computer printout which comes taped to the hard drive itself. It references bad tracks by cylinder and head number. It may be needed to format the hard disk drive.

DISK SPEED

The hard disk tends to be the slowest component in the computer (unless you are using a floppy disk). The CPU operates in millions of cycles per second. Memory chips are accessed in millionths of a second. Most memory chips work at a rate of 150 and 80 nanoseconds (10^{-9} seconds). The hard disk, in contrast, operates in thousandths of a second. Hard disk access times range from about 80 milliseconds (thousandths of a second) to 15 milliseconds. This means that memory and the CPU can operate 1000 times faster than the hard disk.

When a software program is running, frequent access of the hard drive means that the CPU is often waiting for the hard drive. If the hard disk is used infrequently, this indicates processor- and/or memory-intensive application. These applications require more or faster memory, a faster CPU, or a coprocessor. If the hard disk is the slowest factor, there are several ways to speed up its operation.

SETUP AND BUFFERING

There are two techniques that can be used to speed up a hard disk. One technique relies on changing the setup of the disk. Physical adjustments can be made with the proper software. They work by making sure that the hard disk is operating at the fastest speed possible. The other technique relies on buffering, which makes use of intermediate storage. The basic idea is to retain data in memory that can be accessed more quickly than the hard disk. One type of buffer holds the data that have just been read. Then, if the data need to be read again, they can be quickly retrieved from the buffer.

Smart buffers anticipate disk accesses and attempt to read in data before they are needed from the disk. For example, if the first sector of data is requested from a file, it is logical that the rest of the file may soon be requested. By reading the whole file into a buffer memory, future requests for the information in that file can be satisfied without additional accesses of the hard disk.

There are three physical adjustments that can be made to a hard disk and the data stored on it.

HARD DISK INTERLEAVE

The data on the hard disk are grouped into units called *sectors*. Disk interleaving involves the order in which the sectors are physically addressed on the disk. The fastest way to place the sectors sequentially on the disk is 1, 2, 3, and so on.

This sequence is not always used for all hard disks. Many low-cost disks are not fast enough to address the sectors arranged this way. The hard disk must read sector 1, and then save it and find sector 2. During this time, the hard disk is spinning and the rotation can be enough to allow sector 2 to spin past the read/write head of the disk before it is found. Then the disk platter must spin all the way around before it can read sector 2, slowing disk access.

Alternating the sectors accessed, skipping every other one, has the effect of counting by 2. The sectors are accessed in the following way: sector 1, skip to some other number, sector 2, skip, sector 3, and so on. This is the interleave referred to earlier. Counting by 2 provides a 2 : 1 interleave. Other interleaves, such as 3 : 1, 4 : 1, and 6 : 1, are also used.

The right interleave can produce dramatic increases in performance. Hard disks are usually formatted with a standard interleave. If the disk operates best at the standard setting, no adjustment to the interleave is required. To change the interleave, you can perform a low-level format of the hard disk.

A low-level format erases all of the data on the disk, so you have to perform a complete backup first. There is also software that performs nondestructive, low-level formatting. This software preserves the data during reformatting. SpinRite is an example of a program that performs interleave adjustments.

DEFRAGMENTING THE DISK

This procedure also goes by other names, such as *compressing* and *optimizing*. Under ideal conditions, files are stored in physically adjacent clusters on the disk. These clusters are groups of sectors. For example, sectors 1 through 4 might make up cluster 1. The number of sectors in a cluster increases with the size of the disk. Sectors are the smallest unit that can be directly accessed on a hard disk. You cannot get a single character without first reading the entire sector.

It is often inconvenient to work with sectors because of their large number. Therefore, they are grouped into clusters. By storing the parts of a file in adjacent clusters, you can retrieve them more quickly. However, DOS rarely store files in an ideal fashion, so a single file ends up being stored in parts. DOS splits the file where some free space is available. These fragmented files are stored in random clusters throughout the disk, as shown in Figure 1-6C.*

A defragmenting or optimizing program restructures the data on the disk so that each file is stored in adjacent clusters. In small documents, you may not notice much difference, but if you use large files, the improvement can be dra-

*In order to run a program or access the data in a file, DOS always starts at the root directory. It looks at each file entry, which includes the subdirectories. As the number of files and subdirectories increases, the access time increases. For example, to find the document DATA1, DOS starts the search at the root directory, looking for the proper subdirectory. It searches through every file entry, including data files, programs, and subdirectories. The more there are, the longer it takes. Once the proper subdirectory is found, DOS switches to that subdirectory and starts looking for DATA1. When files are stored in random clusters, the disk becomes fragmented.

matic. Several commercial optimizers are available, including Norton Utilities Advanced Edition and PC Tools 6.0.

The defragmenting process of organizing, subdividing, and deleting files to maintain a logical and efficient directory structure needs to be done on a regular basis to be useful.

MEMORY BUFFERS

Another method of hard disk speedup uses RAM memory as a disk buffer. There are a variety of memory buffering solutions. One of the simplest is included with DOS. This is the BUFFERS statement used in CONFIG.SYS, FASTOPEN and RAM disks.

The memory buffering solution likely to produce the greatest speed increase is disk caching. DOS 4.0 or higher versions include a disk caching program called *smart drive* (SMARTDRV).

When the computer is powered on, or booted, DOS checks the file CONFIG.SYS for special device drivers and system settings. A typical CONFIG.SYS file might contain commands for a mouse, memory manager, or disk caching program. CONFIG.SYS should not be confused with AUTOEXEC.BAT, which is a special batch file containing DOS commands that are executed when the computer boots up.

One of the special settings in CONFIG.SYS is the BUFFERS statement. By specifying the number of buffers for DOS to reserve, reserved areas in memory are created where DOS can keep sectors that it has read from or written to the hard disk. These buffers are only one sector in size, but they provide gains in speed by reducing the number of hard disk accesses. The ideal number of buffers varies, but BUFFERS-20 works in many cases. Increasing the number of buffers reduces the amount of memory available for programs. Each buffer occupies about 512 bytes, so 20 buffers use over 10K of memory.

FASTOPEN

This is another technique included in DOS versions 3.0 and higher. It is invoked in the AUTOEXEC.BAT file. FASTOPEN speeds up the search through directories for a file. Each time a file is opened, DOS keeps its directory information and cluster numbers in memory. Subsequent accesses to that file do not require a search through the directories of the hard disk, since the file's location is in memory. DOS still needs to read the file's contents from the disk.

FASTOPEN is particularly useful when the same files and/or directories are used repeatedly. When FASTOPEN is set up, either in the AUTOEXEC.BAT file or from commands, there should be enough file entries so that previous entries are not replaced by new ones too quickly. If the number of entries is set at 2, DOS remembers only the last two files opened. A good number for file entries is 50.

FASTOPEN should not be used with disk optimizers. Disk optimizers can change the physical location of files on the disk, and they can conflict with the file locations that FASTOPEN retains in memory. FASTOPEN should not be used after optimizing without rebooting. A reboot should always be done after optimizing.

RAM DISKS

A RAM disk is a simulated disk located in RAM memory. When a RAM disk is created, a portion of memory is reserved as a disk, complete with its own drive letter, directories, and files. Files can be copied to or from a RAM disk much more quickly than those on physical disks. RAM disks are only temporary, since their contents are lost when power is turned off. Files on the RAM disk must be copied to the hard disk before shutting down the computer. A disk cache offers an alternative to RAM disks.

DISK CACHES

A disk cache retains information in memory, but that information already exists on the hard disk. If the power goes out, the contents of the disk cache are lost, but the files are still on the hard disk. The disk cache is similar to a buffer, but the cache is usually more efficient than DOS buffers. Disk caches act as smart buffers that anticipate the data most likely to be read next.

A disk cache is usually larger than DOS buffers. It requires a large memory, and it should not be used with only 640K of memory. The use of a cache reduces the amount of memory that programs can use. Any performance gain from the cache can be offset by a slow-down in program execution if there is not enough memory for both.

Disk caches like Super PC-Kwik or SmartDrive, the latter comes with MS-DOS 4.0, should be used with extended or expanded memory. Disk caches make use of the extra memory.

HARD DISK CONTROLLERS

Several models of hard disk/floppy controller are currently used in the 286 system. Functions and performance are the same; the only difference is the arrangement of the drive connectors, jumper plugs, and chips. The hard disk/floppy controller can interface two hard disk drives of different capacities, and up to two floppy disk drives in the 286-compatible computers. One type of drive interface is based upon the Seagate Technology Model ST506 hard disk drive.

Communication with the host is accomplished via the host I/O bus. The hard disk/floppy controllers usually have two separate and independent sections (hard disk and floppy) with their own data separator and write precompensation functions. The hard disk section also has a sector buffer. In addition, registers for control and status functions are included for each control section. Bus interface logic, I/O buffering, and timing are shared functions. Commands to one controller do not inhibit host communication to the other.

Controller types include run-length limited (RLL) modulation and modified frequency modulation (MFM). These are the techniques used to encode data on the hard disk. If a second hard disk is connected to the controller, it must use the same encoding technique.

There are also several standard hardware interfaces for disk controllers, including the small computer systems interface (SCSI). In the SCSI, specification devices on the bus can be either initiators or targets or both. Targets include disk drives, and the initiator is the computer. Operation can be asynchronous, where the target and initiator exchange a request/acknowledge handshake for each byte of data transferred. In the synchronous mode, the target and initiator send data bytes at fixed intervals without handshaking.

SPECIAL CONTROLLERS

Some controllers allow you to keep your current drives and add internal or external drives. An example of this board type is the Compaticard from Microsolutions. The Compaticard is a replacement for a standard XT-class floppy disk controller. It can support up to four floppy drives, including double-density floppy drives and high-capacity drives.

INSTALLATION

The kind of machine and the type of controller used will determine how to install the card. If the machine's existing floppy drive controller uses an expansion slot and serves no other function, it can easily be replaced with a high-capacity controller, since most high-capacity floppy drive controllers also work with double-density drives. If the existing floppy drive controller circuitry shares a card with other components, such as a hard drive controller, or if it is built onto the system board, you may not want to remove it. However, you may be able to disable the floppy drive controller with the proper jumper change. Generally, you can disable the existing controller or configure it as the secondary controller. Designating a controller as secondary is equivalent to disabling it, since most systems will ignore a secondary card if no drives are connected to it. Normally, you can designate an existing card as a secondary controller by setting a jumper or DIP switch on the system board. In some systems, the existing floppy drive controller cannot be disabled or made secondary. In a system with this limitation, you need to designate the additional card as the secondary controller. The documentation does not always refer to primary and secondary controllers but instead to the I/O port addresses they occupy. Set the primary drive controller to use addresses 03F0–03F7 hexadecimal (hex) and the secondary to use addresses 0370–0377 hex.

Many times it makes no difference whether a drive is connected to a primary or a secondary controller, but problems can occur with some disk utilities that interact directly with the drives via the I/O ports. For example, a program that performs drive diagnostics can try to verify that the hardware is functioning properly by communicating with the floppy drive controller using a machine language IN or OUT instruction. If the program tries to communicate with the primary controller and the drives are connected to a controller at the secondary address, the diagnostics test may not function.

The installation of a controller may require you to set a jumper or switch on the card to configure it as the primary or secondary floppy drive controller. Then, install the card in an expansion slot in your machine and connect the single ribbon cable connecting the drives of the controller.

A problem can sometimes occur between drive-control cables and the floppy drive controller. Some XT-class machines may have mixed gender connectors. The cable connector on your new card must be compatible with the one on your existing cable or you must replace the cable. After the installation of high-capacity floppy drives and a controller, you may have to install a device driver in the CONFIG.SYS file to help the card control the drives. Compaticard as well as some other boards require this type of software.

COMPACT DISKS FOR THE PC

A compact disk (CD) is like a phonograph record; there is one track all the way around. The track uses shallow holes known as *pits* to represent binary digits, either 0 or 1. The way the light deflected from the surface is interpreted by the scanning electronics as 1s or 0s.

Compact disk read-only memory (CD-ROM) works the same way as an audio compact disk does. Since there is one continuous track instead of the concentric tracks found on a hard disk, CD-ROM have a slower access time than hard disks. The typical access time is about 500 milliseconds, or approximately half a second. The access time for hard disks is about 18 to 28 milliseconds.

In a hard disk, the read/write function is performed by a mechanical head, which can crash against the disk platter, damaging the platter's surface. Head crashes are not possible with CD-ROMs, since a laser beam is used to read from the disk. Another difference is that CD-ROMs are read only. You cannot write information to the disk, as you can to a hard disk.

XTs and ATs are available with built-in CD-ROM drives. Because CD-ROM has the ability to pack so much information on a disk, its main strength lies in its ability to access large amounts of data. The average storage capacity of one CD-ROM disk is 550 Mb, which is equal to the information on a stack paper about 30 feet high. An entire encyclopedia can be put on one disk, and the information can be found with a few keystrokes. Some optical storage devices allow you to write to the disk.

OPTICAL OPTIONS

CD-ROMs are not the only type of optical, laser-based technology. CD-ROMs are good at disseminating large amounts of information, but there are optical mass-storage options as well.

WORM DRIVES

WORM (write-once/read-many) drives were the first optical technology to become widely available. With a WORM drive, you can write to it once, but you cannot change or erase what is already written. WORM drives are useful as an archiving medium. They are good for backing up large amounts of data that do not need to be changed or edited. WORM drives use a cartridge that can be taken out and stored. They are useful for storing scanned images, as well as records and data that need to be stored for a long time. Paper records can deteriorate after a few years. Data stored in a digital format will print out with the same quality in the future.

REWRITABLE OPTICAL DISKS

Rewritable optical disks can be written to as well as read from. Rewritables allow storage much like hard disks, but the technology is newer and more expensive. WORMs and rewritables are similar, but the media are different. The WORM medium cannot change once it is written too. Rewritable disks use a medium that can be modified a number of times for rewriting. Rewritable subsystems are being used for online storage of computer-aided design/manufacturing files with up to 5 Gb of data storage. Both WORMs and rewritables offer the advantage of portability. The disk cartridges can be removed from the drive. Both offer high-capacity storage, typically ranging from 600 Mb to several gigabytes.

Rewritables have a faster access time than WORM drives. A rewritable averages 65 milliseconds, while a WORM averages about 90 milliseconds.

For applications that do not require rewritability, write-once optical technology has a cost advantage over rewritable optical technology. Rewritable drives cost about twice as much, on a per-megabyte basis, as write-once drives.

Standards for rewritable optical drives are more advanced than those for write-once drives. If cartridge interchangeability is a requirement, ISO standards can be used for evaluating rewritable optical drives.

Write-once drives are not the only alternative to rewritable optical drives. For some applications, rewritable optical drives compete with traditional hard drives and even tape drives.

TAPE DRIVES

Tape drives are another option for data storage and retrieval. There are several tape technologies, including half-inch and quarter-inch cartridge drives, 4-mm digital audio tape (DAT), and 8-mm helical-scan tape drives. All tape technologies suffer from low access speeds:

- Quarter-inch cartridge: 30–40 seconds.
- Half-inch cartridge: 24–44 seconds.
- 4-mm DAT: 20 seconds.

Transfer rates for rewritable optical drives are also faster than those for most tape cartridge drives.

REWRITABLE OPTICAL MEDIA

There are three methods of producing rewritable optical media: magneto-optic, phase-change, and dye polymer. Most drives use magneto-optic technology. Phase-change media is the second generation of rewritables, with dye polymer media still in development.

Magneto-optic technology uses a magneto-optic recording layer consisting of an amorphous alloy of rare earth elements. To record data, the laser beam strikes a spot on the sensitive layer and causes the spot's magnetic field to change polarity. A lower power beam of the same laser then reads the information as binary codes. The rewriting sequence uses three steps: erase, write new data to the disk, and verify the new data. The main advantage of magneto-optic media is long life. Magneto-optic media have been operated at 7 million write-erase cycles without degradation. In contrast to the phase-change and dye polymer methods, data written using magneto-optical technology are permanent unless intentionally erased, and there is no physical change to the medium. Most magneto-optic media manufacturers claim a medium life of 10 years.

Phase-change technology provides another approach to rewritable optical media. In the phase-change technique, a laser beam changes the structure of a light-sensitive layer from a stable amorphous state to a crystalline state. The surface reflectivity changes from low to high, and the laser beam reads this change. To erase data, the laser beam changes the light-sensitive layer back to the stable amorphous state. Phase-change media last about as long as magneto-optic media. But unlike magneto-optic media, phase-change media permit direct overwriting without having to erase before rewriting. Phase-change media are less expensive than magneto-optic media but more expensive than dye-polymer media.

In the dye polymer process, the laser beam's heat changes the topology of the disk surface. The laser beam creates a slight ripple or rise on the surface of the sensitive layer. This defect scatters the light, resulting in less reflection than that of the pure surface. To erase data, the beam reduces the defect. This is the least expensive method of producing rewritable optical media. However, media life is relatively low at a few hundred thousand write-erase cycles. Manufacturers have been concentrating on magneto-optic technology.

COPROCESSORS

A math coprocessor is also called a *numerical processing unit*, a *floating-point unit*, or a *math chip*. It is designed to do floating-point arithmetic. The main processor of the computer is a chip called the *central processing unit (CPU)*. It is a member of the Intel family of microprocessor chips: 8086, 8088, 80286, 80386, or 80486. All of these chips operate with the binary digits called *bits*. The CPU moves these bits around in blocks of eight called *bytes* or in multiple bytes called *words*.

A single byte can represent a keyboard or screen character, but it can also be used to represent an integer or a whole number between 0 and 255. A 16-bit word can represent an integer between 0 and 65,535 or between −32,768 and 32,767. The 80386 and 80486 CPUs can work with blocks of 32 bits, called *double words*, which can hold integers between 0 and about 4 billion.

The CPU can operate quite fast when it needs to perform simple arithmetic operations such as addition, subtraction, multiplication, and division on bytes and words. Suppose that you want to multiply 10 by 30; the CPU can quickly compute the answer because both numbers can be represented as bytes and the result can be represented as a word.

Decimal points are handled by using additional instructions employing more internal instructions that the CPU sees as sequences of bytes. These instructions are similar to the long multiplication and division techniques we use for hand calculations. Floating-point numbers allow the decimal point to occur at any position. The math chip is designed to handle only floating-point operations. It does not replace the CPU but works in parallel to it; this is why it is called a math *coprocessor*.

MATH PROCESSING

Almost every MS-DOS computer contains a socket to hold a math coprocessor. The exceptions are a few laptops and some early Tandy 1000 models. The 80486 microprocessor has floating-point arithmetic built in.

The other chips use a separate math chip as an option. During many operations, the math chip is not used. But when a floating-point operation is needed, the coprocessor takes over, doing the arithmetic with much greater speed than the CPU routines used for floating-point math. In an AT clone with an 8-MHz 80287 math chip, floating-point multiplication can take as little as 12 microseconds (12 millionths of a second). To perform the same function without a coprocessor could take up to 100 times longer. This is still only 1.2 milliseconds (thousandths of a second) and is not a problem if you multiply numbers only occasionally. But if you need to perform many multiplications, the time quickly builds up.

In many scientific, engineering, and statistical applications, a coprocessor is required. Graphics painting programs show no improvement, but many drawing programs are speeded up. Different versions of the same program can respond differently to the presence of a math chip. A program written in GW-Basic, for example, does not benefit from a coprocessor. If you transfer this program to Quick Basic, it will use the coprocessor as much as possible. Most of the programs written with the latest compilers that perform floating-point calculations will use a coprocessor if there is one. They will also run, but more slowly, without a math chip. Some products, such as AutoCAD Release 10, won't run unless a coprocessor is available. A coprocessor will never slow down any application.

The Intel math coprocessors follow a simple numbering system. The last digit of the microprocessor is changed to a 7. An XT-class 8086 or 8088 computer uses an 8087 coprocessor, an 80286 machine uses an 80287, and an 80386 uses an 80387. For battery-powered 286 laptops, there is a low-power 80C287 coprocessor, and for 80386SX systems there is the 80387SX math chip.

Coprocessors, like microprocessors, are rated according to the fastest speed at which they will operate. If you install one that is too slow, it will perform incorrectly and may damage itself or other components in the computer. If you install one that is too fast, you will usually be wasting money, since it is the computer, not the chip, that determines how fast a coprocessor runs. There are exceptions; some 286 systems allow a math chip that is slower than the CPU, and some 386 systems allow a math chip that is faster than the CPU.

Some coprocessors, like those made by Integrated Information Technology (IIT), accept instructions in assembly language for the matrix operations used in many engineering and scientific applications.

INSTALLATION

A coprocessor can be installed without problems if you observe a few precautions. Leave the chip in its static-free package until you are ready to put it in the computer. If possible, work in a room without a rug. After the machine is open, and again each time you move around the room, touch the metal power supply box at the rear of the computer to ground yourself before touching anything inside the PC.

A socket is usually available for the addition of a compatible math coprocessor chip. The 286 machine has a socket for an 80287 math coprocessor chip. This chip is specialized to do floating-point arithmetic extremely fast compared to the CPU. If the computer is used for a large number of calculations, it should have an 80287.

Locate the 80287 chip socket on the system board. The 80287 chip has a notch in one end. If you look closely at the system board under the 80287 socket, you will see a printed white outline of the notched chip in the correct orientation. The 80287 must not be installed backward, or else it will be ruined when the system boots up.

There are 20 pins along each side of the 80287. These pins must be perfectly aligned with the 40 holes in the 80287 socket. When you install the 80287, check both ends carefully to be sure that no pins are hanging over the ends of the socket.

Most 80287 coprocessors are sold as 40-pin DIP chips, which means that they are rectangular, with a row of 20 pins along each long edge. However, some 80286 portables have a socket for a square CMOS coprocessor. In order to find the coprocessor socket, you may have to unplug and remove expansion cards or hard or floppy drives, or move cables out of the way.

Once you have found the socket, make sure you know which way the math chip is supposed to face. A diagram is often printed on the motherboard to show the correct orientation.

For the actual installation, ground yourself again, remove the coprocessor from its protective wrapping, and slowly but firmly press it into its socket. Make sure that it faces in the correct direction and that all the pins go into the socket. On some computers there is a switch on the motherboard that must be set to let the CPU know that a coprocessor is present. Some 286 and 386 machines require a disk or ROM-based setup program to run to tell the CPU that a math chip has been installed. There are also diagnostic utilities that will test for and report the presence of a coprocessor.

386 COPROCESSORS

The first motherboards for 386 machines had sockets for 80287 math chips. The more recent model 386 computers have a socket for a 387 coprocessor. The 387DX is used for standard 386 machines and the 387SX for SX machines. The 387 chips are square. Some 20- and 25-MHz 386 motherboards allow a 33-MHz 387 chip.

Other 387 chips include IIT's 3C87 and special chips made for scientific and engineering calculations, like the 387-compatible Fasmath coprocessor made by Cyrix and the Weitek 31167, or Abacus, chip. Some computers have Weitek sockets and a few applications can use a Weitek chip, but the chip is not socket- or software-compatible with the 387 machine.

2

Automation and Control Applications

This chapter illustrates typical measurement and control applications that can be handled using an industrial PC. In many cases, the lower-cost 286 machines can be used; other applications may require a computer that uses the 80386 or 80486 microprocessors. The important variables are considered for these applications.

Probably the most common application involves some method of data acquisition. Another important area involves using the PC as a development tool. Direct digital control has always been considered the ultimate control technique. Environmental concerns are addressed, as well as standardization and training needs.

We will trace the early PC applications, followed by PC data acquisition and the use of PCs as development tools. Direct control has always been considered the ultimate control technique. The present and future status of PCs in this important area is discussed.

In many industrial applications, the environment is harsh and this becomes an important concern. An important area is standardization in computers. PC training is also important, and it is discussed in this chapter. The PC can be valuable in control system integration. This is explained, as well as the usual issues regarding inhouse software development. We will cover programming languages and packaged software.

We will also tackle the problem of how to develop PC automation projects. We will show how to determine a development strategy and how to establish the system architecture. First, we will show how to identify the process and match it to the plant's requirements. Next, we will consider the need to identify I/O and interface requirements and then determine the best architecture for the application. This is followed by the system design and specification step, which includes control panel configuration, communications, and considerations such as start-up support.

The data acquisition problem requires a knowledge of A/D conversion and signal conditioning. You will learn how to use alarms, when to use networking, and batch control's special considerations. Training and maintenance problems will also be explained.

Statistical control is a powerful technique and is of great benefit in many applications. We will explain how this technique works and examine the use of statistical distributions, control charting, $\bar{X}$-R charts, CUSUM charts, and how to use statistical data.

The use of programmable logic controllers with PCs is another powerful technique. We will show how to combine both functions to the best advantage and how to perform diagnostics.

Industrial applications for the PC have grown well beyond their initial beginnings in office and business automation. These initial applications centered on the areas of word processing and spreadsheet analysis. Today's industrial applications include engineering computation and analysis, production, quality control, and computer-aided design.

MEASUREMENT AND CONTROL

One of the most promising application areas is measurement and control. This includes the use of PCs in the laboratory, collecting data and monitoring the status of processes and equipment. The introduction of data acquisition software for PCs has made it possible to apply microcomputers in many measurement and control applications.

Because of the digital interface of PC buses, analog data acquisition requires the use of an analog-to-digital converter to transform measurement data into a PC bus-compatible format. Once the data have been acquired, they can be readily transformed, analyzed, stored, or displayed.

A major application of PCs in industrial plants has been in configuring control systems using computer-aided design (CAD) techniques. Another major application has been data acquisition, where PCs are used either as a front end to programmable logic controllers (PLCs) or process controllers, or with direct connections to a process through I/O boards or I/O processors.

Most PC systems have not had the raw speed or I/O channels to track and control complex processes in real time, so they have been used primarily for relatively simple processes. As the definition of PCs changes to include powerful 386 and 486 systems, these computers can be expected to be used in many types of more complex monitoring and control applications.

INDUSTRIAL CONFIGURATION

PCs are available in four different configurations for industrial applications:

1. Tabletop or desktop.
2. Portable, battery-powered.
3. Rack-mounted.
4. NEMA enclosure per the National Electrical Manufacturer's Association.

A typical configuration for control devices in U.S. plants is rack mounting. NEMA enclosures are generally required by plants with corrosive environments.

Most application software has been developed for the IBM-PC family because of its greater market share and availability in industrial versions designed for plant environments. These computers are often provided with special features such as removable filters and sealed cases.

Industrial versions of the IBM-PC family are available from several suppliers in different forms. IBM's Model 7552 Industrial Computer uses a passive backplane and battery backup for protection against transients. Other industrial PC versions are available from General Electric, Hewlett Packard, and AT&T.

THE 80386 AND 80486 MICROPROCESSORS

A major development in PC growth was Intel's development of the 32-bit microprocessor, the 80386. Because of its greater computing power and speed, the 80386 has the potential capability of operating in many real-time and multitasking applications. The full potential of the 80386 microprocessor may not be realized under current versions of DOS, but the higher operating speeds that some systems offer can increase throughput considerably. DOS-based programs can run about twice as fast as the highest speed IBM-AT. This means that many processor-intensive chores, such as number and data manipulations, will take about half as much time. Additional hardware and software advances can potentially double this performance again.

EARLY PC APPLICATIONS

During the late 1970s, a number of companies were marketing the first PCs. These relatively low-powered units from Apple, Tandy, and Commodore were directed primarily at the home market, with some applications for small business in the areas of word processing and accounting. During this period, the use of PCs in measurement and control was not a consideration. The same microprocessors used as the CPUs in these PCs were found in many dedicated control and acquisition hardware products.

Apple started a trend in applications like data acquisition when it began adding extra slots on the computer chassis for expanding the total memory and adding peripheral devices. Several companies then began to offer data acquisition boards for simple control and measurement applications.

The IBM-PC, introduced in 1981, also featured expansion slots that could be used for control interfaces. During this same period, several companies began offering add-on boards for functions such as analog I/O and IEE-488 interfaces. Most of these initial board products were designed for Apple PCs. Today, a number of companies provide add-on boards for control interfaces. Most of these are designed for the IBM-PC family, but there are also products for other small computers.

PC DATA ACQUISITION

There are many different ways to implement automated data acquisition systems with PCs. This area has received support from many companies, including Hewlett Packard, Fluke, and Burr-Brown. Most vendors have adopted the IBM-PC format for their analog interfaces, mostly due to the acceptance of the IBM-PC family in scientific and engineering applications.

There is a growing library of IBM-PC-based data acquisition products. The newer data acquisition hardware and software products offer higher performance, lower cost, and greater ease of use compared to earlier offerings. There have been major improvements in speed, the number of channels, resolution, ruggedness, and processing intelligence. Many of these systems compete with specialized test, measurement, and control systems.

Declining hardware costs and more powerful software are providing many new applications for PCs, including replacements for oscilloscopes, data loggers, and chart recorders. These stand-alone units have a negative image because of their inflexibility, while PCs can perform a number of tasks and never be idle.

PCS AS DEVELOPMENT TOOLS

One major application of PCs is in engineering development, which includes the development of control and measurement applications and programs for performing calculations in support of engineering. PCs are an important part of constructing fully integrated data-handling systems. These systems often parallel the physical processes being controlled, with the PC acting as one of the fundamental building blocks in the control chain.

PCs can be used as an operator interface for PLC applications. They can be used to perform simple functions such as changing set points, starting and stopping motors, and controlling valves. A PC with the proper graphics software can be used as a replacement for control and graphic panels as well as annunciators.

PCs have been used with PLCs for such applications as automatic crane control and oven control. The ruggedness and quick scan time of the PLC are combined with the data storage, manipulation, and reporting capabilities of the PC.

The PC has the ability to perform precalculations on data from instruments and controllers in the field. In the chemical industry, PCs are used to control gas chromatograms using data from remote instruments. PCs have also been used to digitize pressure waves in the testing of explosives.

DIRECT CONTROL CONSIDERATIONS

PCs have been used in such applications as controlling furnaces in petrochemical operations. Applications are divided between monitoring/development and direct control. Only half of those applications that used PC systems for direct control employ them for any type of time-critical operation.

In the past, most PCs were too slow for time-critical applications. PC applications for direct control were in the laboratory or the pilot plant, where the flexibility of the PC was an important consideration. This flexibility is particularly useful in those parts of processes which are subject to frequent major adjustments, such as start-up environment.

Dedicated electronics are used for those control applications where users do not wish to write software to get the level of control needed. There may also be special speed or reliability requirements that an all-purpose machine cannot provide. Dedicated products may exist that readily meet these requirements with an installed base of users.

ENVIRONMENTAL CONCERNS

PCs were designed for an office environment. Floppy disks and their drives are susceptible to damage from dust or other particulates. PCs normally do not have any means for program loading and maintenance other than floppy disks. A diskless PC can be loaded from a network or modem connection.

Environmental problems include fumes that can destroy floppy disk heads over time. Other electronic devices may also be degraded, but disk drives are usually the major concern. Moisture, dust, and corrosive chemicals are the major environmental concerns, and disk drives are the most sensitive component.

Electromagnetic interference (EMI) can also be a problem. Even though most electronic components are shielded, the higher speeds of the newer PC systems makes them more susceptible to the more penetrating high frequencies.

Shock and vibration can also damage PCs, although these problems are usually easier to solve. In some applications, disk drives can be replaced with magnetic bubble memory packs. These provide offline storage, although they are very expensive compared to magnetic disks. If a network is used, data can be sent over the network for storage.

A reverse cooling system is used in some industrial PCs. In this type of cooling system, air is drawn in through a filter and returned out through the disk openings. The opposite path is used in most office machines. The use of a filter reduces exposure of the internal components to corrosive materials and particulates.

Another solution is isolation from harmful environments. This requires installation in control rooms, enclosured offices, or a location away from the processes the PC is intended to monitor and/or control. This involves a tradeoff between getting far enough away from the harmful environment to keep the system operating without excessive maintenance and being close enough for the system to get the data it needs in a timely manner.

Cathode ray tube (CRT) screens and mechanical keyboards are also susceptible to damage in the industrial environment. Enclosed CRTs are available for washdown and explosive hazards areas. The standard keyboard can be covered with a flexible plastic membrane for protection. The plastic cover remains on the keyboard during use. However, it can change the key pressure and may need to be replaced periodically.

STANDARDIZATION IN COMPUTERS

Standardization is critical in many applications. Without standards, there tends to be greater flexibility, but this can also result in extra costs. The lack of standards can result in a number of problems. Most companies attempt to establish some form of standards for computer hardware and software. Although many PCs are available, most standardize on either IBM or an IBM-compatible model. This may not always provide the best PC available, but it does reduce problems due to the proliferation of brands and models. But *compatible* is not the same as *identical*. Components may be functionally compatible but not mechanically compatible. So, parts may not be exchanged among different computers.

Software compatibility was a problem in many early versions of the IBM-PC. Some software did not run on specific machines, and the data were not transportable. Most of the problems of software compatibility have been minimized by the use of more recent DOS and BIOS versions.

In some cases, users already have a variety of machines, and they prefer to apply the extra effort needed to make different brands and models work together. Some of these applications may have been using a particular machine for a particular function in a satisfactory manner, and this may be judged to be more important than the machine's ability to work well or easily with other machines at that site. Another difficulty is parts availability. As these machines age, not only do they require more maintenance, but replacement parts become more difficult to obtain.

PC TRAINING

In most applications, the use of PCs simplifies training requirements. It has been found that PC utilization requires little or no increase in training. Most training requirements involve increased programming skills and the need to understand the process better. PCs are actually excellent training devices with the proper software, and after users become familiar with the basic techniques of using computers, they can quickly adapt to revised procedures if these are properly presented.

THE PC IN CONTROL SYSTEM INTEGRATION

In many application areas, the trend is toward fully integrated, large-scale systems. While this represents the ultimate system to many, and while the value of a fully integrated monitoring and control system is high, integration is not often easy, nor is it always necessary. Users must take more time to understand automated processes in detail. More effort by people with broader knowledge is needed to get useful results from integrated systems.

Data sharing is basic to integrated systems, but problems facing PCs and other types of computers in this area can be formidable. Data structure and communications standards are two forms of the same problem.

The ability of PCs to collect data and automate inexpensively is well known, and their ability to network with each other certainly aids the integration process. The major problem is to network PCs efficiently with other, often diverse computer systems. Although many new products are appearing to correct this situation, most of them are not supported as well as the computers themselves.

Thus a major problem for integration is efficient communication between automation layers. The proper communications hardware can improve access to data collected by individual PCs, but software is critical to accessing this information. If it does not work in a manner that facilitates connectivity, then the automation system may not meet specifications. Standard data structures and software architectures are needed which can be augmented without complete rewriting.

Complete integration involves a number of complex issues related to the interaction of many diverse operations. PCs tend to generate a great deal of information that may need to be screened.

Communications, data storage, and applications programming can be both the strengths and the weaknesses of the PC. They are strengths in that the PC can perform in these areas, but their weakness is that the effort required to get PCs to perform certain functions can be prohibitive.

INHOUSE SOFTWARE DEVELOPMENT

Software development costs may exceed hardware costs for some types of control and monitoring applications. Software development costs in some automation applications, such as discrete manufacturing, can exceed estimated project costs by a factor of 10. In other applications, costs may not be a significant factor. Much software is available for control and monitoring applications, and in many cases it can be adapted or modified at a reasonable cost. This can be done in house for simple changes; major modifications can be done by the vendor.

Lead times for software development are another problem, since it can be difficult to estimate the time required to complete a particular program change. A wide variety of software development/cost difficulties can occur.

Often it is wise to avoid inhouse software development altogether. Customized programs can be contracted from outside, or packaged software can be purchased for many control and monitoring applications. There may be some loss in flexibility, but there may also be some gains, such as the ability to keep the packages properly maintained and updated.

Software development is needed for those applications where improved flexibility and the ability to optimize performance for each application are necessary. This is often the case in research or laboratory projects where requirements change frequently.

PROGRAMMING LANGUAGES

Software development costs often depend on the language. Standardizing on one or two programming languages is important. The most popular languages for control applications are BASIC and FORTRAN. Once useful code is written and running in a particular language, a trend has been established for more code to be written in that same language.

FORTRAN and BASIC are not the fastest, or the most powerful, or the most elegant languages. But they have been in use for some time, and much control code has been written in these languages. Many users are familiar with FORTRAN and BASIC, and they are comfortable writing programs in them.

BASIC, in its many versions, is by far the most widely used language. It meets the needs of many applications, but it does have some disadvantages. The language was developed before the use of structured programming, so it does not handle repetitive loops as well as other languages. Another problem with BASIC for control or monitoring applications is its slow speed. It is slow even in the compiled version (usually BASIC is interpreted) and even when the BASIC programs are used to call assembly language subroutines. However, slow execution is compensated for by fast I/O calls. Another drawback is that complex programs are difficult to develop with BASIC, and various versions are required for different machines. This hinders the development of complex programs on larger machines. Nevertheless, the language is still a good choice due to the number of drivers and compatible I/O systems available for it, which exceed similar offerings for other languages.

FORTRAN was originally designed for complex calculations; later, other capabilities were added, and it became a general-purpose language. The more recent ISA extensions of FORTRAN provide the language with real-time capabilities. But there is still the problem of portability. Programs written in FORTRAN for use on one operating system often must be rewritten if they are to run on another system.

Among the other languages, C and Pascal allow considerable improvements in programmer productivity, especially for larger programs. Some versions can also handle real-time multitasking applications. C can be difficult to learn. However, it has many useful features and is truly portable.

PACKAGED SOFTWARE

PC software continues to grow in many areas of control and measurement. These packages provide relief from the high cost of inhouse software development. Spreadsheets are the most widely used packages. They are easy to set up and provide quick results for a number of data analysis applications. Data base packages are widely used for everything from parts lists and maintenance inventory control to storage and the manipulation of data. They are more difficult to set up than spreadsheets and are vulnerable to crashes. But once they are in place, they are easy to use and provide valuable information due to their ability to cross-reference related data items.

Graphics packages for the PC are also widely used. The graphics capabilities of the smaller machines tend to be limited due to their slow processing speed and limited memory. Most 286 machines display data in a usable format. Charts and graphs are easy to create for assisting in such tasks as the production of flow diagrams. Word processing software is used wherever text editing is done, as in specification creation and revision. Statistical packages, as well as project management and scheduling packages, are also widely used.

Custom programming may be needed in some applications, but there are many ready-made packages that are easily configurable to specific control applications.

Future trends are likely to focus on high-level application languages for such applications as ladder logic and flowcharting. These languages will function in a CAD-like programming environment as nonprocedural languages.

Ladder logic was in use before digital computers were available. A nonprocedural language allows the user to list the desired result rather than the desired procedure. When these languages are used in conjunction with artificial intelligence tools, they should be able to provide significant reductions in the time required to write code.

PROBLEM: HOW TO DEVELOP PC AUTOMATION PROJECTS

The use of PCs can provide an ideal framework for industrial control and automation. Industrial applications of PCs offer the benefits of plantwide efficiency, improved quality, and increased productivity.

The first step in achieving these benefits is to automate the process. Whether starting a new process or modernizing an existing plant, process automation at some point involves the integration of multivendor control components.

A multivendor system may be composed of a combination of off-the-shelf controls, automated equipment packages, or custom control systems. There are a number of critical decision areas in the design and start-up phases during the implementation of a multivendor automated production process. These include

1. Selecting the development strategy.
2. Establishing system architecture.
3. System design and specification.
4. Communications design and specification.
5. Start-up support and training.

These essential areas are addressed in the typical development flowchart shown in Figure 2-1.

PROBLEM: DETERMINING THE DEVELOPMENT STRATEGY

Successful system development and integration must begin with a well-developed strategy. Development and integration strategies primarily focus on four important questions:

1. Who needs the information?
2. What types of information do they need?
3. How is the information delivered?
4. What are the operator-machine interface requirements to ensure efficient use of the information presented?

These questions must be answered for each level in the corporate structure, beginning at the plant floor and continuing through supervisory, engineering, accounting, and management levels. These questions must be *fully* answered to establish control system design criteria.

PROBLEM: ESTABLISHING THE SYSTEM ARCHITECTURE

A logical procedure for establishing system architecture is as follows:

1. Identify the process.
2. Identify company or plant requirements.
3. Identify I/O and interface requirements.
4. Determine the best architecture for the application.

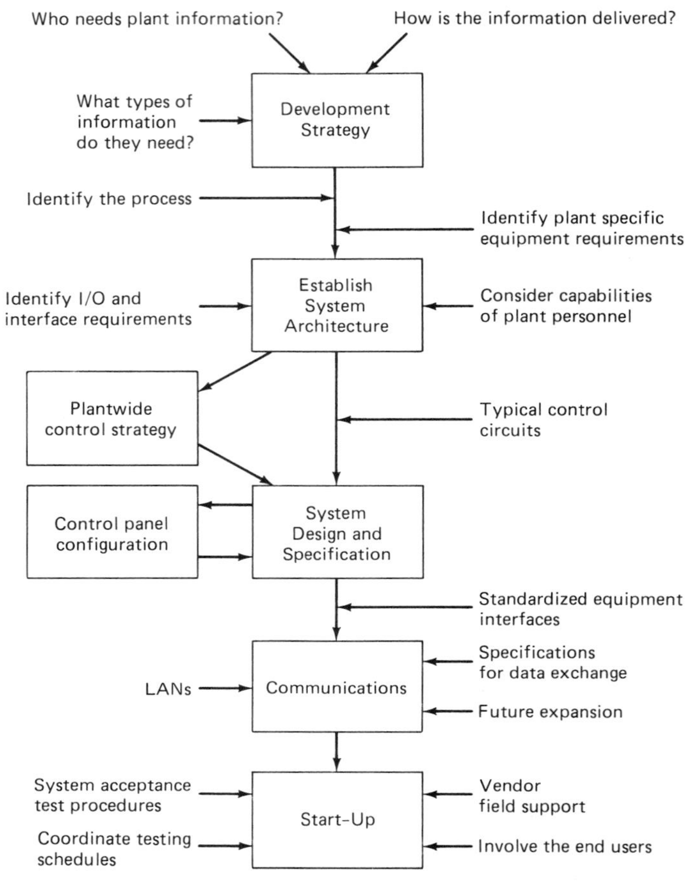

Figure 2-1 Typical development flowchart.

Most problems that occur during the process of establishing system architecture can be traced to one of these four steps.

Step 1: Identify the process

Identifying the process is the first step in engineering a control system. The engineer needs to know what process actions are discrete, continuous, or a combination of both. Often the complete process has not been established when the control system engineer begins design work. Therefore, the engineer needs to keep in mind that the process is evolving, and the control system should be flexible enough to accommodate changes as the process is further defined.

It is important for the people who will be using the plant information to be involved with the project. Plant operators have valuable process experience to offer, and gathering their input at this stage can be beneficial in achieving final system acceptance. Many times production requirements conflict with engineering needs. Therefore, meetings with production supervisors, plant engineers, plant maintenance personnel, management, and other key employees should be scheduled early in the project. System questions that need to be asked at this stage include the following:

1. How is the process to be defined?
2. Are centralized control stations required?
3. What part of the process will be controlled from them?
4. How automated should the system controls be?
5. What types of control panels are desired (hardwired, discrete devices or a computer-based system with CRT graphics)?
6. What higher-level functions, such as production control, scheduling, recipe management, inventory, and data reporting, are required?

To answer these questions, develop a system process and instrumentation diagram (P&ID). The P&ID will show the schematic flow of product through all manual and automatic physical devices. Schematic location of key instruments and control points, including functional interconnection of instrumentation, should also be shown. Use control loop diagrams in conjunction with the P&ID to show additional interconnection details for sensors and controls. A well-developed P&ID will provide a single, controlled source of information and establish a basis for detailed designs, installation drawings, and programming.

Consider using a computer-modeled process simulation to aid in identifying system limitations and to optimize system performance. The ideas of all disciplines involved should be considered. Keep in mind, however, that the project budget may not accommodate all their requests. It is necessary to sort and blend all ideas to produce a design that is both practical and cost effective.

Step 2: Identify the plant requirements

Important issues to consider when dealing with corporate and plant personnel are the following:

1. What are the capabilities and responsibilities of the user (operator)?
2. What is the ability of plant personnel to maintain a new or higher technology system?
3. Will resources be committed to attain necessary levels of competence?

If the commitment to train engineering, maintenance, and operations personnel will not or cannot be made, an advanced system may not be appropriate for the project.

Any corporate or plant-standardized equipment must be identified. Standard equipment helps reduce spare parts inventory and reduces the costs of maintenance and operator training. It is also necessary to determine if the existing equipment technology is adequate and appropriate for the application. If necessary, current-technology equipment must be identified to replace outdated equipment.

These issues must be an integral part of the decision process for establishing the system architecture.

Step 3: Identify the I/O and interface requirements

Identify I/O from existing systems that will be required by the new control system. Determine what hardware and software requirements are necessary to interface this I/O with the new system. To aid in defining new I/O, establish standard I/O requirements for motors, valves, and other field devices used in the process.

Any devices that will need intelligent interfaces should be identified. Examples of these devices include direct digital controllers (DDCs), single-loop digital controllers (SLDCs), and identification scanners. Identification of intelligent interfaces is necessary at this stage of design to determine the types of communication used and how to integrate them into the system.

When interface intelligent I/O is needed, it is desirable to use previously developed software drivers. When a predeveloped driver is not available, it may be necessary to have one developed. Estimating development costs can be difficult, and it is necessary to schedule enough time to develop and test the driver.

Step 4: Determine the best architecture for the application

After all of the needed information has been obtained and evaluated, it is time to determine what architecture meets the project requirements. Decisions can be made on the number and function of (PLCs), (DDCs), control stations, data acquisition systems (DASs), and higher-level supervisory and information processing computers. Always consider the plantwide control strategy before making any decisions.

At this stage, a system block diagram (Figure 2-2) can be developed. This diagram is essential to identify the control and communications architecture. Show each control device and its function on the diagram. Peer-to-peer communications, gateways, LANs, and any other required communications networks can then be added.

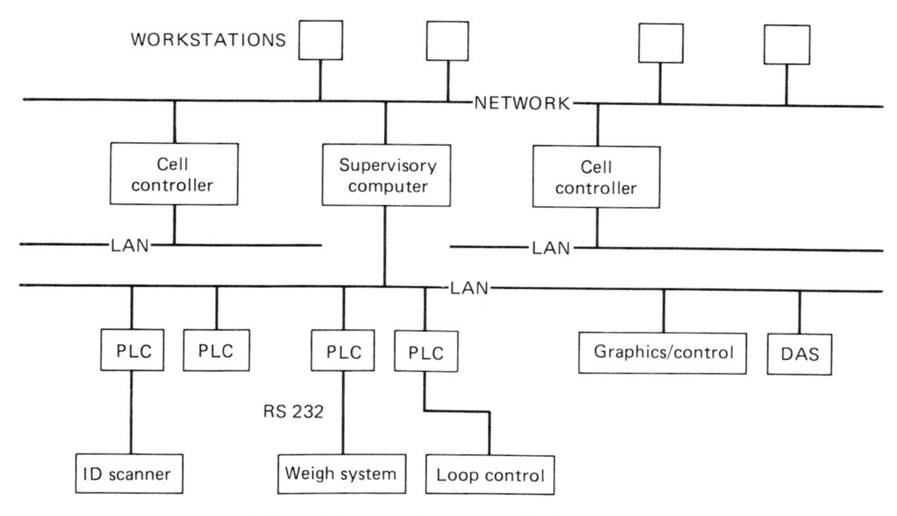

Figure 2-2 Typical system block diagram.

Try to maintain an open systems architecture. This allows a multivendor system to be integrated using industry-standard LANs and provides for future expansion. Consider the possibility of future expansion, and identify these areas on the block diagram.

Step 5: System design and specification

Establish equipment standards for the system. If the process requires more than one subsystem vendor, the type or brand name of major equipment should be specified. Consider the type of equipment used and determine if programming support is necessary. Often the type of equipment is determined by the plant-standardized equipment, helping to reduce spare parts inventory and costs associated with operator and maintenance training.

In addition to equipment standardization, it is advisable to establish typical control circuits for common devices such as motors and valves. Typical circuits aid in identifying system I/O during preliminary design and costing phases, and provide plantwide consistency for future training and maintenance.

Step 6: Determine control panel configurations

Rules for control panel layout, configuration, and display philosophy are needed to ensure plantwide consistency. Rules should be established for color codes, graphics screen or panel layout, and the type and brand of control devices. Color codes include color definition of pilot lights, static and dynamic equipment graphics, and alphanumeric variables and messages. Keep graphics displays simple, and limit the number of displays and colors used for effectiveness. Specifying standardized equipment is recommended.

Determine who will use the system, and define their operating requirements and the operating environment. For instance, the environmental conditions of the control system located on the plant floor will be more severe than those of most office equipment. Consider that the process operator may not be familiar with or trained in using a QWERTY-type keyboard; an alternative input device may be more appropriate.

Following these recommendations will help reduce training time per system; the system will be simpler to operate; and the user will perform more efficiently by having easier access to required information. Subsystem vendors must comply with these specifications to ensure a common philosophy for operator-machine interfaces throughout the plant.

COMMUNICATIONS

Determine peer-to-peer communications

Research the performance specification of the communications link or LAN considered for use to determine the most efficient means of applying the network. A thorough knowledge of the network scanning method and the data transfer capacity is essential in optimizing system performance.

Consider grouping communications data in a common area of the control processor's memory map to maximize the quantity of data that can be transferred in one processor scan. When implementing larger-scale systems, consider dividing the process among two or more networks to help minimize communications delays, to provide for system fail-safe operation, and to enhance overall system performance.

Determine the interface specifications

Interface specifications should be written to govern the passing of data between control stations. The following suggestions may be useful:

1. Establish a master station which will initiate supervisor control functions. This provides a single level for maintaining process communications and systems data exchange. For example, if one station initiates a request for a receive download, it should initiate a command for sending the completed batch data.
2. Use validity checks and data buffering. Operators will gain confidence in the system if the information is accurate and repeatable. False signals will

mislead operators and create problems. Buffer the input of dynamic data to prevent the use of erroneous information.
3. Check for valid input ranges of the data. When applying PLCs, use specific words or bits solely for the communication of information. This will permit filtering of the data with additional logic as required before using the data within the program.
4. Establish a common data format for all stations. The programming required to change data formats for one device can quickly become unwieldy and time-consuming. It is also important to include reserve memory for future additions.
5. Provide a preliminary interface specification with the system specification. It should be as comprehensive as possible, based on what is known and what is expected.

START-UP SUPPORT

Start the system by testing from the lowest control levels to the highest. Preparing a system acceptance test plan and a detailed test procedure is recommended and will help document the testing progress. The following steps are recommended for system checkout:

1. Test hard-wired control circuits (motor and valve controls) first.
2. Verify I/O, including soft control stations.
3. Check all automatic sequencing.
4. Validate the software. This involves testing interlocks and communications (proving and improving software is an ongoing process and will continue well after start-up).

After the initial testing of each system is satisfactory, testing of the interprocess communications and data handling can proceed. The use of a logical sequence of test steps is recommended. For example, do not try to check automatic sequences before I/O has been verified or try to test communications before each system has been proven. This only complicates the troubleshooting process, requiring additional time and costs to find and correct problems.

The availability of existing production equipment also needs to be considered. When integrating existing production lines, it is essential to schedule tests that will have minimal impact on production output or quality.

The basic rules for the design of an automated system are as follows:

1. Research and define the development and integration strategy.
2. Involve the end users in the development of control system requirements.
3. Identify the I/O and interfacing requirements.

4. Develop system P&ID programs.
5. Consider the plant's ability to operate and maintain the system.
6. Try to maintain an open systems architecture.
7. Establish system standards for factors like hardware, control panel configuration, and data exchange techniques early in the program.
8. Use data buffering and validity checks for software-based systems.
9. Test systems from the lowest levels of control to the highest.
10. Involve plant operations and maintenance personnel in the start-up of process systems.

DATA ACQUISITION

Collecting, manipulating, and interpreting data are critical in many automation and control applications, since measurements and calculations are the foundation upon which results are predicted and improved. The available data collection methods are numerous and depend on the processes they measure.

Planning a data acquisition strategy involves a number of choices. There are stand-alone, communication-based, smart, plug-in, single-channel, and multichannel data loggers, in addition to PC-based systems. Data collection technology evolved from the early strip chart recorders. Today's microprocessor-based units provide good I/O capacity and are often used with PCs at the highest level of monitoring, collection, and analysis.

DATA LOGGERS

A data logger is a device which collects analog signals produced by temperature sensors, pressure transducers, flowmeters, and other instruments and converts them into digital form. What the logger does with the data from that point on depends on the particular application.

Data loggers are differentiated according to whether they record and store data in memory for later analysis or on paper for immediate analysis. Hard copy loggers provide the advantage of a permanent record for easy reading and archiving. In addition, many of them can also record data trends, making them competitive with analog recorders. Internal data storage units, on the other hand, are more compact and economical. They can be stand-alone or can download stored data, via an interface, to a host computer. Most loggers are actually a combination of internal memory and hard copy devices, with an emphasis on one capability or the other.

Today's microprocessor-based loggers have taken on many forms, but they all share some commonalities. Following are the basic characteristics of all (or at least most) data loggers, whether they are out in the field or plugged into a host computer.

A/D Conversion

The is usually the first operation in a data logger. This characteristic is an advantage compared to traditional analog recorders, in that digital data are much more accurate than readings taken from a line graph.

Signal Conditioning

Another feature which has increased the versatility of the data logger is its ability to measure specialized signals such as strain, pressure, or flow through the use of a conditioner. The function of a signal conditioner is to amplify, translate, linearize, isolate, and/or filter out signals. If, for example, a logger is measuring temperature via a thermocouple, it may not be able to detect the tiny electric current which is being transmitted; only major changes in temperature should be picked up and recorded. A signal conditioner placed between the thermocouple and the logger increases measurement accuracy by accepting the signal, amplifying it, and sending the stronger signal to the logger.

Signal conditioners are especially useful in applications dealing with common mode voltages, isolation, and ground loops. Modules are available with high isolation for each channel,which protects expensive equipment and reduces noise and reading errors.

Multichannel Inputs

Users often need to measure multiple signals simultaneously in order to make and interpret correlations between measurements. Data logger manufacturers have responded to this need by providing an expanding number of input channels with high resolution and accuracy.

Many loggers are intelligent. Their built-in intelligence frees the operator from slow, error-prone manual analysis by automatically interpreting the data—giving flow rates, differential temperatures, and so on.

Alarming

In applications where measurements change very little, data loggers can record at long intervals and monitor for alarms. Records of these conditions are generated using a minimum amount of paper.

Networking

A network of loggers is the ideal solution for users who want to collect data from a number of different locations and channel them to one single collection point. Satellite loggers can collect data from locations throughout the plant and feed them into one unit for storage and recording. This is an obvious advantage to the operator, who would otherwise have to travel from station to station to retrieve the data.

Interfacing

With the help of a PC, the data logger becomes an important part of many data collections schemes. Collected information can now be fed via interfaces such as plug-in data acquisition boards, RS-232 and IEEE-488 cables to the PC, which can not only manipulate, interpret, and show data in a variety of forms but can also perform control-oriented tasks at the same time.

Computer interfaces allow data loggers to assume many forms. Portable units can be carried to any location, while dedicated units remain in a fixed place adjacent to the host computer. Since data acquisition PC boards receive analog measurements directly and convert them to digital form, they can be considered actual data loggers.

BATCH CONTROL

A batch operation often involves mixing different raw materials to produce a product. A simple control system may consist of stand-alone controllers, indicators, and manually operated hand switches to open and close remotely operated valves. A key parameter to control is the batch temperature, which leads to a consistent, higher-quality product. Several control strategies can be used. First, there is the central computer with dumb I/Os. This requires on-site, specialized personnel to maintain the system, and tends to be expensive in terms of both equipment and personnel. Next, there is the distributed control system. However, these systems can be expensive, slow, or too inflexible to meet many application needs.

A control strategy that combines the supervisory and data collection abilities of a PLC and the operator interface abilities of a PC may be the most cost-effective solution. Such a system can be modular. It does not intimidate the operators, and it offers a variety of rugged I/Os which can be chosen according to application needs. It is also less expensive than other options.

A single PLC can perform all logic, proportional, integral, and derivative (PID) control, and math functions, while a PC could be used for programming and documentation of the PLC, as well as for operator online interface functions such as graphics, alarm trends, and reports. PID control can be housed in a separate hardware platform, such as stand-alone PID controllers linked to the PC. This provides a simpler implementation of a PID algorithm. However, the calculation of the set point would be performed in a separate device (the PLC), creating a failure link in the added communication path.

The control system may also be required to meet certain performance criteria. System speed is an important parameter. A keyboard command can take longer than 2 seconds to travel to a PLC output point, back through an input point, and onto the screen. These 2 seconds include PLC program scan time, I/O image table update time, and input and output response time.

SYSTEM FUNCTIONS

A typical control system consists of an Allen-Bradley PLC 5/15 and two industrial IBM ATs. System debugging, PID loop tuning, and commissioning should take no more than a few days. The PLC and the two PCs are connected on a data highway, with the PCs being functionally redundant. Since reliable, clean power is often critical, an uninterruptible power supply (UPS) can be added, enabling the operator to switch over to a safe mode should the main power fail, as well as avoiding power disturbances. Without a UPS, a careful study is required to analyze the effect not only of power loss on the control system, but also of the response of the control system on sudden power restart.

Each PC can be equipped with an operator-oriented Mylar keyboard instead of the standard QWERTY keyboard. This simplifies the operator interface and makes the keyboard "spill-proof." A printer can also be added to produce hard-copy process alarm messages and system diagnostics.

Critical to quality improvement in many areas, such as resin production, is the reactor's temperature and feed flow control, both of which can be performed in the PLC. Since reactor temperature is critical, several temperature probes should measure this parameter and be individually connected to the PLC.

The PLC program can perform averaging as well as two-out-of-three selection in case of a temperature element failure. On failure detection, the failed element is isolated and alarmed.

In the case of feed flow, load cells under the raw material tanks can monitor the weight change. This is then converted to outgoing flow, with a time function providing feedback for the flow control of raw materials.

PLC programming and documentation is a key requirement in implementing this control strategy. The quality of the PLC program depends not only on the program itself but also on the degree of documentation. Without good documentation, future program changes and editing can become difficult.

Graphics are easier to develop on screen through a basically menu-driven configuration and a library of symbols. An important consideration is the potential color blindness of some operators. Labels can be added to the color changes indicating on/off status (using the words ON and OFF).

The graphics are structured as follows. The first is an index of all graphics on the system. The second is an overview of the complete process. Next are dedicated graphics such as password-protected PID tuning. The last graphic is a system diagnostic which monitors the health and operation of the PLC, the two PCs, and the communication between them, a powerful time-saving tool during system troubleshooting.

Graphics can be developed in close collaboration with the operators. This approach reduces training time and gives the operators a sense of ownership of the system.

Alarms should be individually time-stamped when they occur, when they are acknowledged, and when they return to normal. All alarm points are displayed on the screen through the graphics, as well as printed out by the alarm printer.

Trends are available online for incoming real-time, live data, as well as for information retrieved from disk and displayed on the screen to show historical trending.

DATA ACQUISITION NETWORKS

Computerized monitoring systems, or data acquisition networks, gather data from sensors at hundreds of remote locations, interpret them, and display them at a central location. The networks monitor all types of plant conditions, from environmental hazards to equipment breakdowns. In short, because they help plant managers quickly gather, interpret, and act on information, data acquisition networks help prevent hazards and production problems.

Monitoring systems are not restricted to large plant operations. As both PC hardware and software become more sophisticated, monitoring systems become viable and cost-efficient for production facilities of virtually any size. While the majority of these systems are designed for manufacturing plants, there are also applications for refineries, offshore drilling platforms, ocean-going oil tankers, mines, and sewage treatment facilities.

As monitoring networks become more sophisticated, features such as remote calibration of sensors make maintenance easier and operation smoother.

One of the most important advantages provided by computerized monitoring is linkage. Remote sensors, analyzers, and relays can be spread out over several miles in various environmental conditions.

These devices can be linked in a data acquisition network to provide operators with information from anywhere in the plant. All of the plant's operations, ranging from process control to production tracking, the monitoring of environmental hazards, and even the location of employees and equipment, can be monitored by a single operator at a central location.

Conventional hard-wired systems typically can monitor only up to a few dozen sensors. Yet, depending on the installation, these systems may require a sprawling mass of meters, alarms, and other equipment, in addition to (potentially) miles of wiring between the sensing devices and the central readout station. At the central readout station, individual meters and indicator lights are required for each point that is monitored.

A computerized monitoring system can provide high-speed, centralized monitoring of thousands of sensors, along with comprehensive graphics displayed on color monitors. Pictorial color indicators can be used to pinpoint trouble spots in a large industrial plant or an underground mine.

Computers can be used to identify and monitor employee exposure to hazardous gases and calculate time-weighted averages. The control station operator can activate preprogrammed controls to shut down specific operations.

The software can be used to establish control schemes which allow precautionary actions to take place based on the number of sensors in alarm, sensor values in conjunction with other events such as the time of day, or as the result of a calculation with sensor input as one of the variables.

The data acquisition network can be programmed to dial emergency phone numbers or allow the plant manager to access the system from home, using a PC and a modem.

MONITORING SYSTEM STRUCTURE

The structure of a typical monitoring system works something like the human body's central nervous system, where nerve endings trigger impulses through the body's spinal cord to the brain. Similarly, monitoring systems typically involve three key elements: the central computer or brain; remote terminal units or field data stations; and the sensors themselves. This section will give you a better understanding of how the system works and how each link in the network operates.

Central computer or processor

The central computer acts as the network's brain. It monitors all points along the system and reports on them in the control room on a CRT screen, where reports on sensor activity are generated and control actions are taken.

The control room, in addition to housing the computer, keyboard, and monitor, uses a printer for report generation and a UPS for continuous power. The computer continuously scans the remote sensors for specific inputs regarding information such as fluid levels and equipment status. The monitor can show individual sensor readings and trends regarding liquid levels, gas pressure, temperature, and other environmental conditions in plain English or in graphics on one or more screens at the central station. Also, detailed layouts of any section of the facility under surveillance can be displayed.

The operator can receive warnings about deteriorating plant conditions or emergency conditions. The computer may be programmed to send messages as specific as the cause of an alarm condition and what action should be taken to control it.

Network spinal cords

The next link in this computerized system is a series of remote terminal units (RTUs). These are microprocessor-based subsystems about the size of a common household electrical outlet box. They continuously process the signals sent from remote sensors and relay the information to the central computer. If the link to the computer is broken, the RTU can sound an independent audible or visible alarm, or it can initiate preprogrammed control action if set points are exceeded.

Some RTUs provide local readout of sensor values and digital I/O status. Large computerized networks may include hundreds of RTUs, each of which may be connected to a variety of analog and digital I/Os.

System nerve endings

The final link in the computerized monitoring system is composed of sensors, relays, and instruments which may function in countless combinations. A system may be cost-effective with as few as 10 to 20 sensors. Many systems, however, have the capability to monitor thousands of I/Os. The computer has continual contact with all of the remote monitoring and control points, which may be located anywhere in the facility.

A common class of safety sensors measure for oxygen deficiency, combustible gas, toxic gas, or a combination of these three parameters. The sensors may be simple diffusion devices or sophisticated instruments, such as nondispersive infrared analyzers and other single-point and multipoint monitors.

Sensors used to measure air velocity, motor bearing temperature, voltage, current, and pressure can also be used in computerized systems to help locate and repair equipment breakdowns. Various types of relays to control alarms, lights, switches, and other electrical contacts may be activated.

TRAINING AND MAINTENANCE

Modern data acquisition networks are not only more powerful than their predecessors, they are more compact and easier to maintain, install, and operate. Because they are simple to install and maintain, they are practical for smaller plants with limited computer equipment and operating personnel.

Most of today's data acquisition networks are equipped with menu-driven programs. This makes it easy for an unskilled operator to learn how to use them. Workers who are unfamiliar with computerized monitoring systems do require some training time.

Most plant workers who are already familiar with the facility's operations can learn how to use a data acquisition network with minimal training.

Maintenance is also easier. Regular calibrations to verify sensor accuracy are normally required. Remote-controlled wireless calibration systems can provide this function almost automatically. They allow a single worker to calibrate a sensor quickly. With a wireless system, the sensor can be calibrated simply by pointing a hand-held, remote control device from a distance of several feet.

The remote control device communicates with the individual sensor's transducer assembly using an infrared beam of light. The sensors also have their own liquid crystal display (LCD) readout to display sensor values.

FLEXIBILITY IN CONTROL

Data acquisition networks provide more in-depth information than conventional systems, making it easier to track equipment breakdowns and safety-related emergencies. For example, combustible gas sensors may be set to signal the computer whenever they detect gas concentrations above a particular limit. The system can then trigger relays to activate or deactivate pumps that start backup ventilation systems and other actions.

A single keystroke may command the computer to find the source of an alarm and display it on the monitor. On command, the computer can display sensor-by-sensor readings from locations throughout the facility, providing additional information for analysis and decision making. Additional monitors can be located throughout a facility for easy access by selected plant personnel. The source of equipment breakdowns can also be found quickly so that the downtime required for repair is kept to a minimum.

Data acquisition networks with centralized monitoring and control of environmental hazards offer an efficient way to enhance worker protection while providing more flexibility and control over production processes and environmental hazards. A relatively small amount of hardware is required to put the system online, yet data acquisition networks can be expanded easily by adding software, sensors, relays, and/or terminals. This results in a system that improves both productivity and protection for workers in plants, mines, and other facilities.

STATISTICAL CONTROL

Statistical process control (SPC) and statistical quality control (SQC) are statistically based techniques that can be used to analyze data. The results of the analysis can be used to improve the process under study. SPC and SQC do not measure quality directly. They can be used as analytical tools to indicate if there is some correctable condition which is affecting quality. This condition can then be corrected to improve the quality of the process. SQC involves the gathering of data for analysis at a later time, while SPC typically refers to online data gathering and immediate analysis.

How SPC works

The basic idea of SPC is that, in any product, there will always be variations from one sample to another due to inherent variations in the process. Variations other than these inherent ones are due to assignable causes which can be corrected. Statistical analysis of the quality measurement should be able to reveal the assignable cause variation. If there are no assignable cause variations and the only variations are inherent, the process is in a state of statistical control. If there are variations that need to be corrected, the process is not in a state of statistical control.

SPC requires samples of the product and the data resulting from the measurement of these samples. Charts are created from the data to assist in the analysis. It is assumed that inherent variations will affect every measurement and will be stable over time. Any variations which are greater than the short-term variations and which are present over a longer term are due to assignable causes and are thus correctable.

SPC attempts to control product quality in reference to a standard or target, as shown in Figure 2-3. The desired result is to reduce product variability so that even when product quality is acceptable, all of the product will be as close to the target as possible.

Figure 2-3 shows typical process control limits compared to an SPC target. In part (a), the limits for a particular composition have been set at 59% and 61%. If the composition is between these limits, the product is considered acceptable; otherwise, it is unacceptable. In part (b), the objective is to control to the target of 60%. While the composition may be acceptable within the limits, it is desirable for it to be held as closely to the target as possible.

SPC distributions

SPC assumes a normal distribution. If enough samples are taken, then these samples of a product will follow the standard bell-shaped normal distribution curve. More of the samples will always be grouped close to the center of the range of the distribution than at the extremes. If enough samples are taken, any group of measurements will have some deviation or spread between the highest and lowest measurements. Certain values will deviate from the mean (average)

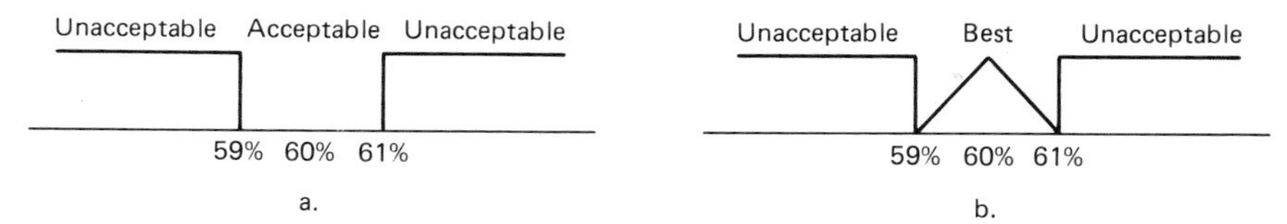

Figure 2-3 Process control limits and SPC control targets. (a) Limits between 59% and 61%; (b) Target set at 60%.

by larger amounts compared to other measurements. The standard deviation of a set of data is a statistical measurement of the dispersion of the numbers. For example, suppose that we measure the following group of numbers:

22
24
25
26
28

The results are as follows:

Mean = 25
Range = 6
Standard deviation = 2.36

Now, if the fourth measurement is higher by one, and the second measurement is lower by one, the data set becomes

22
23
25
27
28

This data set has the same mean and range, but the standard deviation increases to 2.55.

CONTROL CHARTING

Measured data can be plotted on a chart to show variations above and below a mean. This is a type of trend recording. Suppose that upper and lower control limits (UCL and LCL) are used at three standard deviations above and below the mean. If the process data cross either limit, they can be considered to be out of statistical control. This chart is not very sensitive to small variations.

$\bar{X}$-*R* CHARTS

A variation of the basic control chart is the $\bar{X}$ chart (known as an $\bar{X}$ a *bar chart*). In an $\bar{X}$ chart, the product is periodically sampled, usually with four measurements taken. These individual samples are plotted, and the set of measurements is called a *subgroup*. The individual samples are averaged to produce the average X, which is $\overline{X}$. The individual samples are often compared to determine their range, or R, which is combined with the $\overline{X}$ chart.

The $\overline{X}$-R control chart is plotted like an $\overline{X}$ chart, except that instead of using the individual samples, the average X is plotted. The UCL and LCL are determined by adding and subtracting a constant (which depends on the sample size) multiplied by the range R from the grand mean of the average X.

First, the subgroup of samples are collected for each period. The range and average value of each subgroup are computed.

For each period, the variation of the individual measurements within a subgroup is due to inherent process variations. This variation can be considered measurement or instrument "noise." Variations from one subgroup to another which are more than three standard deviations than the subgroup variations are likely due to assignable causes.

In the $\overline{X}$ and R charts, if the process is not in statistical control, then the measured value will fall outside of either of the control limits. Also, there are some general trends which can indicate out-of-control values. For example, a certain number of consecutive values on the same side of the mean can be defined as out of control.

Variations in Subgroup Data

The subgrouping of four to five measurements is an important part of the $\overline{X}$ chart (Table 2-1). These are averaged to provide a point on the chart. The

TABLE 2-1 Sample Data Set Showing Subgroups

Period	Data Set: Subgroup of 4 Samples				Computed Range	Average $\overline{X}$
1	99.282	99.285	100.206	100.197	0.92	99.743
2	98.976	100.404	99.647	101.658	2.68	100.171
3	99.611	99.511	100.590	102.929	3.42	100.661
4	98.142	97.744	99.840	98.153	2.10	98.470
5	98.244	97.613	100.839	99.390	3.23	99.022
6	98.793	99.642	101.179	99.809	2.39	98.856
n	99.260	100.71	101.782	101.365	2.52	100.779

Average range = 2.50 Average $\overline{X}$ (grand mean) = 100.10

For $\overline{X}$ chart
UCL = 100.10 + 0.75 (2.50)
LCL = 100.10 − 0.75 (2.50)

For R chart
Range control limit UCL = 2.30 (2.50)
In many cases, the LCL is zero.

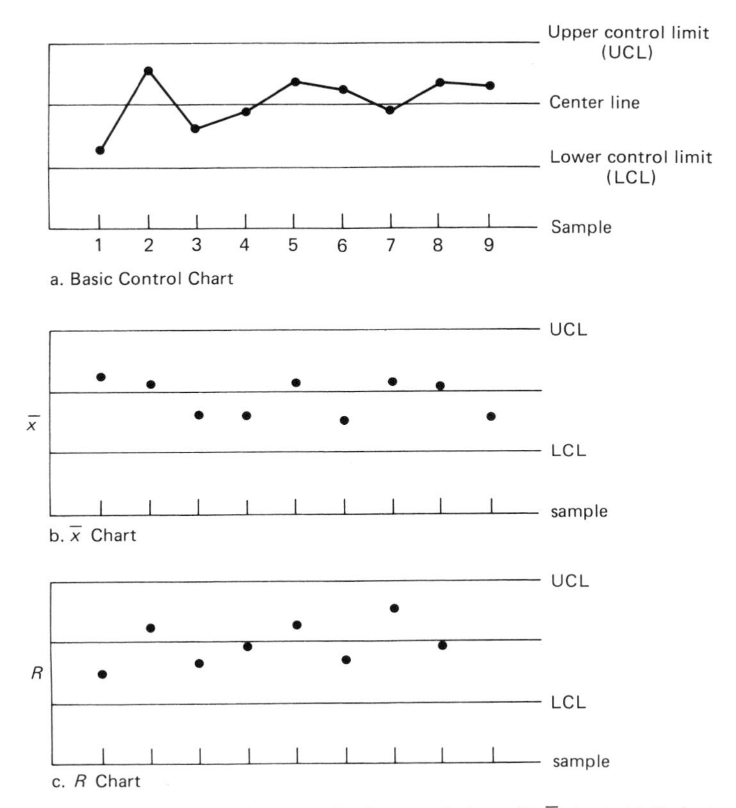

Figure 2-4 Control charts. (a) Basic control chart; (b) $\overline{X}$ chart; (c) R chart.

natural expected variation is shown in the average range of the subgroup; the differences between subgroups should show the variation due to the assignable cause.

In cases where the subgroups do not show any inherent variation, they are omitted and each measurement is plotted on the control chart. The limits are determined by the variation between the measurements. Figure 2-4 shows examples of control, $\overline{X}$, and R charts.

CUSUM CHARTS

Another type of control chart uses the cumulative sum (CUSUM). This technique accumulates the most recent data to provide prompt detection of out-of-control situations. CUSUM charts provide tighter control than $\overline{X}$ charts and are particularly useful in detecting a drift of the mean from the target. The CUSUM algorithm calculates two cumulative values. The amount of measurement above and below the target are calculated.

USING SPC DATA

The control charts provide information about the process which can be used to improve such factors as yield. The chart data can indicate if the variability is due to assignable and correctable causes. They can also show when the variability changed, pointing to operator problems or procedural differences among operators. In order to use SPC, we must have the capability of measuring process parameters at certain designated rates, storing the measurements, and processing them. SPC can be used to produce alarms when the measured variable is out of the statistical control range. Trends can be tracked for the following conditions:

1. When the $\bar{X}$ points increase or decrease for a certain number of samples.
2. When any $\bar{X}$ is outside of the UCL or LCL.
3. When the accumulated value above or below the median is too large.

SPC APPLICATIONS

The use of SPC involves not only control charting methods but also process capability studies and problem solving. Quality circles can meet regularly to discuss problems and areas for improvement. Problem-solving techniques include cause-and-effect analysis, histograms, process control charting, and brainstorming. A multidisciplinary approach is needed, with members from various areas such as quality, manufacturing, and engineering participating.

The purpose of SPC is to control the relevant in-process variables. This often requires machine capability studies to be performed on those operations where critical characteristics are involved or problem areas are suspected.

An IBM computer combined with a DataMyte statistical software package can be used to control process variables. The software can generate online statistical data including histograms, capability studies, $\bar{X}$ and R charts, sigma charts, P charts (% defective), and NP charts (number of defects). Using statistical methods, control charts of individual machine operations can be developed to monitor critical manufacturing operations.

Online statistical control charts can be used to monitor process performance according to the statistical process control measures. Control charts may be collected daily, analyzed, and stored. Machine capability can be analyzed over time to review tool and machine wear.

There are major differences between statistical analysis in the discrete parts area and the continuous process industries. In the typical discrete parts application, only a small number of parts are sampled for testing. Ignoring certain effects such as shift and temperature changes, most of the parts selected for sampling can be considered statistically independent. This means that the difference between a set of two parts taken from the production line should be about the same as the difference between another set of parts taken from the same line.

In many continuous chemical processes, there is a certain amount of backmixing of the product stream due to storage of the product in vessels which are part of the process. This storage tends to smooth any differences in the characteristics of the material. Process measurements may be continuous or sampled at a rate that is faster than the holding time of the product. This eliminates the statistical independence of the samples.

Another major difference between discrete manufacturing and continuous processing is the degree of feedback control. In most continuous processing plants, the product variables are measured and based on some control algorithm; a certain proportion is looped back to control the process. The control algorithms tend to adjust the controlled variable toward a target or set point.

There are some applications where key product quality variables cannot be directly controlled. These include batch processes where the product continues to change after the last control action has been taken. This is known as the *end point control problem.* A similar problem occurs when key quality variables cannot be measured online and must be measured every few hours in the laboratory.

Secondary variables such as impurities are another problem area. An example is a distillation column that separates the product from some other chemical. The control loops on the column measure and maintain the proper proportion of the major constituents of the product. Other minor components can come from impurities in raw materials and contaminants from the materials used in construction of the system. These cannot be directly controlled.

In some industries, such as paper and plastics, flat sheets of material are produced using extrusion or similar processes. The material moves at a relatively high speed. The critical parameters are measured by scanning or sampling the sheet characteristics. SPC can be applied directly to these industries.

BATCH PROCESSING

SPC can be used for batch processes, but there are several factors which must be carefully evaluated. One of the most important considerations is the use of subgroups and the number of samples. Suppose that a batch consists of a well-mixed liquid; then the only difference between samples of the same batch will be measurement error. If several measurements of the same batch are made, the control limits will be very close to the mean. Since little variation in measurements is allowed, any variation from batch to batch will be outside the control limits, and the process will be seen to be out of statistical control.

Another problem is statistical significance. If we use one measurement per batch, there must be a significant number of batches before the SPC results can be used. The number of samples needed will vary, but there is little confidence in the analysis of SPC data for fewer than five or six batches. If analysis of the batch control is needed after only a few batches, SPC cannot be used.

Many continuous processes show short-term variation in measurement due to measurement error or noise. The actual process changes can be filtered out by mixing within vessels during the process. There can also be longer-term variations due to small oscillations in the process or changes in feedstock. These may be insignificant when compared to the product specifications, but they can be much larger than the short-term variations.

In many batch processes, there can be short-term variations between adjacent batches which appear significant in terms of the product specifications. Since the variations are short-term, they will not be considered in the control chart.

SPC is not always applicable to certain processes, but in many cases, it can provide knowledge about the process that is not obtainable more easily through other methods. The knowledge gained through SPC may be used to improve the process operation.

Statistical analysis of data taken from a process can often help to locate the cause of any variation. Some variations, such as measurement noise, are random and can be identified by a statistical test for randomness. One technique is to use serial correlation, whereby each measurement is correlated with the previous sample. Pure random or white noise has a low serial correlation. Random noise which has been filtered by passing through the process will exhibit high serial correlation. This type of testing can separate measurement noise from process fluctuations. Standard SPC control charts can be used to detect trends which are not apparent because the variation is less than the noise.

Statistical techniques can be used to analyze process variations in order to locate their cause. These methods include digital filtering and correlation analysis techniques. Knowledge of the cause of the variation is needed to remove it if the end result of the variation is harmful.

PLCS AND PCS

When the PC first appeared in the plant environment, many people greeted it with skepticism. For many years, the PLC had been the only processor of any significance in harsh duty control applications. The idea of an office PC doing anything useful, let alone surviving, on the plant floor was hard to believe.

But then the industrial PC appeared. Industrial PCs are built to thrive in rugged factory environments, and they can run graphics-oriented software that a PLC cannot handle.

Once PLCs and PCs were tied together, it became clear that each one's strengths complemented the other's weaknesses. PLCs shine in their I/O capabilities. Their I/O interfaces are thoroughly protected against the idiosyncrasies of real-world digital and analog signals, with isolation and filtering being the rule rather than the exception.

PLC I/O cards are designed to be mounted in racks so that diagnostic lights are easily visible and cable connections are rugged, accessible, and organized. I/O electronics can be repaired or updated quickly, minimizing production downtime. The simplicity of the PLC's lock-step input-process output scan cycle also minimizes I/O priority and synchronization problems.

In contrast, typical PC analog and digital I/O cards are better suited for laboratory applications, where electrical noise and catastrophic mishaps are less likely. Due to the PC's small card size, relatively little room is available for cable connections, so wiring tends to be routed through dense D-type connectors which are not suited for rapid troubleshooting and modification. An industrial PC allows rack access to PC I/O cards, which aids in changing I/O configurations, but the physical structure of the PC cards does not lend itself well to the diagnose-and-swap methods of plant maintenance electricians.

The PLC is an optimized computer. Its CPU, memory, and I/O architectures are tailored for solving the logic equations of control systems and for handling the movement of data through the control system. Ladder logic programming, as used in PLCs, allows engineers and technicians with little or no computer experience to solve a wide range of data acquisition and control problems using relatively simple techniques.

The more diverse nature of PC peripherals requires that the PC processor have an I/O architecture much more complex than that of the PLC. The PC is less optimized for control tasks. Its greatest advantage is its ability to adapt to a wide range of applications, including word processing, structural analysis, accounting, data acquisition, and process monitoring and control. The PC's versatility tends to limit its effectiveness in an application compared to a computer optimized to its task. The PC is generally not as simple to use as the PLC in controlling industrial processes and machines, particularly where the control logic programming is subject to adjustment by nonprogrammers.

The PC outperforms the PLC in several important areas. PLCs suffer in the area of operator interface. They concentrate solely on control logic and I/O transfers. In the past, PLCs used simple devices, such as indicator lights and push buttons, on their I/O lines. These light consoles were often colorful and dynamic but conveyed only rudimentary information. Machine operators and maintenance technicians were often required to consult a system manual to interpret the light patterns and numeric codes. This contrasts sharply with the powerful graphics and text capabilities of the industrial PC.

The PC, with the proper software, can provide a superior operator interface. Operator instructions and context-sensitive menus can pop up when needed.

Historical trend charts can be used to show not only what the system's status is now, but also what conditions led up to that status. Alarm and diagnostic messages can be presented in the operator's own language and can become increasingly insistent if they are not promptly acknowledged.

Since its instruction set is optimized for control logic, the PLC does not have the facilities required, for many control tasks, except for simple analysis of its data. Its limited math functions, sparse memory, and lack of mass storage prevent the PLC from performing the more sophisticated computations required for techniques such as statistical process control and predictive failure analysis.

The PC's analytical abilities, on the other hand, are excellent. Since it has mass storage and a wider instruction set, the PC can easily offload chores from the PLC. With the use of data base and spreadsheet formats, the industrial PC can process raw historical data and summarize the results for office-bound computers for further analysis or review.

The availability of standard network interfaces for the PC makes it especially easy to transfer data from the plant floor to the rest of the organization. This is discussed in Chapter 11.

PC-PLC INSTALLATIONS

Many installations can combine the capabilities of the PLC and the PC. The units can communicate with each other using serial data communication links. The link can be as simple as a programming port or as complex as some specialized interface adapters. Many different protocols can be used, some of them proprietary to the PLC manufacturer.

Once communication has been established, the PC can augment the PLC's data acquisition and control functions in a number of ways. The PC can be used as the operator interface for a single PLC or for a network of related PLCs. The operator has access to clear, context-sensitive views of relevant aspects of the control system. System operational parameters can be selected using menus, icons, and range-checked data input fields. The PLC's operations concentrate on the control and data acquisition task, exchanging raw data with the PC as necessary.

Since the PC has mass storage capability, it can store raw or processed data that have been collected by the PLC. In certain industries, such as the food and pharmaceutical industries, process data must be gathered and stored to meet regulatory agency criteria. The industrial PC can streamline this process in a cost-effective manner. Data collection can be a secondary operation or it can be an integral function of the PC running a supervisory/cell control software program.

In the past, special programming units or terminals were used to enter and debug PLCs' ladder logic programs. These terminal units were usually removed from the PLC during normal system operation.

Today, instead of using a dedicated programming terminal, there are PC programs which emulate all of the functions of the programming terminal. The PC has the ability to provide additional documentation, reporting, and formatting functions. When the PC is installed as the operator interface and performs supervisory control, it can double as the PLC programming terminal with the proper software.

Maintenance-oriented functions can also be performed on PLCs by the PC. This may be done by tracking actual cycle counts and component in-service elapsed time against a maintenance log data base. The PC can be used to report when parts are due to be changed prior to their predicted breakdown.

Data from the PCs can be merged plantwide to produce daily or weekly job schedules for maintenance technicians. The PC can be used to perform sophisticated periodic or continuous predictive failure studies. The data can include combinations of mechanical or electrical parameters of production machinery used as early-warning indicators. The PC can then notify the operator of service needs when the indicators show sufficiently degraded system reliability.

As an example, the PC can analyze trends or patterns of lubricant viscosity, sound and vibration spectral density, motor drive current, or other parameters. Maintenance software for the PC is available from a number of companies. Most maintenance management software programs do not communicate directly with the PLC. They operate on data placed in shared disk files by an operator interface or supervisory PC program.

DIAGNOSTICS

An important part of service is diagnosis of control system faults. This is an area where the PC's analytical and mass storage abilities are both needed. In most cases, the PLC program logic can even detect many faults in its own circuitry, especially in the more intelligent I/O interfaces. The control system's resident PC is able to make the cause of the fault, and the recommended solution, clear to the system's human operator, using the graphic interface.

The PC can translate the PLC's fault indication. This is usually a single bit or numeric code in the PLC's memory. The PC can convert this bit or code into a message that can range from a simple phrase to the appropriate section of an online repair manual, complete with pictorial diagrams. The PC takes the output from the PLC's diagnostic logic and, using its mass storage and graphics display capability, attempts to get to the root of the problem. The PLC usually detects the symptoms rather than the causes of system failure.

The PC can also be used to minimize the effort required to isolate the actual cause of the fault. Artificial intelligence (AI) techniques, including rule-based expert systems and neural networks, have the capability of associating the unit's current symptoms with historical data. The PC can then present a list of fault causes in the order of their likelihood.

The system can store each failure and repair history, and adapt future diagnoses to maximize the effectiveness of the repair instructions. Control system diagnostics can be handled by non-real-time AI programs which analyze the data placed in the PC's disk files.

Another use of the PC is to simulate, in real time, a programmable controller's I/O. When the programmable controller is connected to the simulator PC, its control logic can be tested without using the actual I/O devices and machinery. The logic can be tested under both normal and abnormal conditions, with no threat of damage to the equipment. Potentially damaging situations can be detected using this technique.

I/O CONNECTIONS

Some manufacturers offer PLCs in mechanical/electrical configurations which are compatible with popular industrywide, computer standards. PLCs are available for the VMEbus. A major advantage of the VMEbus is its ability to support multiple processors. The benefit of multiple processors is that they allow the division of operations among several smaller, cooperating processors.

For example, the VMEbus PLC can coexist with a VMEbus PC/AT processor. The use of the standard IBM-PC/AT computer architecture allows the use of the vast array of PC/AT software, along with the available VMEbus I/O interfaces for control and monitoring.

The PC and PLC can also share data over the VMEbus backplane at high speed. It may be process data, ladder programming logic, or other information, and it may be exchanged between the processors without the bottlenecks of serial transmission.

SOFTWARE FEATURES

Some PC software packages allow the PC to emulate a PLC in a conventional way. They combine the ladder logic functionality of a PLC with the documentation, debugging, and operator interface features available from a ruggedized PC. These packages typically use PC bus analog and digital I/O cards and may offer as an option a ladder logic coprocessor board. You can also design your own PLC functions with some programs.

Other PLC software uses the same functions as a PLC, but uses techniques like decision-based flowcharting procedures to implement the control system's logic. This format provides a more readable and understandable user interface to operators who may not be completely familiar with PLC ladder logic diagrams. Another technique is to avoid repetitive analysis of all the logic equations, as with a standard PLC, and to process only those portions of the logic that are necessary at any given time to provide the solution.

A PC can be used with an I/O scanner card that provides access to a PLC manufacturer's standard I/O subsystem racks. This allows electricians to remain familiar with the PLC I/O cards while maintaining system control with a PC processor. Allen-Bradley and Texas Instruments provide I/O scanners that plug into the PC bus. Allen-Bradley also manufactures an I/O scanner for the VME-bus.

PLC BATCH APPLICATIONS

Industrial PCs can raise the productivity of many batch operations. The use of the PC and PLC can result in improvements in batch productivity and profitability. Batch control configurations can be more fully integrated for loop, logic, and sequential control. Loop and logic control have different processing requirements. Loop control requires a consistent, predictable update time (.3 to 1 second), while logic control requires the fastest possible update time even though it may vary from one scan to another (10 to 50 milliseconds).

The ideal batch controller control configuration would use independent logic and loop processors on the same backplane for data transfer. Programming languages should be based on industry standards such as the Instrument Society of America (ISA) diagrams for loop control and relay ladder logic or logic control. For small-batch processes, the multiprocessor controller should require fewer than 16 loops and 256 discrete I/O steps. A PC should be used as the operator interface to take advantage of low-cost hardware and software, while providing a low-cost, single operator window.

A PLC with internal PID provides excellent logic control and good sequential control but is difficult to use with loop control. A PLC with a single loop controller provides easy-to-use loop control but lacks control integration (connectability). A distributed control system has excellent loop control, operator interface, and connectability, but it also has a high cost.

PLCS WITH PID CONTROL

PLCs were originally conceived as a microprocessor-based replacement for logic systems composed of hardwired relays, sequencers, and timers. Most are designed to operate in harsh process environments. With internal PID capability, the entire array of control requirements for the small batch process can be satisfied with a single control device. The PID algorithm is written in ladder logic, and it can be applied to many simple regulator control problems.

With PID control resident, control interactions are easier to implement than in modular systems where these control strategies reside in different hardware sections throughout the architecture. For example, since all the controls are in one common program, the set point for a temperature or pressure loop can easily

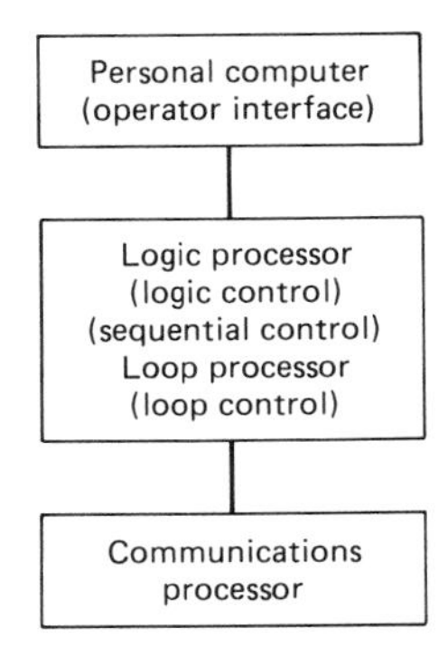

Figure 2-5 Multiprocessor batch controller.

be set by the sequencer logic. Due to the method of implementation of the PID algorithm within the PLCs, the PID algorithm tends to be difficult to use and configure. Setting the values into the PID parameters is usually awkward.

In a small batch process, implementing simple PID loops in the PLC is usually not a problem, since advanced PID algorithms are not needed. An example of a complex control area in this area is the cascading of a reactor temperature loop to the reactor jacket flow loop. The ease of configuring this cascade arrangement depends upon the PLC's implementation of PID.

The PLC's area is logic control. Items such as valve interlocks and alarm logic can easily be implemented in ladder logic. The scan frequency of the ladder logic will depend upon the total size of the program, but it should execute within 10 to 50 milliseconds. Sequential control can also be implemented in ladder logic. A multiprocessor batch controller with the PC and PLC functions is shown in Figure 2-5.

PLCS AND SINGLE-LOOP CONTROLLERS

Single-loop controllers (SLCs) are used for simple closed-loop control applications. These controllers are based on a PID algorithm which determines a control output by comparing the set point and a process variable. The process variable, set point, and output usually are displayed on the controller. Many of today's SLCs provide math, multiple analog inputs and outputs, and communications. The operator interface can be local, using the controller's small display, or remote, using communications and a host computer.

A PLC can be used with SLCs for logic and sequential control. The PLC acts as a master controller and is used for logic and sequential control. The PLC analog outputs are used to set the SLC's set point. The SLCs are used for loop control.

Control integration between the PLC and SLCs is usually minimal. The PLC can set an SLC's set point using analog output, but there is no feedback mechanism to indicate if the SLC is using the PLC's set point. Equipment cost is usually higher for this configuration, since the PLC needs analog I/O in addition to the SLCs. Installation costs also tends to be higher due to the need to mount the SLCs and provide wiring between SLCs and the PLC.

SLCs will continue to decrease in price and increase in functionality. Alphanumeric prompts will make them easier to use, and options like communications will become standard.

DISTRIBUTED CONTROL SYSTEMS

Distributed control systems were originally designed for continuous processes. The controllers were based on the PID control algorithm, and they supported other algorithms such as summers and multipliers. The control needs of discrete and batch control applications were minimal or nonexistent.

Many processes tend to be hybrid, with evolving requirements for continuous, batch changing, and discrete control. Additional equipment may have been purchased to fill these changing needs. PLCs were purchased for discrete control; special platforms were used for batch control; and the distributed control system (DCS) was used primarily for regulatory control. Typically, each of these systems operated independently.

DCS has evolved into a more flexible, integrated control system that provides data acquisition, advanced process algorithms, and batch control capabilities. These functions are usually accomplished with the help of a standard high-level language, such as BASIC or FORTRAN or a proprietory language. Many DCS products used PLCs for discrete control. The operator can monitor and control selected points in the PLC from the DCS operator interface, but the program can be modified only through the PLC's own programming device.

For the small batch process, advanced PID regulatory algorithms are not usually needed. When the logic control requirements are done by the DCS unit controller, it provides a major improvement in control integration over the use of a PLC for logic control.

The DCS configuration provides the ability to monitor and manipulate the process, retrieve historical batch data, configure the system, build schematic displays, develop control programs, and diagnose system failures. Costs are usually much higher compared than those of PLCs with internal PID and PLC/SLC configurations.

Areas of improvement for DCS include providing low-cost entry systems that are upwardly compatible and windows that cover a more complete array of the tasks and tools involved in logic control.

In the future, PC-based operator interfaces will include traditional DCS features such as supervisory control, additional control algorithms, and historical tracking.

PLC TRENDS

Newer PLCs are expected to provide more integrated solutions, easier PID algorithms, and improved operator interfaces. Protocol standards will allow different manufacturers' units to communicate with each other. Although PLCs have been widely used in the process control and automation fields for years as replacements for relays, their capabilities now make it possible to use PLCs for many other purposes.

A number of products are available for batch control, data acquisition, motion/position control, and automation cell control applications. Over one-half of all PLCs sold are used for process control. In these applications, the PLC is connected to a variety of analog and digital sensors and I/O devices, including flow, level, pressure, and temperature sensors.

ANALOG I/O

PLCs evolved from relay replacements to electronic controls that handle analog variables. PLCs offer capabilities such as PID control, floating-point and double-precision mathematics, data manipulation capability, and ASCII/BASIC modules, which allow BASIC programming and interfaces to CRT terminals and printers. These modules plug into the PLC's I/O bus and have direct access to I/O and program variables in the PLC's memory.

There are a variety of analog input modules for temperature, flow, level, pressure, and other sensors and transmitters. Analog output modules are available for control valves, remote controller set points, recorders, and motor drives for process and manufacturing equipment. Analog I/O modules for PLCs include A/D converters, temperature inputs, and D/A converters. A/D and D/A modules are available with up to 16 points and provide 8-, 12-, or 16-bit precision.

DIGITAL I/O

Most process control applications using a PLC involve digital single-bit inputs from devices such as limit switches on valves, on/off signals from pumps and compressors, and alarms from flow or level detectors. Digital outputs in these applications include pump and motor start/stop commands, valve open/close commands (for on/off valves), signals to indicator lamps, and annunciators or status outputs to a graphics interface.

In a machine control, manufacturing, or automation application, digital inputs include photoelectric, proximity, and limit switch sensors, on/off signals from motors, high-speed counter inputs from position encoders, and on/off signals from manual push buttons. Digital outputs typically turn relays on or off. The relays may be used to control items ranging from a small solenoid-actuated gate on a conveyor to a 440-volt motor starter.

In addition to single-bit inputs, other digital inputs include binary coded decimal (BCD) data from thumbwheels, keypads, absolute position encoders, and other devices, ASCII-coded inputs from terminals, and Gray-code data from encoders. Digital outputs include BCD outputs to panel-board displays, terminals and similar devices, and pulse outputs to servo and stepper motors.

CONNECTING DIGITAL I/O

Due to the many types of digital I/O, most PLC vendors offer special modules to handle each category. The module plugs into the PLC's I/O rack and needs no special wiring or programming. Typical digital I/O modules include AC input modules, AC/DC input modules, DC input modules and counters, DC output modules, and AC output modules. Generally, it is not possible to mix I/O types in the same module.

The wiring is not difficult, but the following rules should be followed in routing the wires to the various modules:

1. Twisted pairs should be used whenever possible.
2. Low-voltage dc signals should be separated from ac sources.
3. Proper grounding procedures should be used.
4. Wiring ducts for cables should be employed.

Most of the newer PLCs and I/O modules are much smaller than previous systems. The size reduction is due to modern semiconductor technology and packaging advances that provide more capability and performance in a smaller package. Power requirements are also less, so interference from power sources is lower.

PLCs often use high-density I/O terminations, especially for low-voltage dc signals. These terminations are designed to save space and money, since they allow up to 64 points to be installed in a single module. Additional I/O can be connected to modules mounted in expansion racks or remote I/O racks. These remote I/O racks may be located several feet or a few hundred feet from the PLC and connected with a communications link. Fiberoptic cable can be used to reduce noise interference that can occur when the link passes near high-voltage equipment such as pumps, compressors, or motor starters.

I/O SCANNING

A PLC executes programs quite differently from a computer- or microcomputer-based process control system. A PLC performs a scan, in which it reads all the inputs into an image table, executes the program logic, and then sends all the outputs.

In a continuous process where analog or pulse outputs must be changed regularly, it is not always possible to rely on the PLC's scan time. This is because the time can change from scan to scan, depending upon which inputs are on or off, or because the output signal cannot be ramped in a stepwise manner.

In some larger PLCs, PID control algorithms are handled by the CPU, using special PID functions. The disadvantage of this approach is that the lengthy computations use part of the scan time, slowing down the overall processing speed of the PLC.

SPECIAL I/O MODULES

Some PLCs offer special I/O modules that operate independently of the scan to avoid this problem. For example, PID control can be done by a separate module. This module is connected to the PLC's I/O bus and controls one or more PID loops according to set points received from the PLC's CPU. The loop input, such as pressure or flow, and output, such as 4–20 maDC outputs to valves or heaters, are connected directly to the module.

Motion and position control modules are available to provide a continuous stream of pulses to operate stepping motors. These modules also connect to the I/O bus and take position control set points from the PLC, but they are independent of the scan. Position encoder inputs and motor control outputs are wired directly to the module.

A specialized input module is available to handle high-speed inputs from position encoders or photoelectric sensors. This module can accept inputs at speeds of up to 50 kHz and count the pulses under the control of the PLC.

Interrupt modules are available to generate hardware interrupts. When an interrupt occurs, the PLC's program will transfer to an interrupt subroutine where the input can be processed. If the input requires an immediate output (such as an emergency shutdown), the subroutine can execute an immediate I/O command to send the proper output signals.

Most PLCs allow a partial or complete I/O refresh during a program scan so that critical inputs can be monitored more than once per scan. Typically, the refresh will be done for only a few I/O critical variables that need to be monitored closely.

ASCII/BASIC modules function as small computers connected to the PC's I/O bus. These modules do not connect directly to the I/O but access I/O variables by reading and writing data to and from the PLC's image table. Since they are programmed in BASIC, the modules allow a user to write an advanced control algorithm and to perform statistical analysis, data acquisition, and logging functions.

FAULT-TOLERANT APPLICATIONS

In many process applications, dual computers, redundant highways, or fault-tolerant systems are the techniques used to prevent a component failure from causing the entire control system to shut down. PLCs also provide this capability using a hot backup system. In a hot backup system, dual CPUs are used with an automatic fail-over switching system. When one CPU fails, the switching system will transfer control to the standby CPU in what is called a *bumpless transfer*.

An advantage of a hot backup PLC over a fault-tolerant computer is lower cost. A dual-processor PLC with an automatic fail-over switch can cost as little as one-fourth of the price of a fault-tolerant computer. Some units permit the CPU and I/O modules to be removed and replaced without shutting the system's power off, which provides minimal interruptions when doing repairs and maintenance.

PLCs can be interfaced to distributed control systems. Their primary traditional role has been to execute interlock logic, such as setting all the valves and pumps to the right configuration. With improved computing and analog I/O capabilities, PLCs can be used with PCs for process control tasks such as batch, unit, or distillation column control. A typical configuration is shown in Figure 2-6.

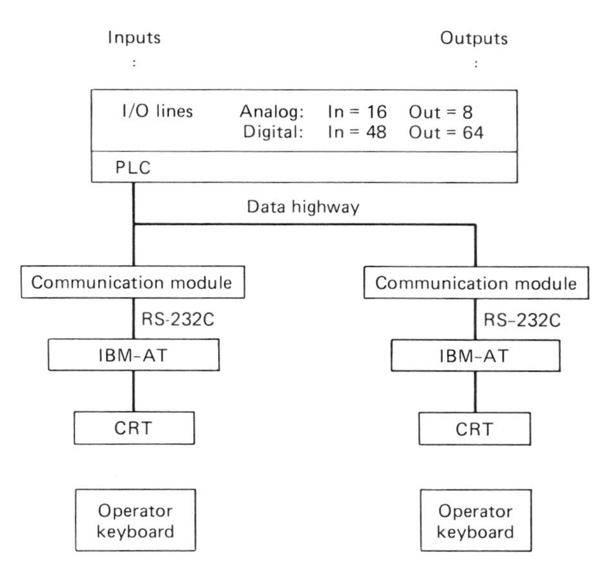

Figure 2-6 Typical PLC-PC configuration.

PROCESS CONTROL APPLICATIONS

An example of a simple process control system is one that maintains the temperature of a processing tank. The temperature of the tank is sensed and converted to a digital signal. It is then compared with the commanded temperature, which is also a digital signal. The difference between the two is the error signal. It is converted back to an analog signal, amplified, and applied to the heat control motor. The motor may adjust the heat control valve to change the flow of steam, forcing the temperature toward the desired value.*

In most process control systems, a major activity is the collection and processing of analog sensor data into digital form. There are many different data acquisition configurations, and a number of considerations are involved in system design.

The data may be stored for later use or transmitted to other locations. The processing can be used to obtain additional information or to produce specialized graphics for analysis or recording. The data may be stored in raw or processed formats. It can be retained for short or long time periods or transmitted over long or short distances.

Data processing can range from simple value comparisons to complex calculations. The system may be required to collect information; convert data to a more useful form; and use the data to control a process, perform calculations, separate signals from noise, and generate information for graphic displays.

The designer has a choice of configuration, components, and other elements of the system.

In a single computer control configuration, a single computer performs all of the supervisory control and process monitoring. The computer monitors a process by reading analog or digital data from the I/O equipment. It may change a control set point in performing supervisory control function.

Both process control and machine tool drive systems are similar in that they are both based on closing the control loop, using the feedback signal from the function or device being controlled. The closed-loop control system compares the actual behavior of the device being controlled with the behavior commanded. If a difference between the two or an error exists, it automatically produces a correction signal which tends to drive the error to zero. If a machine tool drive is commanded to move the machine slide to a certain position until the slide reaches that position, the position feedback will differ from the input command, and the actuator will continue to drive the slide toward the commanded position.

The choice of sensor dictates much of the system hardware. The monitoring and control of motor shafts, for example, may be done with one of three dif-

*This example is greatly simplified. In an actual system, oscillation may be a real-life problem. To minimize it, proportion control, at a minimum, would be used.

ferent position-sensing techniques: shaft encoders, synchros, or potentiometers. Temperature measurement may be accomplished by resistance detectors, thermocouples, or thermistors, and force may be measured by strain gauges or obtained by integrating the output from accelerometers. If the transducer signals must be scaled from millivolt levels to an A/D converter's typical ±10-volt full-scale input, an operational amplifier is required. If a number of sources is used, each transducer can be provided with its own amplifier so that the low-level signals are amplified before being transferred.

When analog data are to be transmitted over any distance, the difference in ground potential between the signal source and the final location can add substantial error to the unamplified signals. The low-level signals can also be obscured by noise, radio frequency interference (rfi), ground loops, power-line pickup, and transients.

Most PC applications fall into one of two basic categories: those with environments like calibration laboratories and office areas, and those with more hostile environments such as a factory floor or vehicles. The latter group includes industrial process control systems where temperature information may be developed by sensors on tanks, boilers, vats, or pipelines that can be spread over miles of facilities. The data may be sent to a central processor to provide real-time process control. Digital control of steel and petrochemical production and machine tool manufacturing occurs in this environment. The vulnerability of the data signals requires isolation, sampling, and other retention or reconstruction techniques. Hostile environments may also require components with wide temperature ranges, shielding, common-mode noise reduction, data conversion at an early stage, and even redundant loops with voting techniques for critical measurements.

In laboratory environment applications, the major concern is related to the performance of sensitive measurements under mostly favorable conditions. In this type of environment we are also concerned with protecting the integrity of the collected data, but the environment is much more favorable.

The multiplex configuration is also a major consideration. In remote multiplexing, remote multiplexer units are located throughout the plant, and the analog and digital signals are sent to the nearest remote multiplexer. A/D converters in the remote multiplexer convert signals to digital words, which are usually 16 bits long.

A control system with a remote multiplexing configuration is sometimes called a total *multiplexing system*. This type of system thus allows both analog and digital signals to flow either to or from the computer and the remote multiplexers.

The installed cost of wiring has been increasing; at the same time, the installed cost of digital systems has been decreasing. The use of remote multiplexing has increased, since it can greatly reduce the wiring necessary in a process installation.

In the past, remote multiplexing was used mainly for process monitoring rather than process control. As the reliability of remote multiplexing improved, it began to be used for control signals as well.

MOTION CONTROL

The PC offers solutions to problems in dynamic control. The development of microcomputers as control elements has resulted in a greater variety of positioning control systems. Early computer controllers evolved from programs stored in a time-shared, general-purpose, medium-sized computer. This was followed by the use of minicomputers as special-purpose controllers and then by the use of today's microcomputers.

A certain amount of commonality can be traced to the hardware configurations and computer programs used in the earlier designs. The basic dynamic characteristics of the actual power drives were usually modeled after a Type 1 servo: double integration with velocity feedback, and velocity and acceleration limiting. Many electrohydraulic power drives fall into this class, as do a large number of electrical power drives.

The typical design approach was to make the model of the controlled variable appear to have the same form from the controller's point of view. This was done mainly through the selection of the velocity loop gains. The result was a controller that could be used with a variety of power drives. While the early controller hardware was composed of a variety of circuit elements, more recent units have been implemented with microprocessors.

These microprocessor-based controllers used four types of functional sections:

1. Interface.
2. Analog output.
3. Status.
4. Processor.

The interface section allows different types of systems to be controlled. These systems may have different characteristics, such as acceleration limits and unique coefficients for the difference equations. The solution of various polynomial equations may be required, and the values of these coefficients are typically stored in ROM. The signal interfacing requirements for a particular system and computer are also handled by the interface section.

The analog output section can contain the buffer amplifiers and the D/A converters necessary for the conversion of the digital commands from the controller into analog signals required for control. Each of the D/A converters may be connected to a separate summing amplifier so that it can be summed with the output of the feedback device. Some of the D/A converters may be used for other functions, such as monitoring position and velocity errors.

The purpose of the status test section is to generate input commands for test purposes. Special commands, such as steps, and sinusoids, can be generated. The coefficients for these functions are usually stored in ROM.

The processor section provides a source of arithmetic capability, temporary storage for the calculations, mode selection, and timing and control functions. In this section module, all of the calculations are normally performed by the CPU.

In a distributed control system, a major goal is to avoid generation of commands from the master computer at an excessively high rate, since high update rates can saturate the bus bandwidth. This fixes the maximum sample rate at about 20 to 30 samples/second; however, this low rate can make it difficult to maintain smooth system operation. A conventional analog system driven at 20 samples/second through a digital-to-synchro converter can result in rough, uneven operation of power drives. This places mechanical stresses on some components and can result in eventual failure. To avoid this problem, an interface can be employed that provides some type of data extrapolation for the digital conversion. The low sample-rate commands will then appear to an analog loop to be quasi-continuous if the control loop is closed properly. If the loop is closed digitally and the digital error is used, a sample rate of 20 samples/second can provide acceptable performance using position feedback data from a shaft encoder or a similar feedback sensor.

Stepping and servo motors are commonly used in control and automation applications that require precise motion to reach a digitally defined position. Acceleration of the operating speed is typically followed by deceleration of the programmed position. An alternative is to use digital interpolation to generate command pulses for a linear velocity ramp. This tends to minimize the travel time to the desired position.

This technique also allows the last motion pulse and zero velocity to occur simultaneously. The actual commands are sent to a stepper or servo motor in a closed-loop configuration. The major objective is to reduce the lag between the command and the actual positions to provide faster positioning.

An open-loop system is sensitive to aging and temperature effects in the analog circuits, which can cause variations in both the speed and the exponential time constant. If the deceleration is started at a fixed distance from the final position, the part being positioned may stop short of the destination or arrive at too high a speed and overshoot the position. A lower final speed can be employed by creeping to the commanded position, but this tends to consume excessive time.

An exponential velocity change can also be used for acceleration and deceleration. A closed-loop system with velocity and position feedback is used to produce the exponential output velocity change derived from a velocity step input command. A servo motor system requires feedback with an exponential characteristic to produce a lag between the commanded and actual positions. This lag provides the deceleration needed to avoid overshooting the commanded position.

A stepping motor control can be used for this technique. Here a voltage-controlled oscillator is coupled to a pulse generator to produce either an exponential rising or falling voltage for the acceleration or deceleration.

Another approach in stepping motor control depends on a linear voltage ramp input to the oscillator producing the command pulses. Positioning time is less than with the exponential method, but creeping is still needed. The following functions are needed:

1. A constant-frequency pulse generator which provides an output frequency proportional to the desired velocity. Each pulse represents one increment of motion.
2. An acceleration data store used to hold a digital number which defines the desired acceleration.
3. A distance data store for the digital number which represents the distance of motion required.
4. An acceleration pulse generator which is used to generate the command pulses. The frequency of the pulses is increased for acceleration and decreased for deceleration.
5. A PC-based controller which monitors the command pulses and determines the acceleration and deceleration periods, as well as when the required distance is reached.

PLCS CAN FUNCTION WITH PCS

The simpler computer operations are generally those dealing with logging or reporting functions. Many computer applications in continuous processing are involved in such tasks.

Advances in microprocessor technology have improved PLCs, so they are used increasingly for the control of processes and discrete-parts manufacturing. The early PLCs were memory-based logic controllers. Their main function was to replace relay control systems. Simple ladder-diagram programming was used; this is still a popular programming technique for PLCs. The next evolutionary step produced stand-alone controllers which performed computer-like functions beyond the capability of relay panels.

The more advanced PLCs of today offer English-language programming using functional blocks that are displayed on a CRT. Many of these PLCs have enough computing power to perform floating-point calculations and to compute basic trigonometric functions. They can also manipulate relatively large blocks of data, and have the capability to do square roots and other transcendental functions. These characteristics allow PLCs to handle many industrial application tasks normally performed by larger computers.

PLCs can be used with PCs as an integral part of factory automation. They can serve as links between machine tools, processes, and order systems. As their capacities and capabilities continue to increase, many of the newer micro-PLCs can take on more control and automation functions. Larger PLCs will be able to handle functions in process control that formerly required dedicated computers.

Motor control is a typical function for PCs and PLCs. Since PCs and PLCs can be programmed to deliver specific numbers of pulses, they can be used to drive stepper motors. Additional I/O boards are usually needed before some PLCs can drive servo motors. The PLC can generate the initial command, while a separate processor on the I/O board handles the control function and performs the continuous position corrections.

Material and part control is another performance area for PLCs and PCs. Many plants can use PCs for motion control. Even though they may not be able to supplant numerical control (NC) equipment completely, PCs can perform many tasks in motion control, such as positioning parts for assembly. The combination of PCs and PLCs offer the following:

1. Easy operation (depending on the PC software)
2. A flexible choice of control method
   a. Logic
   b. Timing
   c. Counting
3. Minimal maintenance effort
4. Easy troubleshooting
5. Rapid, reliable documentation
6. Changeable control sequences

PCs and PLCs can do machine control, testing, and recording of counts, run times, parts produced, and fault conditions.

Both labor and energy are growing more expensive, and manufacturing firms must increase their productivity to remain competitive. In a manufacturing process, one way to achieve a productivity increase is through the use of transfer machines to automate a particular step or series of steps. In the past, most transfer machines were restricted to high-volume processes because of high start-up costs and because they usually had to be dedicated to manipulating a single part. The carousel-type transfer machine with chain plates can be used to accommodate many different manufacturing fixtures. A PLC under the supervision of a PC can be used as the machine's central control mechanism.

A typical transfer mechanism system requires system head controls and indicator lights for system status. Each head on a transfer machine requires a control system which generates a head-cycle-complete signal as a system interface. A head can be classified in one of the following four general categories:

1. Attachment devices or machines, such as eyeletting heads, spot welders, riveters, or nut runners, used to fasten two or more parts together.
2. Feeders or devices that place parts in an assembly. These range from simple pick-offs to complex escapement mechanisms with several axes of motion.
3. Inspection stations, ranging from a single sensor to determine if a part is in place to an automatic tester which examines the completed device.
4. Ejection stations, which range from simple air blowoffs to pick-and-place units that retain the part's orientation.

WHY USE A PLC INSTEAD OF A RELAY BOARD?

The strongest reasons for using a PLC are reliability and maintainability. If a machine makes 40 parts each minute and you allow for 50 production minutes, or 2000 parts, each hour, the machine will produce, on a one-shift, 5-day-week basis, 4,160,000 parts each year.

The life expectancy of a good relay ranges between 200,000 and 2,000,000 operations, depending on the electrical load. Transient suppression circuits will help, but even with these, you can expect to replace each relay a minimum of two to three times each year. Even on a small, eight-relay system, this means about 20 failures a year.

Relays fails in many ways, but seldom is the first failure the final one. Usually a failing relay operates intermittently, it contacts either hanging closed or failing to make. The maintenance department can usually get things moving again by repairing the contacts or replacing the relay. But even with a well-planned maintenance effort, the lost production time for each failure will average about 3 hours or 6000 parts. This represents an annual loss of 120,000 parts. Electronic relays can reduce the failure rate, but if the system uses a PC or PLC under the same operating conditions, the failure rate can be much less.

Built-in error checking LED indicators on inputs and outputs and self-testing techniques can reduce downtime to less than 2 hours for each failure, which represents a large reduction in lost parts.

Microprocessor-based PLCs allow closed-loop, point-to-point servo positioning. The interrupt mechanism is used to divide the operation of the microprocessor between ladder logic scanning and motion control. This allows the PLC to service the control loop frequently enough to provide the positioning needed. These higher-level PLCs are not suitable for most contouring applications, but they can be used on certain types of production equipment, like grinding machines and transfer devices, where precise positioning is required.

Many PLCs can handle simple process equations and correct some process variables in real time. Some high-level PLCs are capable of more complex machine control.

3

Amplifiers and Signal Conditioning

Operational amplifiers are used in many data acquisition applications, and their characteristics are important in understanding many control applications. We will show how to use operational amplifiers, discuss the characteristics of idealized operational amplifiers, and show how both noninverting and inverting amplifiers work.

An understanding of the important gain parameters is required, and the use of difference amplifiers is explained. Some of the more important characteristics of operational amplifiers include input impedances, offset voltage, input noise, common-mode voltage, open-loop gain, overload recovery, settling time, slew rate, and temperature drift effects.

Instrumentation amplifiers are particularly useful in control circuits. We will explain the different configurations used and the application of reference and sense terminals. Isolation amplifiers are another extremely useful operational technology. We will explain their use and develop their utility. Noise considerations become important, along with mutual capacitance and ground loops. The use of shields and floating inputs will be explained, and we will explore cable selection and measuring devices. Amplifier selection tips will be given. Frequent system problems will be explored, and general troubleshooting advice will be given.

OPERATIONAL AMPLIFIERS

The operational amplifier got its name from its early association with computing circuits to simulate various mathematical operations. It is used in many A/D conversion and interface circuits. Applications include scaling and input conditioning, sample hold, precision comparisons, current-to-voltage and voltage-to-current conversion, and active filtering.

An understanding of the operational amplifier is necessary to appreciate its many possibilities. The objective of this chapter is to assist the reader in understanding how and where operational and instrumentation amplifiers are used in typical PC control applications. The chapter starts with a brief discussion of operational amplifier principles, followed by guidelines to definitions and specifications. Instrumentation amplifiers are then introduced and its applications are discussed, as well as grounding and shielding problems.

USING OPERATIONAL AMPLIFIERS

No single operational amplifier can meet the requirements of every application. The concept of an ideal operational amplifier is always useful in the initial analysis of a given application and for synthesizing the basic circuit configuration to perform some mathematically described task. The ideal device is useful for deriving a closed-loop relationship that can be directly applied to many real circuits, with little error in most cases. When the desired circuit configuration has been established, the desired error can be analyzed in terms of the following factors:

1. Static sources such as dc offset errors.
2. Dynamic sources such as gain, common-mode errors, and the response to rapidly changing inputs.

IDEAL OPERATIONAL AMPLIFIERS

The symbol for an operational amplifier is shown in Figure 3-1a. The triangle points in the direction of signal flow. Considering the amplifier as a four-terminal network, the output e_0 is related to the inputs $(e_1 - e_2)$ as shown in Figure 3-1b.

For changes in e_1 that are positive with respect to e_2, the output moves in the same sense or phase, and the terminal at e_1 is marked positive (+) as the noninverting, or reference, input. Corresponding variations in e_2 cause the output to move in the opposite sense, or 180° out of phase. This terminal is marked negative (−) as the inverting input. The inverting input terminal is also known as the *summing point*.

The idealized properties generally assumed for the ideal performance of the operational amplifier are as follows:

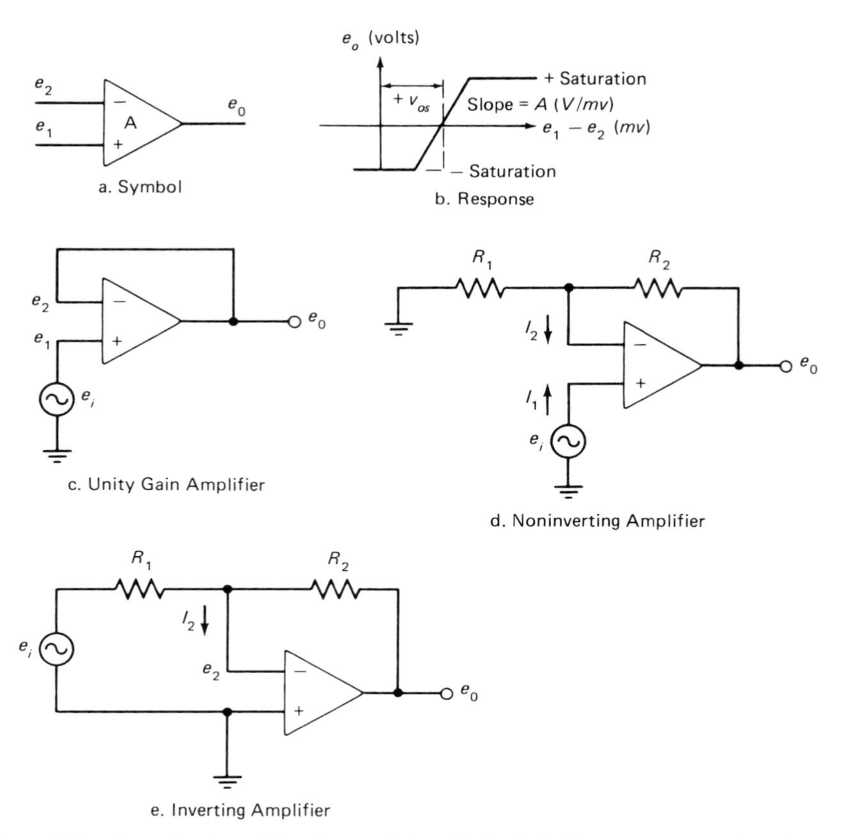

Figure 3-1 Operational amplifier characteristics. (a) Symbol; (b) response; (c) unity gain amplifier; (d) noninverting amplifier; (e) inverting amplifier.

Ideal Value	Typical Value
Open-loop gain $A \rightarrow \infty$ dc to high frequencies	100 k V/V
Voltage offset $V_{os} = 0$ volts	+1 mV @ 25°C
Bias currents $I = I_1 = I_2 = 0$ amperes	10^{-14}–10^{-6} A
Input impedance $Z_{in} \rightarrow \infty$	10^5–10^{11} ohms
Output impedance $Z_o = 0$ ohms	1–10 ohms
Common-mode rejection $CMRR \rightarrow \infty$	60–120 dB

Based on the ideal values, the operational amplifier acts like a two-level, high-gain comparator if used in the open-loop configuration. They can be used as comparators in this open-loop circuit configuration. The closed-loop configuration offers stability, along with the other benefits of operational circuitry. It is necessary to employ negative feedback to obtain the required stability. With negative feedback, the operational amplifier circuit acts as a control loop dedicated to maintaining the error voltage between the inputs close to zero. In other words, if the output voltage is to be within its desired range, $e_1 - e_2$ must approach zero. The relation between the inputs and output of Figure 3-1a is

$$e_1 - e_2 = \frac{e_0}{A}$$

As

$$A \to \infty \; e_1 - e_2 \to 0$$

and

$$e_1 \cong e_2$$

In the nonideal case, $e_1 - e_2$ is actually

$$e_1 - e_2 = \frac{e_0}{A} + \frac{e}{CMRR} \quad (1)$$

Neglecting the *CMRR* term, in order for

$$e_0 = 5 \text{ V}$$

if

$$A = 100{,}000$$

Then

$$e_1 - e_2 = 50 \; \mu\text{V}$$

The simplest feedback circuit is the ideal unity-gain follower of Figure 3-1c, in which all of the output voltage is fed back to the negative input. The input signal source is connected directly to the noninverting input, and the output will follow.

Summing voltages around the loop,

$$e_i + (e_2 - e_1) - e_0 = 0$$

since

$$e_2 - e_1 \to 0, \qquad e_0 = e_1$$

The gain is unity while the input impedance is infinite. The output impedance is zero, since we are considering an ideal amplifier. In practice, the major performance-limiting factors are the common-mode voltage range and error and the bias current if the source impedance is high. Changes of bias current with common-mode level can also be important.

THE NONINVERTING AMPLIFIER

In Figure 3-1d, the entire output voltage is not fed back. Instead, it is divided through an attenuator to reduce the feedback voltage:

$$e_0 = \left(\frac{R_2}{R_1} + 1\right) e_i$$

The gain is ideally determined only by the passive components used. It is positive and equal to or greater than unity. The voltage follower is the special case for

$$R_2 = 0$$

The input impedance is infinite for

$$I_1 = 0$$

and the output impedance is zero. However, $R_2 + R_1$ is now part of the load on the operational amplifier output. The same basic limitations as for the follower apply, except that any common-mode problems are reduced as the gain is increased.

THE INVERTING AMPLIFIER

The basic circuit for the inverting amplifier is shown in Figure 3-1e. In this circuit, the noninverting, or reference, input is connected directly to ground and the input signal is connected at R_1, which forms part of the negative feedback network. In the noninverting circuit, e_2 follows the input signal e_1 through the common-mode range. In the inverting amplifier, e_2 still follows e_1, but it is constrained to zero within the gain error of the amplifier (e_0/A). This makes e_2 act like a virtual ground.

Summing currents at e_2 gives

$$\frac{e_i - e_2}{R_1} + \frac{e_0 - e_2}{R_2} - I_2 = 0$$

Since

$$e_2 = e_1 = 0 \qquad \text{and} \qquad I_2 = 0$$

$$\frac{e_i}{R_1} + \frac{e_0}{R_2} = 0$$

$$e_0 = \frac{-R_2}{R_1}(e_1)$$

The gain is determined only by the passive components used to set the amount of feedback. But in this circuit the output is an inverted version of the input. The input impedance seen by the signal source is

$$Z_{\text{in}} = R_1$$

The output impedance remains zero:

$$Z_{\text{out}} = 0$$

This circuit has several unique characteristics.

VIRTUAL GROUND

Voltage e_2 is essentially at ground potential.

VOLTAGE-TO-CURRENT TRANSCONDUCTANCE

The current through R_2

$$\frac{e_0}{R_2}$$

is determined by

$$\frac{e_i}{R_1}$$

independently of

$$R_2, e_0$$

R_2 may be nonlinear (a diode) or may contain storage (a capacitor), which produces integration, or an n-terminal network (an attenuator).

CURRENT-TO-VOLTAGE TRANSRESISTANCE

If a current source is connected at the negative terminal, the current will flow through R_2 to the output and develop an output voltage $-iR_2$, which depends only on the values of i and R_2. The virtual ground at this point is an ideal load for current source. The output resistance is reduced by the loop gain.

SUMMATION

Since the negative terminal is maintained at virtual ground, the currents developed by current sources, voltages, and resistors will flow through R_2, developing an output voltage that is dependent on the sum of the currents. The summing point acts as a virtual ground, or current sink, only when the amplifier's output is linear. If the amplifier is saturated, it is no longer amplifying and must be treated using an equivalent circuit for the new mode of operation.

In normal operation, there is little voltage difference across the differential input. The differential input impedance has a negligible effect on the circuit. If the amplifier is used in a charging mode, with a capacitance, to form a differentiator, integrator, or charge amplifier, a resistance should be used in series with the amplifier's input terminal for protection against transient charging currents.

GAIN PARAMETERS

There are five types of gain that can be defined for operational amplifiers. These will now be discussed.

Open-loop gain (A) is the relationship between the output voltage and the differential input voltage. The output voltage is specified for a certain value of load resistance. A lighter load will result in a higher open-loop gain, depending on the output impedance of the operational amplifier.

Feedback ratio (beta) is the total amount of voltage fed back from the output to the amplifier's input. It is a function of the whole circuit from output back to input, including stray circuit elements and the input-impedance characteristics of the amplifier. Its inverse (1/beta), which is often called the *closed-loop gain*, is the ideal closed-loop gain.

The *loop gain* ($A \times$ *beta*) is the fraction of open-loop gain that is available for error correction. The magnitude of the loop gain can be used to represent a figure of merit for a particular circuit, and the magnitude-phase relationship can be used to predict the closed-loop stability.

Closed-loop gain is the gain for signals in series with the positive input terminal or, if opposite polarity is used, for a signal in series with the negative input. Closed-loop gain is

$$\frac{1}{\text{beta}} \cdot \frac{1}{1 + \cfrac{1}{A \times \text{beta}}}$$

Offset voltage, drift, and noise voltage are usually referred to the input, and they are amplified by the closed loop gain. Because of this, it is sometimes called the *noise gain*. When $A \times$ beta is much greater than 1, the closed-loop gain is 1/beta.

Signal gain is the closed-loop transfer relationship between the output and any signal input to an operational amplifier circuit. For example, suppose that the amplifier is connected as an inverting summing amplifier with gains of 1 and 2 for the two inputs being summed, $A = 100{,}000$, and beta $= 1/5$. Then the closed-loop gain is approximately 5:

$$A \times \text{beta} = 20{,}000$$

An offset of 100 millivolts referred to the amplifier's input becomes 0.5 millivolts at the output. An offset of 100 millivolts referred to the actual signal inputs becomes 0.5 millivolts, for a gain of 1, and 167 volts, for a gain of 3.

DIFFERENCE AMPLIFIERS

To analyze an operational amplifier connected differentially, the equivalent circuit of Figure 3-2 can be used.

If e_{CM} is zero, the circuit acts as a subtractor, while for e_1 or e_2 zero, the circuit reduces to a bridge or differential amplifier, with or without the common-mode voltage e_{CM}.

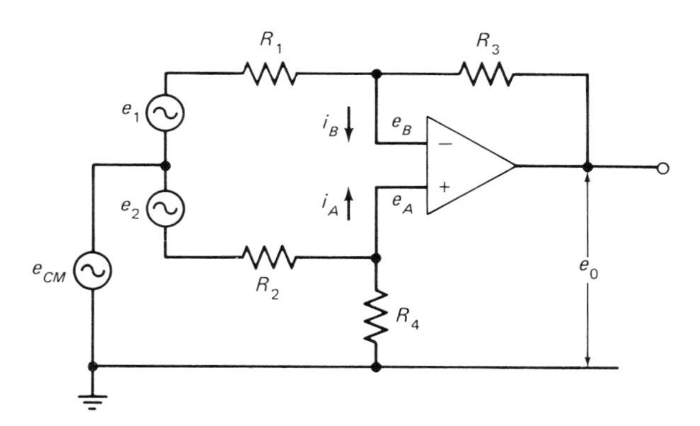

Figure 3-2 Difference amplifier.

Using superposition and the usual assumptions for a ideal operational amplifier, if

$$\frac{(R_4)}{(R_2)} = \frac{(R_3)}{(R_1)}$$

then

$$e_0 = (e_2 - e_1)\left(\frac{R_3}{R_1}\right)$$

Thus it is important that $R_3/R_1 = R_4/R_2$ if common-mode errors due to e_{CM} are to be eliminated. R_1 and R_2 should include the internal impedance of the sources e_1 and e_2. The source impedance of e_{CM} is important only in determining the actual amount of common-mode signal level at e_A. The actual common-mode signal seen by the amplifier is e_A, not e_{CM}.

Any common-mode error is due to resistance-ratio mismatch. The common-mode response of the amplifier is dependent on this mismatch. If the circuit is linear, an adjustment in the matching of resistors may be used to compensate for the amplifier's common-mode error.

OPERATIONAL AMPLIFIER APPLICATIONS

All of the operational amplifier circuits described in this section use resistive feedback elements. In general, any complex linear or nonlinear feedback element, passive or active, may be used (provided that the configuration is stable). Typical control and measurement applications include the following:

1. Voltage-to-current and current-to-voltage conversion.
2. Computing sums and differences.
3. Scaling, rectifying, and filtering.
4. Modulating and demodulating.
5. Classifying.
6. Peak following.
7. Logarithmic functions.
8. Multiplying/dividing.
9. Extracting square roots.
10. Integrating/differentiating.
11. Oscillation.
12. Phase shifting.

The number of possible useful circuits using operational amplifiers continues to grow as new uses are thought up for these devices.

CMRR

The ideal operational amplifier responds only to the difference voltage between the inputs. It produces no outputs for a common-mode voltage, which occurs when both inputs are at the same potential. Due to slight differences in the gains between the plus and minus inputs, common-mode input voltages are not entirely subtracted at the output.

When the output error voltage is referred to the input by dividing by the closed-loop gain, it reflects the common-mode error voltage between the inputs. The common mode rejection ratio (*CMRR*) is defined as the ratio of common-mode voltage to common-mode error voltage. *CMRR* is sometimes expressed in decibels.

$$CMR = 20 \log_{10} (CMRR)$$

The common-mode voltage error can be a nonlinear function of the common-mode voltage, and it also varies with temperature. This is especially true for field effect transistor (FET)-input amplifiers. An average value of *CMRR* is usually specified. The incremental *CMRR* for a large common-mode voltage can be less than the average *CMRR* specified, but it can be much greater for lower voltages. Most *CMRR* specifications apply only to dc input signals, while *CMRR* decreases with increasing frequency.

MAXIMUM DIFFERENTIAL INPUT

Under most operating conditions, the feedback loop maintains the error voltage between inputs very close to zero. However, in some circuits, such as voltage comparators, the voltage between inputs can become large. The maximum differential input voltage defines the maximum voltage which can be applied between inputs without causing permanent damage to the amplifier.

FULL-SCALE RESPONSE

The large-signal and small-signal response characteristics of an operational amplifier differ. An amplifier does respond to large-signal changes as fast as small-signal characteristics indicate, mainly due to slew rate limiting in the output stages. Full-scale response can be specified in two ways; full linear response and full peak response.

Full linear response is the maximum frequency at unity closed-loop gain, where a sinusoidal input signal will produce full output at rated load without exceeding a specified distortion level. This specification does not relate to any gain reductions with frequency. It refers to distortion in the output signal. The generally accepted values for distortion levels range from 1% to 3%. In some applications, the short-term distortion which is caused by exceeding the full linear response can be ignored. A more serious effect is the dc offset voltage that can be generated when the full linear response is exceeded. This is caused by rectification of the asymmetrical feedback waveform or overloading of the input stage, with large distortion signals at the summing junction.

Some amplifiers for high-frequency applications provide a full output swing beyond the distortion-based limit discussed above. A linear waveshape is not important in the use of these devices, and they are specified for the maximum frequency at which they provide a full output swing. This is called the *full peak response* or *full power response*.

INITIAL BIAS CURRENTS

The bias current is the current required at either input from an infinite source impedance to drive the output to zero, assuming that there is no common-mode voltage. In a differential amplifier, bias current is present at both the negative and positive inputs. Most specifications refer to the higher of the two, not to the average. In a single-ended amplifier, the bias current refers to the current at the active input.

The initial bias current is the bias current at either input measured at +25°C, the rated supply voltages, and zero common-mode voltage. Internal or external compensation can be used to reduce the initial bias current. An external compensating resistor can be connected to zero the initial bias current. Compensating initial bias current usually has only a small effect on the bias-current temperature coefficient. The bias current of FET amplifiers will increase by a factor of 2 for each 10°C rise in temperature.

INITIAL DIFFERENCE CURRENT

The difference current is the difference between bias currents of a differential amplifier. The input circuit of a differential amplifier is usually symmetrical, so that the bias currents at both inputs tend to be the same and tend to track with changes in temperature and supply voltages. The difference current is about one-hundredth of the bias current at either input, provided that any initial bias current has not been compensated.

DRIFT

Parameters like offset voltage, bias current, and difference current will change as components age. The time drift for amplifiers does not accumulate linearly. For example, the voltage drift for an amplifier might be specified as 1 microvolt per day, but the cumulative drift might be specified for 30 days not to exceed 5 microvolts. The drift accumulation can be extrapolated by multiplying the specified drift per day by the square root of the number of days. For drift per month, divide by 5.5 to obtain drift per day.

INPUT IMPEDANCES

The differential input impedance is the impedance between the two input terminals, usually measured at +25°C and assuming that the error voltage is close to zero. The dynamic impedance can be represented by an equivalent capacitance in parallel with the resistive component of the impedance.

The common-mode impedance is defined as the impedance between each input and ground and is usually specified at +25°C. In most circuits, common-mode impedance at the negative input is not significant, except for the capacitance it adds at the summing junction. Common-mode impedance at the plus input R sets the upper limit on the closed-loop input impedance for the noninverting configuration.

The dynamic impedance can be represented by a capacitance in parallel with the resistive component. The usual range for this capacitance is 5–25 pF for the plus input.

The common-mode impedance is a nonlinear function of both temperature and common-mode voltage. In a FET-input amplifier, the common-mode impedance is reduced by a factor of 2 for each 10°C temperature rise.

INITIAL OFFSET VOLTAGE

The offset voltage is the voltage required in series with the input from a zero source impedance to drive the output to zero. The initial offset voltage is usually defined at +25°C and rated supply voltages. In many amplifiers, the initial offset can be set to zero with an external resistance.

INPUT NOISE

Input voltage and current noise characteristics are specified much like offset voltage and bias current characteristics. Drift can be considered to be noise which occurs at very low frequencies. The main difference in measuring and specifying noise as opposed to dc drift is that the bandwidth must be considered. Two noise specifications are usually given. The low-frequency random noise for a bandpass of 0.01 to 1 Hz is usually specified as peak-to-peak, while wideband noise in a bandpass of 5 Hz to 50 kHz or more is usually specified in root-mean-squared (rms) values.

MAXIMUM COMMON-MODE VOLTAGE

In a differential-input amplifier, the voltage at both inputs can have values above or below ground potential. The common-mode voltage is defined as the voltage above or below ground when both inputs are at the same voltage. It can be specified as the maximum peak common-mode voltage which will produce less than a certain error at the output, usually 1%. The common-mode voltage establishes the maximum input voltage for the voltage follower circuit.

OPEN-LOOP GAIN

The open-loop gain (A) is defined as the ratio of a change of output voltage to the error voltage applied between the amplifier inputs required to produce the change. Gain is usually specified only at dc, but in many applications the frequency dependence of gain is important and is specified as the open-loop gain response.

OVERLOAD RECOVERY

This relates to the time required for the output voltage to recover to the rated output voltage from a saturated condition. In some amplifiers, the overload recovery increases for impedances greater than 50K ohms. Most specifications apply for low impedance and assume that overload recovery is not degraded by stray capacitance in the feedback network. The overload recovery is usually defined for a 50% overdrive.

RATED OUTPUTS

The rated output voltage is the minimum peak output voltage which can be obtained at rated output current before clipping or excessive nonlinearity occurs. The rated output current is the minimum value of current that can be supplied at the rated output voltage.

SETTLING TIME

The settling time is the time required, after an input step change, for the output to reach and remain at the final value within a band of specified magnitude. This time includes an initial delay, a period of slewing, a period of recovery where an overshoot may occur, and a final relaxation period to the defined band.

SLEW RATE

The slew rate is the maximum rate of change of output voltage for a large input step change.

TEMPERATURE DRIFT

Parameters like offset voltage, bias current, and difference current change or drift from their initial values with changes in temperature. This is the major source of error in most applications. The temperature coefficients of these parameters are defined as the average slope over a specified temperature range. Drift is a nonlinear function of temperature, and parameter changes are greater at extremes of temperature.

One method of specifying this parameter involves subtracting the measured offset values at the upper and lower temperature extremes and dividing this difference by the temperature excursion. This can be misleading if offset drifts in the same direction at the two extremes. It is possible to have no difference in the two endpoint measurements, yet there may be a large difference in the center of the range. In this case, the specified drift is zero.

A better technique involves adjusting the amplifier to zero at room temperature and ensuring that the offset at any temperature does not exceed the value of the specified drift rate.

UNITY-GAIN SIGNAL RESPONSE

The unity-gain small-signal response is the frequency at which the open-loop gain becomes unity or 0 dB. The small-signal response is used, since in general it is not possible to obtain a large output voltage swing at high frequencies because of distortion due to slew rate limiting or signal rectification. If the amplifier has a symmetrical response on both inputs, this parameter may be obtained from either the inverting or the noninverting circuit. Some wideband amplifiers restrict high-speed use to the inverting circuit.

INSTRUMENTATION AMPLIFIERS

In many control and measurement applications, it may be necessary to retrieve millivolts of analog data from volts of common-mode interference. In some applications, it may be necessary to isolate the amplifier's input galvanically from its output and the power source. This may be done either to protect the amplifier from high voltages or to protect the measured unit, such as a hospital patient, from stray leakage current. In other cases, it may be done to obtain a better common-mode rejection.

The devices available for these applications are known as *instrumentation amplifiers*, which includes the subclass of isolation amplifiers. These amplifiers often contain operational amplifiers, but they are distinguished from the latter in being committed devices with a more fixed configuration. They also have a definite set of output-input relationships and are designed to meet specific parameter requirements like high *CMRR*, low noise, and drift. Instrumentation amplifiers generally have a moderate bandwidth and often employ a limited range of gains, such as 1 to 1000, which can be programmed by a single resistor.

The instrumentation amplifier is a committed-gain amplifier that normally uses internal high-precision feedback networks. Its main features are low drift, good linearity, and noise rejection. It is often used for extracting and amplifying low-level signals in the presence of high common-mode noise. These devices are popular transducer amplifiers for thermocouples, strain-gauge bridges, current shunts, and biological probes. When used as preamplifiers, they are useful for extracting small differential signals superimposed on large common-mode voltages. Wideband units are popular for data acquisition applications.

A wide variety of instrumentation amplifiers are available, ranging from units with digitally programmed gains and autoranging for low-level multiplexed systems to simple, low-cost instrumentation amplifiers. Low-cost amplifiers encourage the growth of per-channel amplification in measurement applications.

Circuit Configurations

Figure 3-3 shows some popular circuit approaches. All of these use only one resistor for adjusting the gain. Most commercial units employ feedback sense and reference terminals for lead compensation, current-output sensing, and adjustable offset reference voltage. The basic subtractor shown in Figure 3-3a uses only one operational amplifier. It does not have good source unbalance characteristics due to low input impedance and the need for critical resistance matching. A FET-input amplifier can be used with large values of resistance, but noise and bandwidth become more of a problem.

Buffers can be added as shown in Figure 3-3b. The additional operational amplifiers can be FET-input types for signal inputs having a high source impedance. Matched input followers can provide low drift and high *CMRR*, provided that the main amplifier's drift is low and the resistances are well matched. Resistance matching can be relaxed and the bandwidth improved for high-gain applications by using the buffered inputs with gain shown in Figure 3-3c. The buffers provide unity gain for the common-mode signal and increase overall the common-mode resistance by the differential gain of the first stage. The use of separate followers with gain would not allow this improvement, since they would amplify differential and common-mode signals equally. Matched amplifiers will help *CMR* and drift stability. The two-stage amplifier circuit shown in Figure 3-3d has high input impedance and saves the cost of an amplifier, but it also increases dependence on resistance match for high *CMR*.

These circuits are sensitive to resistance in series with these terminals, which must be matched to prevent common-mode errors. A differential current-feedback configuration is sometimes used to provide higher impedance at the sense and reference input terminals, which allows a wider range of applications.

Applications

In data acquisition systems, instrumentation amplifiers are used for pre-amplification and signal conversion. Since they are designed to respond only to the

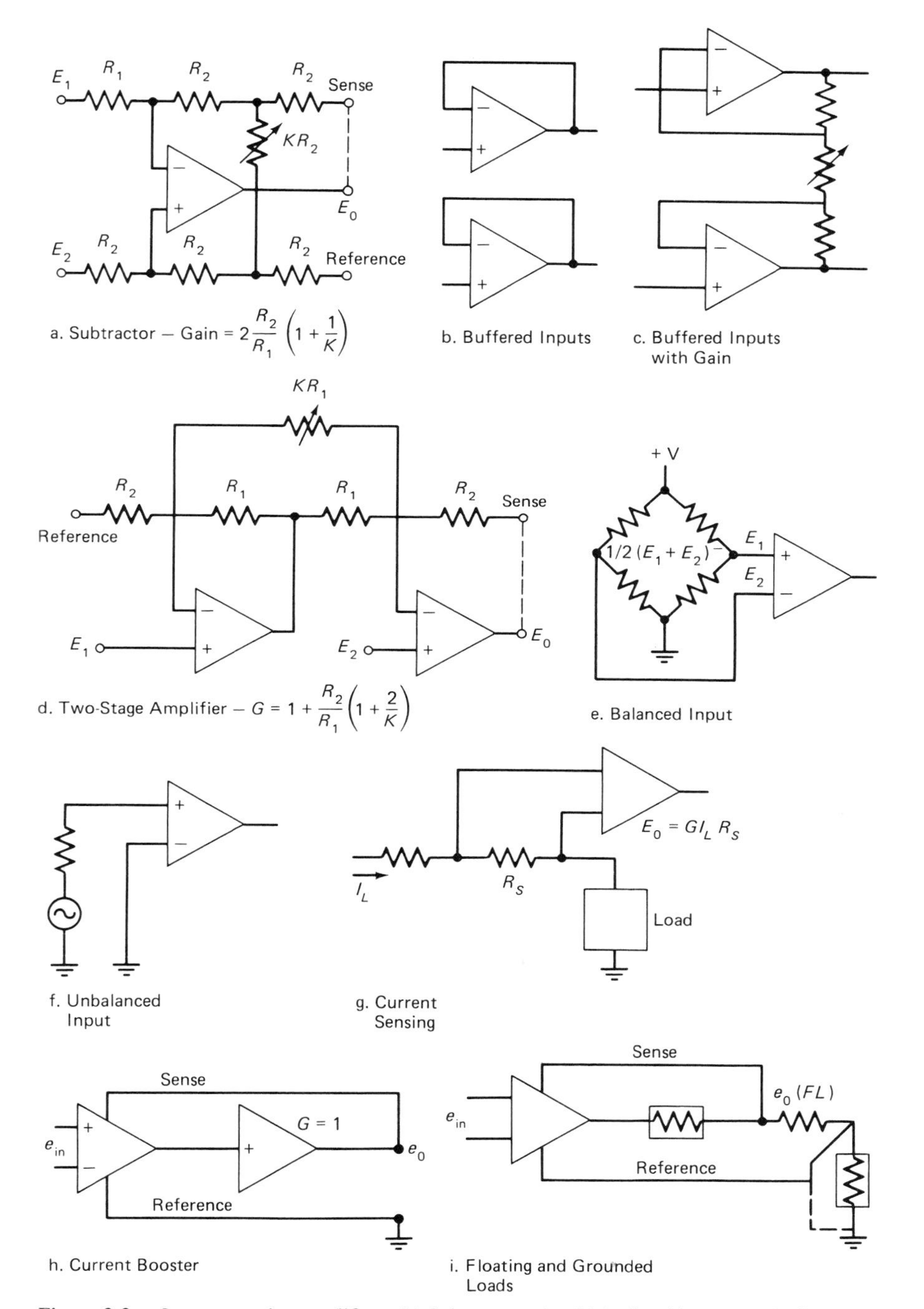

Figure 3-3 Instrumentation amplifiers. (a) Subtractor gain; (b) buffered inputs; (c) buffered inputs with gain; (d) two-stage amplifier; (e) balanced input; (f) unbalanced input; (g) current sensing; (h) current booster; (i) floating and grounded loads.

difference between two voltages, they can be used in both balanced and unbalanced systems.

A balanced system implies that the output of the signal source appears on two lines and that these lines are balanced with respect to the source resistances and output voltages in relation to either ground or the local common-mode level. This is illustrated in the bridge circuit of Figure 3-3e.

In an unbalanced system, symmetry is not part of the configuration, as shown in Figure 3-3f. In these systems, the instrumentation amplifier is used to reduce the effects of ground potential differences.

Since instrumentation amplifiers can measure voltage differences at any level within a certain range, they are often used for current measurement. In this application, they are used to measure and amplify the voltage that appears across a low-resistance shunt which is often placed in the high side of the circuit (Figure 3-3g).

Reference and Sense Terminals

In a typical application, the sense and reference terminals are connected to the specific points in the circuit where the output is to be accurately maintained. This is done to eliminate voltage drops in the output signal or ground lines. In a differential amplification circuit, it is important to consider the possibility of feedback problems and to take precautions to avoid dynamic problems such as overshoots, ringing, or oscillation.

The sense and reference terminals can be used, for example, with a current-booster follower. Since the booster is part of the feedback loop, offsets, drifts, and gain errors are nullified by using the sense and reference functions, as shown in Figure 3-3h.

When the sense and reference terminals have a high enough impedance to avoid limit-loading effects, they can be used for driving current to either floating or grounded loads. This is shown in the circuit of Figure 3-3i for a floating load where the reference is grounded and for a grounded load where the sense terminal is connected to the output.

When the reference terminal of an instrumentation amplifier is available for connections, it can be used for biasing. It can be used to bias out dc normal-mode voltages that occur from contact potentials, or it can be used to adjust the bias for relay or comparator trip points.

The reference terminal can be driven by the output of an operational amplifier with either constant or variable voltage. When the amplifier has high input impedance at the reference terminal, it can be adjusted with a voltage divider.

Specifications

Table 3-1 shows some typical specifications for instrumentation amplifiers. In many respects, they are similar to those for operational amplifiers. The major differences include the following:

TABLE 3-1 Typical Specifications for Instrumentation Amplifiers

Gain range	1 to 1000
Nonlinearity	±0.01%
Rated output	±10 V at 10 mA
Frequency response	
Gain = 1	300 kHz
Gain = 1000	300 Hz
Full power response	1.5 kHz
Slew rate	0.1 V/microsecond
Unity gain settling time to 0.1%	100 microsecond
Offsets referred to input	
Unity gain (temperature)	±150 microvolts/°C
Unity gain (supply)	±0.2 mV/%
Input bias current	
Initial, 25°C	0, +100 nA
Vs. temp.	1 nA/°C
Input difference current	
Initial, 25°C	+100 nA
Vs. temp.	1 nA/°C
Input impedance	
Differential or common mode	10^{9} ohms
Noise referred to input	
$G = 1$	15 microvolts p-p, 0.1 to 10 Hz
$G = 1000$	1.5 microvolts p-p, 0.1 to 10 Hz
Voltage noise, $G = 1$	5 microvolts rms, 10 Hz to 10 kHz
Input voltage range	
Linear differential input	±10 V
Max. differential input	±20 V
Max. common mode	±10 V
Reference terminal	
R_{in}	10^{6}
Output offset range	+10 V
Gain offset range	+1.00
Bias current	20 nA

1. *Gain:* An open-loop gain specification is not required, but gain nonlinearity and instability are specified.
2. *Offset:* Offset drift specifications may be given, referred to the input at two gains, A and B. To determine the corresponding specification at any arbitrary value of gain referred to the output, the following formula can used to provide a useful approximation:

$$\text{Drift}|_G = \frac{(A)\ \text{Drift}|_A - (B)\ \text{Drift}|_B}{1000}(G) + \text{Drift}|_B$$

This works well for wide ranges of gains, such as 1 and 1000.

Some instrumentation amplifier specifications may be listed for specific values of gain. Intermediate values can usually be interpolated. Specifications not associated with a gain value are essentially independent of gain. Typical specifications are shown in Table 3-1.

Errors from a number of sources may not be individually significant, but the total error can add up quickly. It is often useful to perform an error-budget analysis to find the most significant terms.

ISOLATION AMPLIFIERS

Some applications require galvanic isolation of the amplifier's input circuit from its output and the power supply. These include medical electronics and applications with high common-mode voltages (500 volts or more) between input and output. Other applications require a two-wire input, without any ground return for bias currents or a high *CMRR*.

The two most popular approaches to isolation amplifiers are transformer and optical coupling. Optical coupling is effective for isolation, since it uses a portion of the electromagnetic spectrum that isolates it from normal electrical current effects. Isolation amplifiers employing transformer coupling are available with galvanic isolation and the following characteristics:

1. A low capacitance (less than 10 pF) between input and output ground circuits.
2. A high *CMRR* (115 dB at 60 Hz).
3. High common-mode voltage ratings to 5k volts and beyond.

These devices are capable of transmitting millivolt signals in the presence of up to 1000 volts common-mode, with unity gain or adjustable gains. They are useful for medical applications where an EKG waveform is the input signal to the data system and where the patient must be isolated from potentially lethal ground-fault currents. They are also used in measurement applications where it is necessary to interrupt ground loops between transducers and output-conditioning circuits.

Most amplifiers of the transformer type have committed gain circuits with internal feedback networks and operate from dc to 2 Hz. They are usually divided into two sections: an isolated front-end amplifier section and a grounded output section. The front end includes a fixed-gain operational amplifier, modulator, and dc regulator circuit which are enclosed in a floating guard shield. The output section contains the demodulator, filter, and power-supply oscillator circuit. Power is transformer-coupled into the shielded input circuits and capacitively or magnetically coupled to the output demodulator circuit.

NOISE CONSIDERATIONS

Electrical noise refers to unwanted and interfering voltages developed either inside or outside a system which degrade the performance of the system. Interfering noise can be handled by filtering, which attempts to reduce the noise after it enters the system. This method can be expensive compared to other methods, but it can be effective for systems that are not too large or complex. When high information rates are used along with low-level analog or digital pulse systems, a different approach is needed.

Filtering can produce excessive deterioration of the desired pulse waveforms, as well as inaccuracies and distortion of analog signal voltages. The best way to achieve noise reduction is to keep the noise out of the system. This can be accomplished with the use of noise-rejecting cables and equipment isolation and grounding techniques.

Not all equipment produces unwanted noise, but when it is connected to other equipment to form a system, the interconnecting wiring can be a source of noise due to ground loops, common-mode returns, and capacitive or inductive pickup of radiated fields. A desired signal in one circuit can be noise in another. The noise can be produced by local circuits in the system or by equipment external to the system. Cables between circuits can radiate the signal they are carrying into adjacent circuits. The amount of unwanted noise is increased by poor cable-to-equipment impedance matching, which produces signal reflections and high-standing wave ratios. The lower the signal voltage level, the greater is the susceptibility to any interference. The following noise sources are most commonly encountered:

1. *Inductive pickup from power sources:* This includes 60-Hz noise from power lines, 120-Hz noise from fluorescent lighting, and higher frequencies from electric arcs (motor brushes and welding) and pulse-transmitting devices.
2. *Electrostatic coupling to ac signals:* The distributed capacitance between signal conductors, and from signal conductors to ground, provides low-impedance ac paths for crosstalk and signal degradation. The internal capacitance of power transformers can also cause voltage fluctuations and electrical transients in power lines to be conducted into electronic systems through the ac power connection.
3. *Common impedance coupling ground loops:* When there is more than one ground on a signal circuit, a ground loop is produced. The loop provides a closed path which can generate 60-Hz noise from circulating 60-Hz currents and degrade the signal.
4. *Inadequate common-mode rejection:* Common-mode signals include common-mode voltages and in-phase signals that appear simultaneously at both differential amplifier input terminals. These signals must be rejected without

changing the useful signal, which is the difference in voltage across the input terminals of the amplifier.

MUTUAL CAPACITANCE AND GROUND LOOPS

Figure 3-4a shows the three most important mutual capacitances between an amplifier and its shield enclosure, Conductor 3. The input-signal reference may be called the *signal reference conductor*, or *zero-signal reference conductor*. It is common to both input and output. Figure 3-4b shows the equivalent circuit for Figure 3-4a.

The three mutual capacitances form a feedback path from output to input, providing a source of instability and noise. Capacitor C_3 is eliminated by connecting the signal reference (3) to the shield, as shown in Figure 3-4c.

Any voltages that exist between separate ground points are called *ground-loop potentials*. These have the ability to cause ground-loop currents when a ground loop exists, which occurs when a current has more than one ground-reference point. This can be an earth ground or current ground. The path 3-4-6-3 in Figure 3-4d provides a ground loop for currents.

SHIELDS

A shielding enclosure protects the contained volume by terminating most of the electric field lines or charges that are supplied to it. These charges are supplied by some source through the ground point for the shield. The motion of these charges becomes a current which flows in the shield. This current must be prevented from flowing into the input lines of an amplifying circuit. This can be accomplished by connecting the shield to only one point, which becomes the signal reference point, as shown in Figure 3-4c. The use of separate grounds can result in induced voltages, as shown in Figure 3-4d.

When the shield is divided into sections, all of the sections should be connected in series and then connected to the signal reference point at the signal earth point. If this is not possible, then the loop currents should be confined to the high-signal level portions of the circuit.

Entry points for power to the amplifier can become noise entry points unless batteries are used. One side of the power line is usually at its own ground reference, which is not the same as the signal reference. Try to avoid bringing the power supply reference into the amplifier circuit because of the potential difference that exists between the power ground and the circuit ground. Although breaks in the amplifier shield can cause unwanted signals at the amplifier reference, this is usually small compared to the pickup from induction along a long line to the signal reference point.

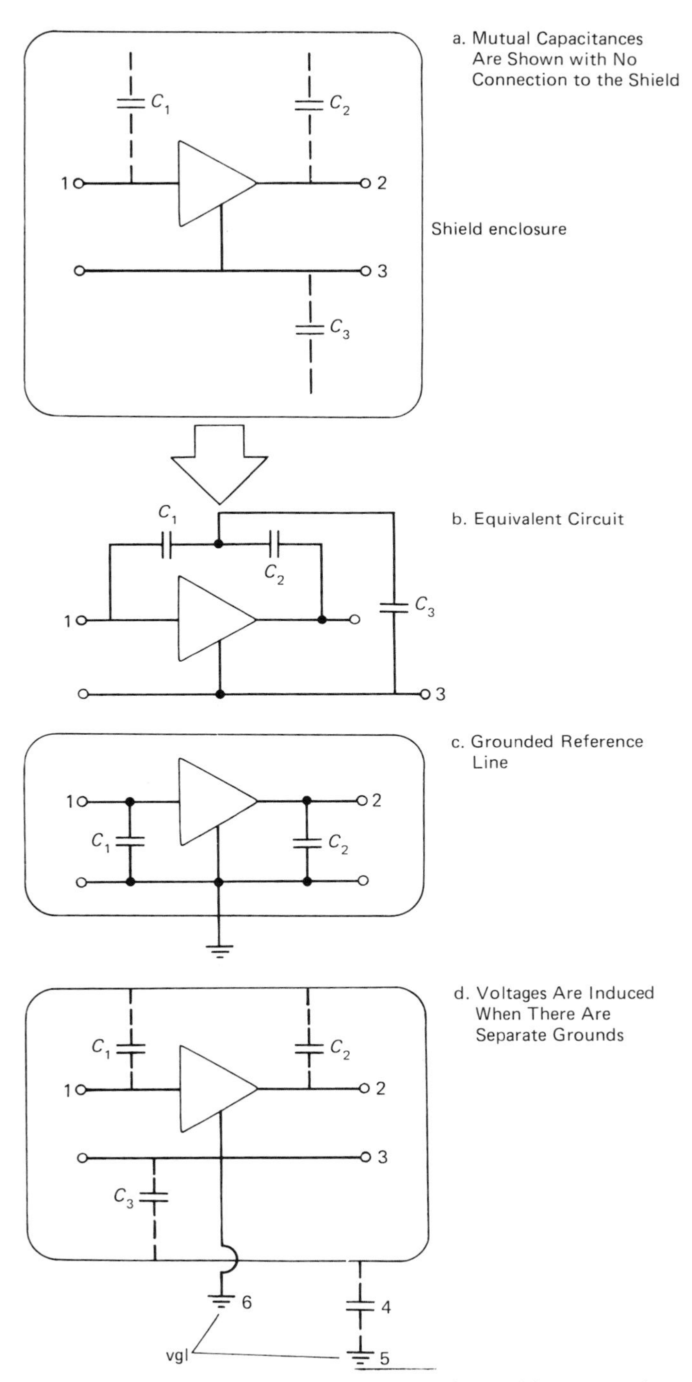

Figure 3-4 Shield grounds. (a) Mutual capacitances are shown with no connection to the shield; (b) equivalent circuit; (c) grounded reference line; (d) voltages are induced when there are separate grounds.

SINGLE-ENDED AND DIFFERENTIAL AMPLIFIER PROBLEMS

A single-ended amplifier requires a reference line that is common to both inputs and outputs with the shield. All current return paths must be connected to only one point, which is the signal reference. Thus, only one ground reference point is used for all devices, including sensors, amplifiers, and output devices. This is often difficult because of the physical layout of the devices.

This problem can be solved by the use of differential amplifiers instead of single-ended amplifiers. The differential amplifier allows the signal to be developed at one ground reference and measured at another. When the amplifier is single-ended, there is only one input line for the signal. The other line is the reference or ground line, which functions as the return path for the signal.

The differential amplifier has two input lines, and the amplifier responses only to the difference between the two input signals. The differential amplifier is not sensitive to common-mode signals, those signals that appear simultaneously at both of its inputs. In addition, the differential amplifier reduces ground-loop problems, since the input and output references can be separated.

A common-mode signal is common to both inputs, but in a single-ended amplifier the signal appears only on one input to the amplifier, with the return being the signal reference point. The desired signal is actually the difference in voltage between the two amplifier input terminals, but only one is used as the zero reference in the single-ended amplifier.

The reactance of this path is usually high for operational amplifiers, exceeding 1000 megohms in many cases. When the impedance of the circuit carrying the unwanted noise signal is 1000 megohms or greater, the circulating ground-loop current is only nanoamps and the noise signal associated with this current should be tolerable for most applications.

FLOATING INPUTS

Floating inputs can be a problem and should be treated with caution. The shield should be connected to the signal reference point. Grounding a shield does not satisfy this requirement if the signal is floating. A ground system may float without earth contact, such as in aircraft.

Differential amplifiers use two signal reference potentials. If one is grounded, the other can assume any potential, as determined by the ambient electrostatic (E) and magnetic (H) fields and parasitic currents. It is better to define this point than to permit it to find its own level by floating. A floating reference can also present a hazard if the floating line comes into contact with an unknown conductor that has a high voltage.

SYSTEMS WIRING AND CABLING CONSIDERATIONS

After purchasing the equipment, it is necessary to determine the best methods of cabling, as well as the type of interconnections and the routing of signals within the system. This should be done properly to achieve a minimum of signal loss, degradation, and noise pickup.

In many cases, the solutions are relatively simple, such as selecting the correct type of cable, eliminating common-mode grounds, and separating long runs of parallel wires. These practices should be implemented at the time of the initial design, since they become much more difficult to after the system is completed.

Most interface problems fall into two classes; signal degradation and noise. Initially, there should be a study of the system's parameters and the noise environment. The following questions need to be answered:

1. What are the signal frequencies and the voltage and power levels?
2. What are the tolerable losses or degradation levels?
3. Is there pickup of noise from direct contact common-mode, ground-loop returns and radiated magnetic fields?

Low-voltage wiring should always be shielded, regardless of the frequency to be transmitted. Coaxial cable is designed to carry high frequencies, but it is also useful for low-frequency, shielded-wire applications, since it is relatively inexpensive and many types of connectors are available. For higher-frequency applications, coaxial cable should always be used for point-to-point wiring, since it has the desired transmission characteristics, flexibility, and cost required for most systems.

Coaxial cable is available in 50-, 75-, and 93-ohm impedance configurations. The 93-ohm cable was developed primarily to meet the need for a low-capacitance instrumentation coaxial cable. This was done by removing some of the coax dielectric and substituting air in RG-59. The distributed capacitance was lowered, resulting in a lower-loss voltage transmission medium, RG-62. Cables and connectors of twinax, triax, and quadrax are also available. They are used in many situations where improved external noise rejection is needed.

CABLE SELECTION

In most control and measurement applications, the signals must move with minimum signal degradation and loss. It is often necessary to reduce unwanted external noise to an acceptable level. Signal degradation in the transmission cable may consist of voltage amplitude reductions, wave shape changes, phase or delay changes, or power losses where power is transmitted.

Interconnecting cables provide the longest transmission path in most systems, and its selection is most important. The following parameters must be considered:

1. The length of the cable run.
2. Heat and other environmental effects.
3. Frequency and power to be transmitted.

These must be compared to the acceptable losses inherent in the cable, the external noise fields and frequencies to be encountered, and the availability of connectors to terminate the cable.

If a cable that is too small in diameter is used, the signal losses will be too high. Fast rise time digital pulses will suffer from leading edge distortion due to the higher resistance skin effect of the smaller coax cables. Longer cable runs compound the insertion loss problem, since the signals must travel longer to reach their destination. Too small a cable can result in excessive signal loss. Incomplete coverage in the outer braid of the cable can cause transmission line losses, as well as cable susceptibility to signal leakage or noise pickup.

During installation, several precautions should be taken:

1. Make sure that the cable does not support equipment.
2. Do not place cables in areas where they will be subject to prolonged heat.
3. Do not bundle cables tightly, since this can cause crosstalk.
4. Do not bend cables beyond the manufacturer's recommended radius, since this can produce cable discontinuities.
5. When routing cables, try to separate signal and power circuits.

Signal cables are usually manufactured with polyethylene. High temperatures or chemical action from oxidizing gases can attack these cables. Teflon cable should be used for these applications.

COAXIAL CONNECTORS

Connectors should be able to make a connection with a low dc series resistance of about 10 milliohms. The ac impedance of a connector is not critical below 300 MHz, since the connector does not contribute to the circuit's performance until its length reaches about 1/20th of the wavelength. This allows 50-ohm connectors to be attached to 75-ohm cables without major problems at lower frequencies. Above the 300-MHz region, all coax connectors should be impedance-matched to the cable used.

TABLE 3-2 Basic Connector Types

Nominal Cable Outside Diameter	Connector Size Class	Quick Disconnect Type	Threaded Type
.3 to .425 in.	Medium	C	N
.10 to .3 in.	Miniature	BNC	TNC
.10 to .242 in.	Subminiature	TPS	SMA

Many cable systems were originally designed for specific problem solutions or as proprietory products. The connector types listed in Table 3-2 have been shown to be especially useful due to their simplicity of design and high overall quality and performance.

The selection of one type of connector over another is usually based on cost or performance. Crimping is used to save time in assembly or in the field where there is a lack of available soldering iron power. One disadvantage of crimping is that it requires an expensive crimping tool that must be properly used. Crimped contacts can also corrode over time, especially in chemical or salt environments. Soldered connections will resist corrosion better if correctly done. Most coaxial connectors can be reused with proper care. This can be an advantage in field locations where spare parts or special tools may not always be available.

MEASURING DEVICES

For monitoring performance and troubleshooting, the devices that perform dc measurements should have at least twice the resolution and accuracy of the units they are checking. Devices that perform high-speed measurements should have a faster response than the devices they are checking.

An oscilloscope should always be used to avoid flying blind. A multimeter cannot see dynamic changing signals or oscillations. Its dc resolution may be inadequate for useful measurements on the high-resolution devices usually found in data systems, and its load impedance may affect the accuracy (if not the actual character) of measurements.

AMPLIFIER SELECTION

In general, the objective of amplifier selection is to use the least expensive device that will meet the physical, electrical, and environmental requirements of the application. This means that a general-purpose device is the best choice for the application if the desired performance requirements can be met. When this is not possible, the limitations are usually imposed by bandwidth requirements, offset, or drift parameters.

In selecting the right device for a specific application, the designer needs to understand clearly the design objectives, as well as the published specifications and the important features affecting the application. The design objectives include the signal levels, desired accuracy and bandwidth, and circuit impedances, as well as the environmental conditions. Two devices may have comparable specifications, which may have been obtained with different measurement techniques.

A checklist of essential characteristics would include the following:

1. Character of the application: differential or single-ended, follower or inverter, linear or nonlinear.
2. Accurate description of the input signals: voltage or current source, range of amplitudes, source impedance, time/frequency characteristics.
3. Environmental conditions: maximum ranges of temperature, time, and supply voltage over which the circuits must operate (to the required accuracy) without readjustment.
4. Accuracy desired as a function of bandwidth, static and dynamic parameters, and loading.

Check the system out in small sections and functional groupings before assembly. After assembly, the system should be thoroughly inspected. Be sure that all connections have been made, that the right devices have been plugged into the right connectors, and that no bent or broken pins exist.

FREQUENT SYSTEM PROBLEMS

Common system malfunctions include the following:

1. The wrong digital code, for example, the use of positive true instead of complementary.
2. The wrong polarity relationship.
3. Grounds not interconnected.
4. Power supply not connected properly.
5. Improper connections for inputs and outputs.
6. Damage due to applying power to devices in the wrong order. In general, power should be applied to downstream units first.
7. Incorrect control logic—polarity, duration, timing, and levels.
8. Uncontrolled overflow in counters.
9. Wrong diode polarity.

Poor overall performance can be due to the following:

1. Common-mode problems in single-ended systems. Use proper grounding or a difference amplifier.
2. Grounding problems, poor ground connections, wrong ground connections (common analog and logic return), or shields returned to the wrong ground or grounded at both ends.
3. Pickup due to proximity of the digital ground plane to analog circuits, or proximity of analog and digital wiring, or stray capacitance.
4. Excessive load capacitance on outputs of voltage D/A converters or other analog devices, causing slow response, poor settling, ringing, or oscillation noise.
5. Improper connection of built-in references; unused bipolar offset references may require grounding in unipolar applications. The external use of an internal Zener voltage reference generally requires buffering.
6. Operational amplifier voltage offset adjustment used for zeroing other offsets, such as system offsets, can result in increased thermal drifts.
7. Logic overloading. Logic outputs can also be used for internal purposes.
8. Nonlinearity when the maximum output voltage range is exceeded.
9. Noise, high differential nonlinearity, and missing codes caused by noise on the input signal or pickup from wiring.
10. A bent pin that did not go into the socket or broke off.
11. Degradation due to lack of filtering, inappropriate converter choice, marginal logic timing, limited slewing rates due to excessive capacitive loading, or stray capacitive coupling to analog circuits.
12. Gain and offset adjustments performed in the wrong order.
13. Excessive thermal drift due to an improper adjustment procedure.
14. Use of an operational amplifier voltage offset adjustment to counteract bias current or system offsets.
15. Loss of monotonicity in code convertors over small temperature ranges.
16. Fast pulses causing rectification effects that produce offsets.
17. Long-tailed responses due to thermal transients in operational amplifier or comparator circuits.
18. Inappropriate capacitor choice. Precision capacitors must have good stability and must use a low dielectric absorption material such as polystyrene, Teflon, or polycarbonate.
19. Excessive drift in low-level circuits due to differential thermoelectric effects in input leads. Differential input leads should always be close together, and their junctions should be similar.
20. When other possibilities have been exhausted, assume that the device is malfunctioning or out of specification, due either to a recent overstress or to an early failure mechanism.

21. Data sheets with errors or insufficient data. If a user finds questionable information on a data sheet, most manufacturers will discuss this and clarify the points in question, especially as it pertains to the application.

Some devices may fail, for no apparent reason when first plugged in. The failure may be catastrophic or slightly in error, permanent or intermittent, affected by mechanical movement and/or shock.

GENERAL TROUBLESHOOTING ADVICE

1. A good initial analysis of the basic problem is needed.
2. Use a conservative analysis of the basic problem.
3. Use double-checking.
4. Make sure that the best available data have been used.
5. Check that the tolerances on resolution, accuracy, and timing are adequate.
6. Check that the connection scheme is proper.
7. Follow the manufacturer's suggestions when appropriate.
8. Use breadboarding to verify problem areas.

General test and troubleshooting procedures include the following:

1. Check the supply and ground voltages at the terminals of pluggable devices, with the devices removed.
2. Perform dc, manual, and low-speed checks before performing measurements at higher speeds. This will show if the system is working properly under some static or low speed conditions.
3. Try to isolate the problem. If more than one unit of a given type is in use, an apparent failure can be checked by substituting another unit. When a similar unit of the same kind has the same problem, it is likely to be a design or system problem. If it is a serious system problem involving a damage-producing condition, the original unit and its substitute may no longer be in condition for further use.
4. Check the grounding, using continuity testing. Use a systematic procedure.

Common-mode and normal-mode noise problems can be solved with differential amplifiers, filtering, and proper lead location. Grounding must be proper, with no ground loops. Only one path is allowed for ground current.

Digital and analog ground leads, and high-power and low-level signal ground leads, must be separated. Use one point where all these ground leads meet, if possible, along with heavy ground conductors, to avoid voltage drops in signal return leads. Also, be careful that the various interconnections of devices do not produce transient overloads during switching.

4

Data Acquisition Components

In a typical data acquisition system, data are collected in analog form. It is then necessary to convert these data into the digital codes utilized by the CPU of the computer. In this chapter the process of conversion is explained, and the digital coding techniques used in converters are presented. Conversion techniques such as voltage-current conversion, A/D conversion, and multiplexing are explained.

Multiplexing is an important part of conversion schemes, and we explore multiplexer requirements, error sources, and noise problems. Sample holds are another important component in conversion. We examine basic sample hold techniques, sample hold errors, and performance specifications. We show how to configure the PC Interface and conclude with a data acquisition design example.

In this chapter we introduce a number of components used in data acquisition systems. By understanding the functions of these components, the user will be better prepared in choosing a data acquisition system or board. This chapter provides basic functional descriptions, definitions of specifications, and the most important features for selection and evaluation.

DIGITAL CODING

Digital numbers in control and measurement applications are represented by the presence or absence of fixed voltage levels. These digital numbers are binary, since each bit or unit of information can have one or two possible states:

true or false
on or off
1 or 0
high or low

In a binary code, the different bits represent different portions or weights of the digital number. The bit with the greatest weight is the first bit in the leftmost position and is called the *most significant bit* (*MSB*). The bit with the least weight is the last bit in the right most position and is called the *least significant bit* (*LSB*).

The resolution of an A/D converter is determined by the number of bits. The coding used is the set of coefficients representing the fractional parts of full scale.

NATURAL BINARY PROGRESS

The most basic digital code follows a natural binary progress. In a natural binary code of n bits

the MSB has a weight of $\frac{1}{2}$: 2^{-1}

the second bit has a weight of $\frac{1}{4}$: 2^{-2}

the third bit has a weight of $\frac{1}{8}$: 2^{-3}

This progression continues to the LSB, which has a weight of 2^{-n}.

The value of the binary number represented is obtained by adding up the weights of the nonzero bits. A 4-bit representation is shown in Table 4-1, with the binary weights and the equivalent numbers shown for both decimal and binary fractions.

Table 4-2 shows the bit weights in binary for numbers of up to 12 bits, the range for many data acquisition boards (for larger numbers of bits, continue to divide by 2).

The weight assigned to the LSB is the resolution. The decibel value is the logarithm (base 10) of the ratio of the LSB value to unity multiplied by 20. Each successive power of 2 represents a change of 6.02 dB.

TABLE 4-1 Natural Binary Code

Decimal Fraction	Binary Fraction	MSB (×1/2)	Bit 2 (×1/4)	Bit 3 (×1/8)	Bit 4 (×1/16)
0	0.0000	0	0	0	0
1/16 = LSB	0.0001	0	0	0	1
2/16 = 1/8	0.0010	0	0	1	0
3/16 = 1/8 + 1/16	0.0011	0	0	1	1
4/16 = 1/4	0.0100	0	1	0	0
5/16 = 1/4 + 1/16	0.0101	0	1	0	1

TABLE 4-2 Binary Weights

Bit	1/2^n (Fraction)	dB	1/2^n (Decimal)	%	ppm
0 (FS)	1	0	1.0	100	1,000,000
MSB	1/2	−6	0.5	50	500,000
2	1/4	−12	0.25	25	250,000
3	1/8	−18.1	0.125	12.5	125,000
4	1/16	−24.1	0.0625	6.2	62,500
5	1/32	−30.1	0.03125	3.1	31,250
6	1/64	−36.1	0.025625	1.6	15,625
7	1/128	−42.1	0.007812	0.8	7,812
8	1/256	−48.2	0.003906	0.4	3,906
9	1/512	−54.2	0.001953	0.2	1,953
10	1/1,024	−60.2	0.0009766	0.1	977
11	1/2,048	−66.2	0.00048828	0.05	488
12	1/4,096	−72.2	0.00024414	0.024	244

In the A/D conversion process, a quantization uncertainty of $\pm\frac{1}{2}$ LSB occurs in addition to conversion errors existing in the converter. To reduce this quantization uncertainty, it is necessary to increase the number of bits. Statistical interpolation techniques can be used during processing or filtering following the conversion. This tends to fill in missing analog values for rapidly varying signals but does not reduce errors due to any variations within $\pm\frac{1}{2}$ LSB.

It is usually easier to determine the location of a transition than to determine a midrange value. Thus, errors and settings of A/D converters are normally defined in terms of the analog values when actual transitions occur in relation to the ideal transition values.

Both D/A and A/D converters have offset errors, since the first transition will not always occur at exactly $\frac{1}{2}$ LSB. Scale-factor or gain errors can cause a difference between the values at which the first and last transitions occur, since these are not always equal to $\frac{1}{2}$ LSB. Linearity errors can also exist, since the differences between transition values are not all equal or uniform when changing. When the differential linearity error becomes too large, it is possible for codes to be missed.

In some codes, the bits are represented by their complements. Unipolar converters use analog signals of one polarity, while bipolar converters use an extra bit for the sign. The unipolar version of natural binary is sometimes called *unipolar straight binary (USB)*.

When the sign digit doubles both the range and the number of levels, the LSB's ratio to full scale in either polarity is

$$2^{-(n-1)}$$

and not

$$2^{-n}$$

The most common binary codes for bipolar conversion are as follows:

1. Sign-magnitude (magnitude plus sign).
2. Offset binary.
3. Two's complement.
4. One's complement.

Each of these codes in 4 bits (3 bits plus the sign) can be compared in Table 4-3.

When the analog signal is given a choice of polarity, the relationship between the code and the polarity of the signal must be specified. In the positive reference column, the analog signal increases positively as the digit number increases. In the negative reference column, the analog signal decreases toward negative full scale as the digital number increases.

Sign magnitude refers to a way of indicating signed analog quantities. It adds a polarity bit to the code and is useful in converters that operate near zero, where the application calls for a smooth linear transition from a small positive

TABLE 4-3 Binary Codes

	Decimal Fraction					
Decimal Number	Positive Reference	Negative Reference	Sign Magnitude	Two's Complement	Offset Binary	One's Complement
+7	+7/8	−7/8	0 1 1 1	0 1 1 1	1 1 1 1	0 1 1 1
+6	+6/8	−6/8	0 1 1 0	0 1 1 0	1 1 1 0	0 1 1 0
+5	+5/8	−5/8	0 1 0 1	0 1 0 1	1 1 0 1	0 1 0 1
+4	+4/8	−4/8	0 1 0 0	0 1 0 0	1 1 0 0	0 1 0 0
+3	+3/8	−3/8	0 0 1 1	0 0 1 1	1 0 1 1	0 0 1 1
+2	+2/8	−2/8	0 0 1 0	0 0 1 0	1 0 1 0	0 0 1 0
+1	+1/8	−1/8	0 0 0 1	0 0 0 1	1 0 0 1	0 0 0 1
0	0+	0−	0 0 0 0	0 0 0 0	1 0 0 0	0 0 0 0
0	0−	0+	1 0 0 0	(0 0 0 0)	(1 0 0 0)	1 1 1 1
−1	−1/8	+1/8	1 0 0 1	1 1 1 1	0 1 1 1	1 1 1 0
−2	−2/8	+2/8	1 0 1 0	1 1 1 0	0 1 1 0	1 1 0 1
−3	−3/8	+3/8	1 0 1 1	1 1 0 1	0 1 0 1	1 1 0 0
−4	−4/8	+4/8	1 1 0 0	1 1 0 0	0 1 0 0	1 0 1 1
−5	−5/8	+5/8	1 1 0 1	1 0 1 1	0 0 1 1	1 0 1 0
−6	−6/8	+6/8	1 1 1 0	1 0 1 0	0 0 1 0	1 0 0 1
−7	−7/8	+7/8	1 1 1 1	1 0 0 1	0 0 0 1	1 0 0 0
−8	−8/8	+8/8		(1 0 0 0)	(0 0 0 0)	

voltage to a small negative voltage. It is the only code in which the three magnitude bits do not have a major transition, such as all 0's to all 1's, at the zero. However, this makes sign magnitude more expensive to implement compared to other codes. The problem is that the two codes for zero require additional hardware or software. Zero errors can be larger with sign magnitude, since the zero level is usually obtained by taking the difference between the MSB ($\frac{1}{2}$ full scale) and a bias ($\frac{1}{2}$ full scale), which are usually two large numbers.

The two's complement code consists of a binary code for positive magnitudes with a 0 sign bit. The two's complement is formed by complementing the number and adding one LSB. The two's complement of

$\frac{3}{8}$ (binary 0011)

is the complement plus the LSB:

$$1100 + 0001 = 1101$$

Two's complement can be viewed as a set of negative numbers.

To subtract $\frac{3}{8}$ from $\frac{4}{8}$
add $\frac{4}{8}$ to $-\frac{3}{8}$
or add 0100 to 1101.
The result is 0001,
neglecting the extra carry,

or $\frac{1}{8}$

The bipolar version of two's complement is known as *BTC*, and the bipolar version of offset binary is known as *BOB*. Comparing the two's complement code and the offset binary code, the only difference between them is that the MSB of one is replaced by its complement in the other. An offset-binary-coded converter can be used for two's complement by using the MSB's complement at the output of an A/D converter or at the output of a D/A converter's input register.

Offset binary is the easiest code to implement with converter circuits. The offset binary code for 3 bits plus the sign bit is the same as the natural binary code for 4 bits, except that zero is at negative full scale, the LSB is $\frac{1}{16}$th of the total bipolar range, and the MSB is on at zero.

An offset binary 3-bits-plus-sign converter can be built up using a 4-bit D/A converter with a 0- to 10-volt full-scale range. The scale factor is doubled to 20 volts and then offset the zero by half of the full range to -10 volts. For an A/D converter, the input is cut in half and then incremented a bias of one-half of the range.

The main disadvantage of offset binary is the bit transition which occurs at zero when all bits may change, for example, from 0111 to 1000. The difference in switching speeds due to electronic components turning on and off can lead to spikes and linearity problems. These linearity errors are likely to occur at major transitions, since the transition represents the difference between two large numbers. Offset binary is more compatible with microcomputer inputs and outputs, since it is easily changed to the more common two's complement by complementing the MSB. Two's complement has a single unambiguous code for zero, but it also has the same disadvantages as offset binary, since the conversion process is the same.

The one's complement code is a way of representing negative numbers. The one's complement is obtained by complementing all bits.

The one's complement of
$\frac{3}{8}$ (0011) is (1100).
A number is subtracted
by adding its one's complement.

The extra carry that is disregarded in the two's complement code causes one LSB to be added to the total in the end-around carry.

Subtracting $\frac{3}{8}$ from $\frac{4}{8}$, we get
$0100 + 1100 = 0001$ (or $\frac{1}{8}$).

The one's complement is formed by complementing each positive value to obtain its corresponding negative value, including zero, which is represented by two codes, 0000 and 1111. Besides the ambiguous zero, another disadvantage is that it is not as easy to implement as two's complement in converter circuits.

In sign magnitude and one's complement converters, there are two ways to treat the ambiguous zero:

1. One of the codes may be forbidden.
2. The $\pm\frac{1}{2}$ LSB zero region can be divided into two regions, 0 to $+\frac{1}{2}$ LSB and 0 to $-\frac{1}{2}$ LSB. Each of these can then have its own unique code.

D/A CONVERSION

The *R*–2*R* ladder circuit is often used in D/A conversion. The basic circuit is illustrated in Figure 4-1a, which shows its use with an inverting operational amplifier.

If all bits but the MSB are off (grounded),
$$\text{output} = (-R/2R)V_{\text{ref}}.$$
If all bits except bit 2 are off,
$$\text{the output} = \tfrac{1}{2}(-R/2R)V_{\text{ref}} = \tfrac{1}{4}\,V_{\text{ref}}.$$

The lumped resistance of the LSB circuit to the left of bit 2 is $2R$. The equivalent circuit looking back from the MSB toward bit 2 is $V_{\text{ref}}/2$, and the series resistance is $2R$. The grounded MSB series resistance is also $2R$, but it has no effect since the amplifier is grounded. The output voltage is then $-V_{\text{ref}}/4$. This line of reasoning can be extended to show that the nth bit produces an output equal to $2^{-n}V_{\text{ref}}$.

The R-$2R$ network can be used to provide an unattenuated, noninverting output by connecting the output to a high-impedance load, such as the input of a

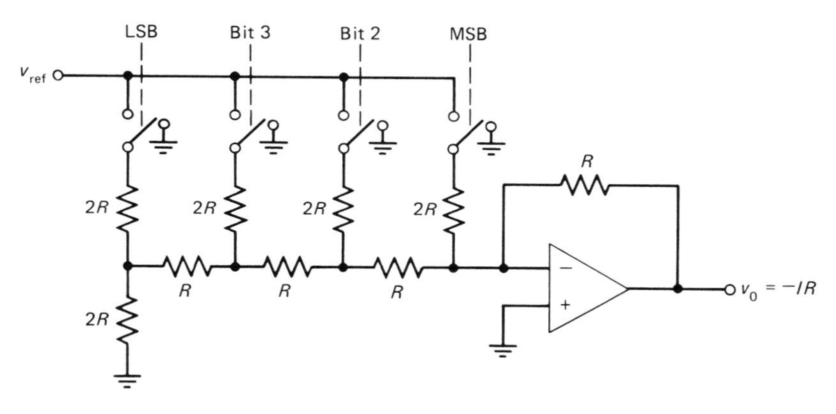

a. Basic $R - 2R$ Resistance Ladder

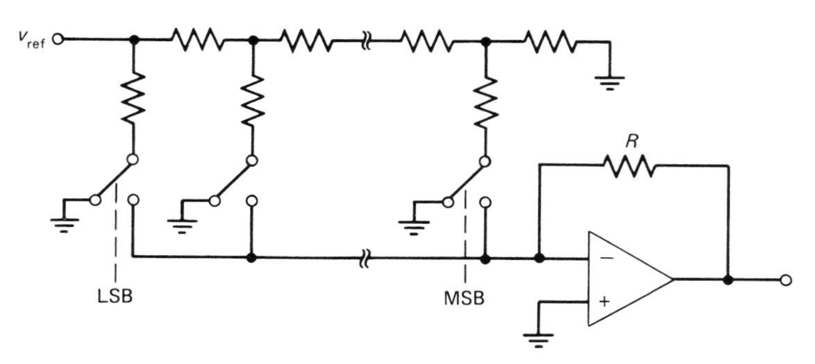

b. Inverter $R - 2R$ Ladder

Figure 4-1 R-$2R$ ladder circuits. (a) Basic R-$2R$ resistance ladder; (b) inverter R-$2R$ ladder.

follower amplifier, as shown in Figure 4-1b. Following the reasoning used above, the MSB output is $\frac{1}{2}$ V_{ref}. The entire network can be considered a generator having an output voltage NV_{ref} (where N is the fractional digital input) and an internal resistance R.

The most common reference in converters is the temperature-compensated zener diode. It is often used with an operational amplifier for operating-point stabilization. In bipolar current-switching D/A conversion with offset binary or two's complement codes, an offset current equal and opposite to the MSB current is added to the converter output. This may be a resistor and a separate offset reference, but it is often derived from the converter's basic reference voltage in order to minimize drift with temperature.

The gain of the output-inverting amplifier can be doubled to increase the output range, from 0–10 to ± 10 volts. If the amplifier is connected for sign inversion, the conversion is negative reference. In a noninverting application, the same values of offset voltage and resistance can be used, but the value of the output voltage scale factor will depend on the load.

Bipolar D/A converters using voltage switching with R–$2R$ ladder networks and offset binary or two's complement coding may use terminals that are normally grounded for unipolar operation (one side of the switches and the LSB termination) and the reference signal in the opposite polarity. If the LSB termination is grounded, the output will be symmetrical. For sign magnitude conversion, the converter's current output may be inverted.

The analog output in a parallel-input D/A converter circuit continually follows the state of the logic inputs. If the conversion circuit is preceded by a register, the converter will respond only when the inputs are gated into it. This technique is useful in data distribution applications, where the data are continually appearing, but it is desirable for a D/A converter to respond at certain times and then hold the analog output constant until the next update.

A D/A converter with buffer storage can be used as a sample hold with digital input, analog output, and an infinite hold time. The register is under the control of a strobe which causes the converter to update. The rate at which the strobe updates is determined by the settling time of the converter and the response time of the logic.

VOLTAGE-CURRENT CONVERSION

A/D converter inputs are normally voltages. D/A converter outputs are often voltages at low impedance from an operational amplifier. Many converters provide an output current instead of a voltage. The conversion process can result in a current output that is linear and free from offsets. An operational amplifier may be used to convert the current to voltage.

Converters that use current or voltage outputs directly from resistive ladders can be considered either as voltage generators with series resistance or as current generators with parallel resistance. They can be used with operational amplifiers in either the inverting or the noninverting mode, as discussed earlier.

The inverting current output connection provides a high internal impedance, and the loop gain is close to unity and essentially independent of the feedback resistance, minimizing any amplifier errors, such as voltage drift.

The conversion relationship of D/A current (D/I) converters is normally positive reference. As the current flowing out of the converter increases, the value represented by the digital code increases. It does not depend on the actual polarity of the converter's reference. If current flowing toward the converter increases as the number represented by the digital code increases, the relationship is a negative reference.

A/D CONVERSION

Most data acquisition boards for PCs use successive approximation conversion. These A/D converters are built around a D/A converter. The conversion process is as follows:

1. When a conversion command is applied, the D/A converter's MSB output (one-half full scale) is compared with the input.
2. If the input is greater than the MSB, it remains on and the next bit is tested.
3. If the input is less than the MSB, it is turned off, and the next bit is tested.
4. If the second bit does not have enough weight to exceed the input, it is left on and the third bit is tested.
5. If the second bit exceeds the input, it is turned off and the third bit is tested.
6. The bit-testing process continues until the last bit has been tested.
7. When the bit tests are complete, a status line indicates that a valid conversion has occurred.
8. An output register is used to hold the digital code corresponding to the input signal.

Figure 4-2 is a block diagram of a successive approximation A/D converter. Internally, the converter operates as follows. When the logic signal is applied to the command terminal, the D/A switches are set to their off state, except for the MSB, which is set to logic one. This turns on the corresponding D/A switch to apply the analog equivalent of the MSB to the comparator.

If the analog input voltage is less than the MSB weight, the MSB is switched off at the first edge of the clock pulse. If the analog input is greater than the MSB, the 1 will remain in the register.

During the second pulse, the sum of the first result and the second bit is compared with the analog input voltage. The comparator is gated by the next

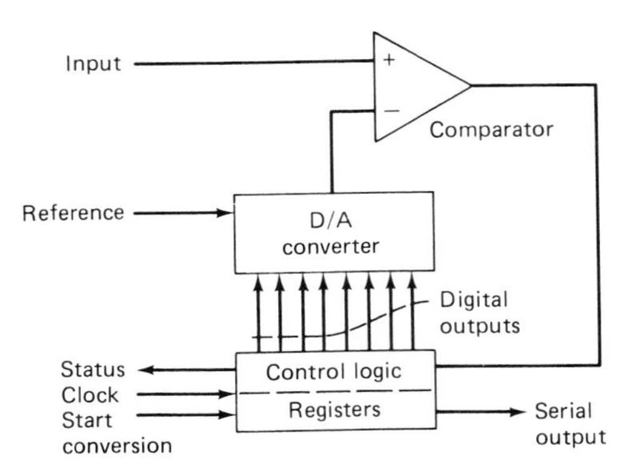

Figure 4-2 Successive approximation conversion.

clock pulse, causing the register to either accept or reject that bit. Successive clock pulses cause all the bits, in order of decreasing significance, to be tried until the LSB is accepted or rejected.

As the converter's full-scale range is adjusted, it will be set with respect to the reference voltage, which can be traced to some recognized voltage standard. The absolute accuracy error is the tolerance of the full-scale point referred to this absolute voltage standard. Offset is measured for a zero input. It is the extent to which the output deviates from zero, and it is usually a function of time and temperature. Nonlinearity monotonicity is the ability to include all code numbers in actual operation. It is the amount by which the plot of output versus input deviates from a straight line. Settling time is the time required for the input to attain a final value within a specified fraction of full scale, usually $\pm\frac{1}{2}$ LSB.

MULTIPLEXING

When more than one analog quantity has to undergo conversion, it is necessary either to time-division multiplex the analog inputs to a single A/D converter or to provide an A/D converter for each input and combine the converter outputs by digital multiplexing. Analog multiplexing is generally favored for achieving the lowest system cost. The factors that are important in multiplexer selection will now be discussed.

The resolution of measurement is important. The cost of conversion rises quickly as the resolution increases. This is due mainly to the cost of the higher-precision components required in the converter. At the 8-bit level, the per-channel cost of an analog multiplexer becomes comparable to that of a converter. At resolutions above 12 bits, the reverse becomes true and analog multiplexing tends to be more economical.

The number of channels controls the size of the multiplexer required, as well as the amount of wiring and interconnections that will be needed. Analog multiplexing can be used for handling up to 256 channels. Beyond this number of channels, analog errors become difficult to minimize.

High-speed conversion is costly, and analog time-division multiplexing may require a high-speed converter to achieve the specified sample rate. Wide dynamic ranges between channels can be a problem with analog multiplexing. Signals of less than 1 volt will usually require differential low-level analog multiplexing. This is expensive, and programmable-gain amplifiers may be needed after multiplexing. An alternative is to use fixed-gain converters on each channel, using signal-conditioning specifically designed for the channel.

Analog multiplexing is best suited to making measurements at distances no greater than a few hundred feet (100 m) from the converter. Analog lines are subject to losses, transmission line reflections, and interference. Cabling ranges from simple twisted-wire pairs to multiconductor shielded cable. The choice depends on the signal level, the distance, and the noise environment.

Digital multiplexing is viable for distances of up to thousands of miles with suitable transmission equipment. Digital transmission systems employ noise-rejection characteristics, which are essential for long-distance transmission.

MULTIPLEXER REQUIREMENTS

The most important requirement for a multiplexer is the ability to operate without introducing unacceptable error at the sampling speed. In digital transmission, the speed is determined from propagation-delay parameters and the time required to achieve adequately settled output on the data bus. Analog multiplexer speed is a function of internal parameters and external parameters such as channel source impedance, stray capacitance, the number of channels, and the cable type. Because of their nonideal transmission and open-circuit characteristics, analog multiplexers introduce static and dynamic errors into the selected signal path. These include leakage through switches, coupling of control signals into the analog path, and interaction with both sources and following amplifiers. Poor circuit layout and cabling can compound these effects and further degrade performance.

Analog multiplexers are usually connected directly to sources which may have little overload capacity or poor settling times after an overload. The switches used must have an inherent break-before-make action to prevent the possibility of shorting channels together. Some multiplexers are deficient in this area. It is often necessary to avoid shorted channels when power is removed, and some commercial devices do not offer this feature.

The channel-addressing lines are usually binary-coded. It is also useful to have one or more inhibit or enable lines to turn all switches off, regardless of the channel being addressed. This simplifies the external logic required to cascade multiplexers. Another requirement for multiplexers is a tolerance of transients and overload conditions, as well as the ability to absorb transient energy without damage.

ERROR SOURCES

Multiplexing introduces static (dc) and dynamic errors into the signal. If these errors are large in relation to the measurement, they will produce undesirable variations in performance. Static errors can come from switch leakage and offsets in output amplifiers. There may also be gain errors due to switch-on resistance, source resistance, amplifier input resistance, and amplifier-gain nonlinearities. Dynamic errors can occur from charge injection of the switch control voltage, settling times of the common bus and input sources due to circuit time constants, crosstalk between channels, and output amplifier settling characteristics.

Gain or transfer ratio errors are due mainly to source (R_s) and leakage (R_e) circuit resistances, which will now be discussed.

R_s is the internal resistance of the source, and R_c is the leakage from cabling and wiring. R_c is mainly controlled by circuit insulation. It should always be several orders of magnitude greater than R_s. Teflon-insulated wiring can be used to achieve this, if necessary. R_s can then be neglected. Good wiring practices and layout are also helpful in minimizing R_c and can be important in high-humidity environments. R_o is the switch-on resistance and is a function of the switch design. Solid-state switches have a predictable, stable on-resistance which ranges, for FET switches, from a few to a few thousand ohms. The very-low on-resistance, large-geometry FETs provide good static performance at the expense of dynamic performance, since they have large gate and drain-source capacitances. The on-resistance of FETs is a function of temperature and increases with temperature at $\frac{1}{2}°$ to 1 °C.

The input resistance, R_i, of the output amplifier is a function of its differential and common-mode input resistances and loop gain:

$$R_i = \frac{1}{A_c/A_o R_d + 1/R_{cm}}$$

where

A_c = closed-loop gain of the amplifier circuit

A_o = open-loop gain of the amplifier

R_d = differential input resistance of the amplifier

R_{cm} = common-mode input resistance of the amplifier

The open-loop gain and input resistance are functions of temperature, input voltage, and supply voltage.

An input resistance of 100 megohms is typical for a unity-gain follower with integrated circuit (IC) amplifiers. Higher input resistances are possible with FET-input ICs.

If R_c is neglected, the dc transfer ratio (output/input) is

$$\frac{E_o}{E_s} = \frac{A_c R_i}{R_s + R_i + R_o}$$

This equation allows variations in the above parameters to be calculated. Errors in the transfer ratio can be due to a high source resistance, but these are systematic errors that can usually be tolerated or compensated for. Variations in transfer ratio due to temperature and voltage generally are more critical, since they cannot easily be compensated for. Gain nonlinearity of the amplifier is common when the loop gain is inadequate. It is usually possible to achieve 80 dB of loop gain with amplifiers used as followers:

$$A_c = \frac{\text{Ideal Gain}}{1 + 1/\text{Loop Gain}}$$

The loop gain is equal to $A_o \times$ beta, where beta is the fraction of output that is fed back.

FET switches have a finite off-resistance and are affected by drain-to-source leakage currents (Ids) and gate-to-source leakage currents . Junction FETs and Zener-diode-protected metal oxide semiconductors (MOSFETs) have leakage paths between the gate and the channel. Unprotected insulated-gate MOSFETs have negligible gate leakage but are not used because they are vulnerable to transients. The leakage currents of the switches in a multiplexer return to ground via the input source resistance and the input resistance of the amplifier. Leakage on the output side of nonconducting channels does not affect the signal path. Leakage on the output side of the channels and the conducting channel into the signal path will cause a voltage error.

Ids(off) and Igs(off) are about 0.1 nA at 25°C for junction FETs and double for every 10°C rise. Insulated-gate MOSFETs have a Igs(off) of about 10–50 times this amount. The gate leakage is in the femtoampere (10^{-15} amperes) range and can be neglected. For both types, 10:1 variations of leakage parameters are possible from device to device.

Leakage is important when working with high source impedances, and

$$R_i >> R_{s1} >> R_o$$

The terms Igs(off) + Ids(off) are usually lumped together on the multiplexer data sheet and called the *channel leakage*. The error voltage is then

$$V_{error(CH1)} = R_{s1} \text{ (Number of Channels } - 1\text{) (Channel Leakage)}$$

For a single 8-channel multiplexer with a channel leakage of 3 NA at .255°C with a source resistance of 50 k ohm, the leakage error would be 1.05 millivolts. For large multiplexers like those with 256 channels, the leakage error is decreased by cascading multiplexers in a submultiplexing fashion, as shown in Figure 4-3. This configuration reduces the leakage and improves the dynamic

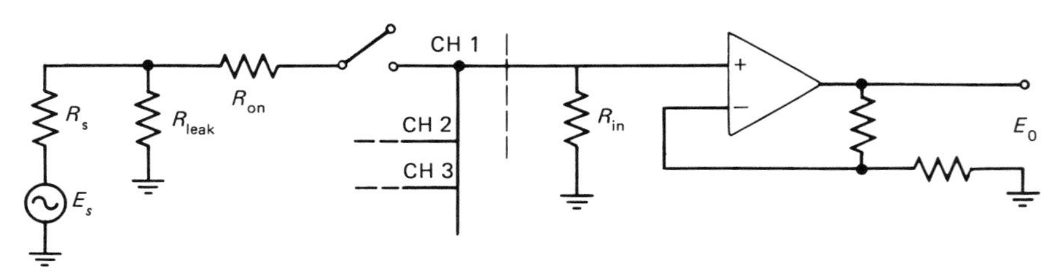

a. Gain Errors

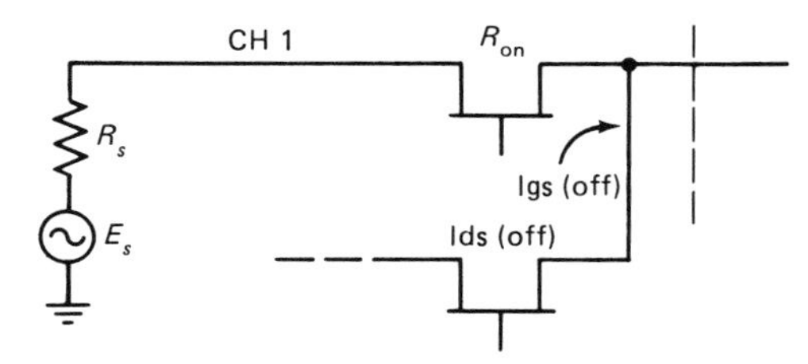

b. Current Errors

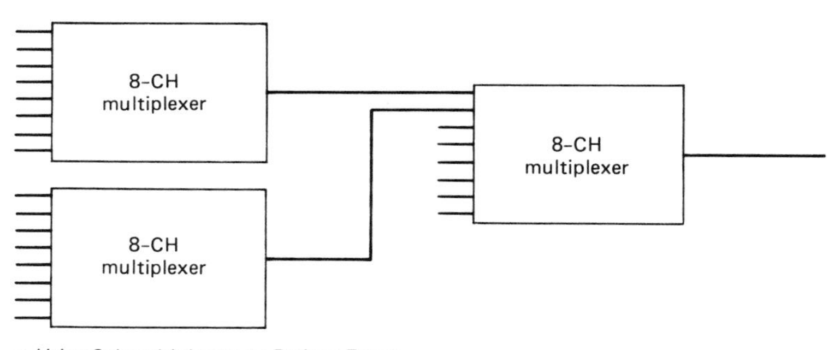

c. Using Submultiplexers to Reduce Errors

Figure 4-3 Multiplexer errors. (a) Gain errors; (b) current errors; (c) using submultiplexers to reduce errors.

performance. The leakage error voltage using this type of connection is

$$V_e = R_s \text{ (Number of Channels/Stage } - 1)$$
$$\text{(Number of Stages) (Channel Leakage)}$$

Multiplexing 256 channels by submultiplexing with 8-channel multiplexers can provide an order-of-magnitude improvement over a simple parallel connection.

The on-resistance of most types of multiplexer switches requires a high-input impedance amplifier to buffer the voltage on the output bus of the multiplexer. This amplifier may be part of the multiplexer, or it may be included as part of a sample hold or A/D converter following the multiplexer, or it may need to be provided separately. If a large dynamic range is needed, a programmable-gain amplifier can be used.

Offset and bias-current drifts, as well as common-mode effects of the amplifier, can also introduce errors. The errors introduced by the buffer amplifier will usually be much less than those from the multiplexer. The use of an amplifier with adequate open-loop gain will lessen errors from gain nonlinearities. Voltage drift is less of a problem in high-level multiplexers operating with low-gain buffers. Current drift and offset are more critical, since the bias current flows through the multiplexer and source resistance. Because the source resistance will vary from channel to channel, it is difficult to compensate for the inverting input of the buffer, so that offset current controls the drift. Usually the bias-current variation with temperature will control the drift in a general-purpose IC. A FET-input operational amplifier tends to set up a static offset:

$$V_{\text{offset}} = I_{\text{bias}}(R_{\text{on}} + R_s)$$

Typically, using an operational amplifier following a multiplexer driven by a 10K source can produce a static offset of 5 millivolts and a temperature variation of 1 millivolt due to current drift alone.

One technique is to ground the input of one channel, which is used as a reference. Measurements of this channel can be subtracted from each of the other channels to provide corrected readings. This compensates for drift in the amplifier, as well as for A/D converter zero-drift and multiplexer leakage errors. It is a most useful technique when measuring low-level signals.

DYNAMIC ERRORS

The output bus of a multiplexer will have a capacitance to ground, and when switched to another channel, the output voltage will not change as fast as the source voltage. Consider the simple case where

$$R_s = 0$$

The settling time is a function of the time constant

$$R_{on}(C_{gs} + C_{ds} + C_i + C_o)$$

C_i is the amplifier input capacitance and may be a function of frequency. C_o is the stray bus-to-ground capacitance, as shown in Figure 4-4. Typically, the gate-drain and gate-source capacitances (C_{gs} and C_{ds}) are 2–5 pF for small-area FETs with on-resistances in the 200- to 500-ohm range. There are wide variations, depending on the construction. Large-area, low-on-resistance FETs have higher capacitances. Stray capacitance depends to a large extent on the physical layout. It is typically 15 pF. The input capacitance of the buffer amplifier is usually about 5 to 10 pF. For an 8-channel multiplexer, the bus capacitance is typically 35 to 65 pF. This gives a time constant of 18 to 34 nanoseconds with $R = 500$ ohms. The settling on time to 0.01% F.S. is 165 to 310 milliseconds. This time will depend on the number of channels connected to the common bus. For a large number of channels submultiplexing should be used, since it will decrease the output bus capacitance and keep the settling time reasonable. If R_s cannot be neglected, then the settling time is controlled by 2. There is a charge transfer through R_{on} with a time constant of

$$(R_{\text{on}} C_{\text{line}} C_{\text{bus}})/(C_{\text{line}} + C_{\text{bus}})$$

and a charging through

$$R_s \text{ by } E_s$$

with a time constant of

$$R_s(C_{\text{line}} + C_{\text{bus}})$$

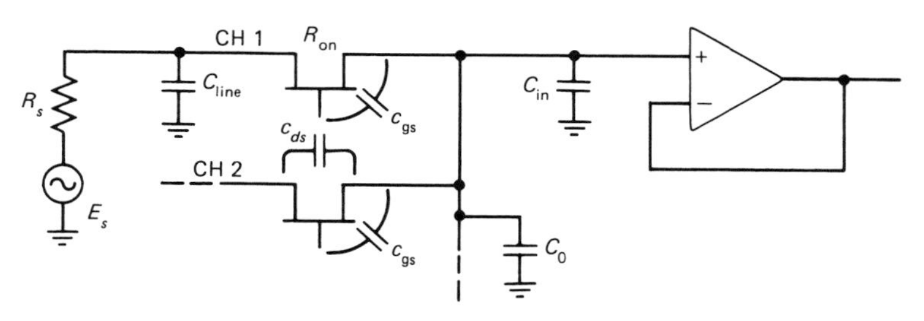

Figure 4-4 Multiplexer settling time.

The first time constant is usually shorter, controlling the initial settling, while the second controls the final settling. If an input filter is used, C_{line} becomes part of the line filter capacitance. Without a filter, C_{line} depends largely on the type of cable used. It is typically greater than 8 pF/ft. When C_{line} is of the same order as C_{bus}, the differential equations of the circuit must be solved to obtain the settling time to a given resolution.

The speed of the output buffer amplifier should be great enough so that it does not affect the settling time. Wideband amplifiers with fast settling time are available for high-speed multiplexing.

When settling time is not a critical parameter, a rough estimate can be obtained from the slew rate and bandwidth of the operational amplifier.

Every time a channel is switched in a multiplexer, some of the switch-control signal is coupled inductively and capacitively into the analog signal path. In FET multiplexers, the coupling is due to the gate-drain and gate-source capacitance of the FET switches. Every time a switch is operated a quantity of charge equal to

$$V_c(C_{gs} + C_{gd})$$

is injected into the bus. V_c = gate-driving voltage

This charge is dissipated through the switch on-resistance and the source resistance R_s. When R_s approaches zero, the charge injection produces a short spike on the output bus each time a switch operates. If the output amplifier response to the transient remains in the linear range (the amplifier is not overloaded or driven into slewing), charge injection will produce only a slight increase in settling time. This is the time required for most of the injected charge to leak away and for the amplifier to recover from the small transient.

If R_s is large, the initial voltage step will be reduced to $V_c(C_{gs} + C_{gd})/(C_{\text{bus}} + C_{\text{line}})$, but the decay time constant will be increased to $R_s(C_{\text{bus}} + C_{\text{line}})$ while the switch remains on. The effective values of C_{gs} and C_{gd} can differ between the switch-on and switch-off conditions due to nonlinear switching. The line capacitance can remain charged, decaying through its own time constant, which is $R_s\,C_{\text{line}}$.

Continuously operating the switch adds to the charge on the line capacitance, and at high sample rates, charge is pumped into the line capacitance more rapidly than it can leak away. This produces an offset voltage in series with the signal source. This is called *pumpback* and can limit the sampling rate in high-speed systems. For high sampling rates, low-capacitance, medium-on-resistance switches will minimize pumpback errors.

Crosstalk is a measure of the coupling between the off channels and the conducting channel of a multiplexer. It depends on the cable and circuit layout used for the multiplexer, as well as the switch on/off impedance. Crosstalk is measured by applying a voltage of known magnitude and frequency at one or more of the off channels of a multiplexer and measuring the output voltage on the bus or at a source with a defined internal resistance (usually 1 k ohm). Crosstalk can be measured at both dc and ac (usually 1kHz) and depends on the affected source resistance.

In addition to the delays required for analog settling, there are propagation and risetime delays associated with the logic that drives the switching elements. Turn-on and turn-off times of switches are measured on each channel with full-scale input voltage. Turn-on time is the delay from application of channel address to the appearance of 90% output voltage on the bus, and turn-off time is the delay from removal of channel address to 10% output voltage on the bus. It is usually necessary to load the bus with a resistive load to make the bus time-constant short with respect to the switching time.

MULTIPLEXERS IN HIGH-NOISE ENVIRONMENTS

When multiplexers operate under conditions of high common-mode interference, guarded two-wire differential or flying-capacitor multiplexers can be used. Normal-mode interference may require techniques such as filtering, averaging, and integrating. These will now be briefly discussed.

The use of low-pass filters to the channel inputs of a multiplexer is a common method of reducing normal-mode interference. The filter characteristics can be customized to the requirements of the individual channel. Filters can increase settling time and pumpback effects, but these are usually small. It is possible to place the filter after the multiplexer, but this is not usually done, since each channel has to charge the filter and this greatly increases the settling time. In a differential system, filters should have balanced impedances in both inputs or be connected differentially.

When passive filtering of each channel is not practical, an integrating A/D converter can provide high normal-mode rejection. This is especially true at frequencies which have periods that are integral submultiples of the integration interval. The rejection is obtained with a conversion time that is usually much shorter than the settling time of a filter that would be required to provide the same rejection.

A rejection level of normal-mode interference of 40 to 70 dB is possible with an integrating converter. Many integrating converters use floating guarded input operation and provide both common- and normal-mode rejection. In a PC system where storage is available, the converter can be used to track the variations in input signal produced by interference, and software can be used to reduce the effects of interference. Multiple samples can be collected on each channel, and the results summed and averaged. The signal/noise ratio will improve as the square root of the number of samples, provided that sampling and interfering frequencies are uncorrelated.

SAMPLE HOLDS

A sample hold is a device having a signal input, an output, and a control input. It has two operating modes:

1. Sample or track in which it acquires the input signal as rapidly as it can and tracks it until commanded to hold.
2. The hold mode is the time it retains the last value of input signal that it had at the time the control signal initiated the mode change.

Sample holds are also called *track holds* if they spend a larger portion of the time in the sample mode tracking the input.

Sample holds usually have unit gain and are noninverting. The control inputs are logic levels. Logic 1 is usually the sample command, and logic 0 is the hold command.

A sample hold goes through the following four states:

1. Sample.
2. Sample to hold.
3. Hold.
4. Hold to sample.

In data acquisition systems, sample holds are used to freeze fast-changing signals. A typical application is to store multiplexer outputs for conversion while the multiplexer is seeking the next signal to be converted. In data reduction applications, sample holds are used to determine peaks or lows or to collect signals obtained at different instants in time. Fast sample holds are often used to acquire and measure fast pulses for analysis.

SAMPLE HOLD TECHNIQUES

The type of storage element used divides sample holds into two major classes. The conventional technique employs a capacitor for storage. The other technique uses digital storage with an A/D converter, a register for storage, and output via a D/A converter. A major advantage of this method is the essentially infinite hold time.

The open-loop follower shown in Figure 4-5a is the most basic circuit in use. When the switch is closed, the capacitor charges exponentially to the input voltage, and the amplifier's output follows the capacitor's voltage. When the switch is opened, the charge remains on the capacitor. The capacitor's acquisition time depends on the series resistance and the current available to charge its capacitance.

Once the charge is acquired to the desired accuracy, the switch can be opened,

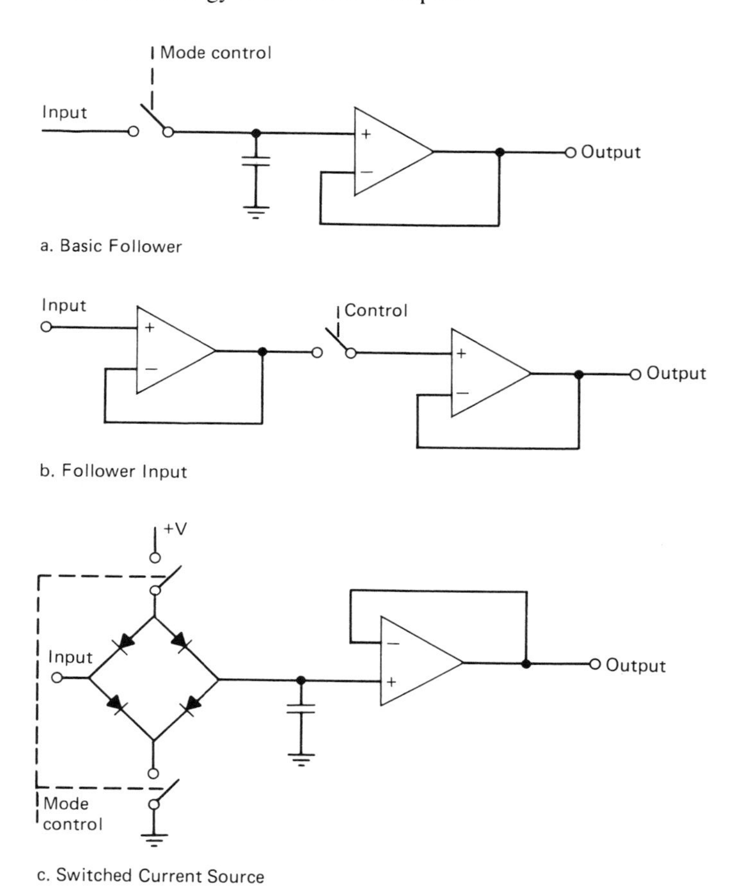

Figure 4-5 Sample hold techniques. (a) Basic follower; (b) follower input; (c) switched current source.

even though the amplifier has not yet settled, without affecting the final output value or the settling time. This assumes that the amplifier's input stage draws negligible current.

The switch is usually a FET, and the amplifier has a FET input. This circuit has the disadvantage that the capacitor loads the input source. This can cause oscillations or limit the current available to charge the capacitor.

An input follower can be used to isolate the source, as shown in Figure 4-5b. Many commercial devices use this scheme. For faster charging at a more linear slew rate, a diode bridge can be used as a current source, as shown in Figure 4-5c. If the bridge and current sources are closely balanced, the current flow into the capacitor stops when the capacitor voltage is equal to the input voltage.

SAMPLE HOLD ERRORS

In an ideal sample hold, tracking is error-free, and acquisition and release occur instantaneously. The settling time is zero and the hold time is infinite, since there is no leakage. Commercial units are rated in terms of the extent to which they differ from the ideal. The most common deviations that occur during the four states of operation are discussed below.

During the sample state, the offset can be defined for a zero input as the extent to which the output deviates from zero. It is a function of time and temperature. The nonlinearity is the amount by which a plot of output versus input deviates from a straight line. Scale factor error is the amount by which the output deviates from a specified gain, usually unity. Settling time is the time needed for the output to attain its final value.

In the sample state, the device acts like a slow unity gain follower. Other specifications typical of such devices include phase shift, slew rate, full power, and small signal bandwidths.

During the sample hold state, the aperture time is often important. This is the time between the command to hold and the actual opening of the hold switch. It is usually made up of two parts: a time delay and an uncertainty due to jitter or variations from device to device.

There is also a sample hold offset—a step error occurring at the initiation of the hold mode. It is caused by a flow of charge into the storage capacitor from other circuit capacitance. Some of this is due to the capacitance of the switch, such as the gate to drain capacitance of the field effect transistors. Switching transients include residual transients from switching. They remain after compensation of the sample hold offset. Settling time is the interval required for the output to attain its final value within a specified fraction of full scale following the opening of the switch.

During the hold period, there is droop. This is a drift of the output at an approximately constant rate caused by the flow of current through the storage capacitor ($d_V/d_t = I/C$). This current is the sum of the leakage across the switch, the amplifier's bias current, and leakage to the power supplies and ground. The first of these is the most important. Sample holds that use digital storage have no droop.

Feedthrough is the fraction of input signal that appears at the output in hold. The major cause of this is capacitance across the switch. There is also some error due to dielectric absorption—the tendency of charges in a capacitor to redistribute themselves over a period of time, resulting in a different voltage level. The change is less than 0.01% for polystyrene and Teflon capacitors, and can be as large as several percentage points for ceramic and Mylar capacitors.

In the sample hold state, the acquisition time is the time that an input must be applied for sampling at the desired accuracy. There are transients or spikes that occur between the sample command and the final settling. These are not important for large changes, but they become important when the spikes are large compared to the actual change. The acquisition time may include the settling time of the output amplifier. It is possible for the signal to be acquired and the circuit switched to hold before the output is settled.

PERFORMANCE SPECIFICATIONS

The problem of determining performance requirements can be simplified by dividing the specifications into three groups:

1. Specifications that determine accuracy under optimum conditions.
2. Specifications that are time dependent.
3. Specifications that are affected by the environment.

The first group includes specifications such as resolution, relative accuracy, differential linearity, noise, quantization uncertainty, and monotonicity. Time-dependent specifications include conversion time, bandwidth, and settling time. Environment-related specifications include gain, scale factor, temperature coefficients, and operating temperature range.

Relaxation of the specifications of the first group can be achieved through the use of signal conditioning. The particular type of signal conditioning used depends on the input signals and the form of the information to be extracted from them. Unwanted signal components can be extracted from the input signals. A differential instrumentation amplifier may be used to reject common-mode signals, bias out the dc offsets, and scale the input, as discussed in Chapter 3.

A relaxation of time-dependent specifications can be achieved by adding a sample-hold amplifier to the system. The use of this amplifier will increase the system throughput rate and increase the highest-frequency signal that can be encoded within the converter's resolution. Without the sample hold, system throughput rate is determined mainly by the multiplexer's settling time and the A/D conversion time. Multiplexer settling time is the time required for the analog signal to settle to within its error budget, as measured at the input to the converter. In a 12-bit system with a +10-volt range, the multiplexer units typically settle within 1 microsecond.

A sample hold can be used to hold the last channel's signal level for conversion while the next channel is selected and settles. If the throughput time is reduced enough, it will approach the conversion time. Pairs of sample holds and A/D converters may be used for alternate conversions to increase the throughput rate even further.

Reduction of the errors due to environment-related specifications can be achieved by allotting one multiplexer channel to carry a ground-level signal and the other to carry a precision reference-voltage level that is close to full scale.

The data from these channels are used to correct gain and offset variations common to all the channels. These can be generated in the sample hold, A/D converter, or wiring.

ERRORS

Errors may be due to the physical interconnections, grounding, power supplies, and protection circuitry. To evaluate the available trade-offs, an error budget is always useful. Errors can be classified into three groups:

1. Errors due to the nonideal nature of the components.
2. Errors due to the physical interconnections.
3. Errors due to the interaction of components.

The first type of error can be determined from the specifications of the components. The second type results from the parasitic interactions that are a function of the interconnections. These are affected by grounding and shielding methods and contact resistance.

Interactions between components in the system are a result of the specifications of the devices. An example of this source of error is the offsets due to series impedances in the signal path. These impedances may come from signal sources or multiplexer switch impedances. Another example is the disturbance caused when a multiplexer switches channels.

The error budget is a useful tool for establishing the performance requirements of the system. The error budget can be used to predict the overall expected error. A worst-case summation, a root sum of the squares summation, or a combination of these methods may be used.

If the system is capable of tolerating a constant fractional error of 1% or less, logarithmic data compression can be used. Medium accuracy in a fixed ratio is substituted for extreme accuracy over the complete full-scale range. These applications can be handled using a PC for processing the data.

The wide use of modular interface boards for PC interfaces requires consideration of their design. Many types are programmable, which allows the user to select one of several voltage ranges by choosing the appropriate jumper connection. Many boards allow these modifications by connecting external resistors. The gain and offset temperature coefficients of these resistors can be sources of error.

To reduce common-mode errors, a differential amplifier can be used to eliminate ground potential differences, as discussed earlier; the signal source may be a remote transducer. The common-mode signal is the potential difference between the ground signal at the interface board and the ground signal at the transducer, plus any common-mode noise produced at the transducer and voltages developed by the unbalanced impedances of the two lines.

The amount of dc common-mode offset that is rejected will depend on the *CMRR* of the amplifier. Bias currents flowing through the signal source leads may cause offsets if either the bias currents or the source impedances are unbalanced. *CMRR* specifications may include a specified amount of source unbalance. The specifications may also indicate an upper frequency for which *CMRR* is valid. At higher frequencies with unbalanced conditions, the series resistance, the shunt capacitance, and the amplifier's internal unbalances reduce the common-mode rejection, producing a normal-mode signal. This error can be reduced by the use of a shielded line for the inputs. In the shielded configuration, the common-mode signal does not affect the capacitance of the input lines to the shield, since the shield itself is driven by the common-mode source.

The shield also provides electrostatic shielding to limit coupling effects to other lines that may be in close proximity. It is important that the shield be connected only at one point to the common-mode source signal and that the shields be continuous through all connectors. Since the shield is carrying the common-mode signal, it should be insulated to prevent it from shorting to other shields or ground. A return path must exist for the bias and leakage currents of the differential amplifier unless it has transformer or optically coupled inputs.

A common signal-conditioning device is the filter. Low-pass filters can extract carrier, signal, and noise components above the signal frequencies. These components will appear as noise if the A/D converter is not fast enough to follow them. A/D converters will often use follower circuits for impedance buffering. If these connections are available externally, they can be connected as active low-pass filters.

CONFIGURING THE PC INTERFACE

A key step in configuring the PC interface is to define the objectives as completely as possible. It is essential to understand what the manufacturer means by the specifications. Product information must be interpreted in terms meaningful to the user's needs, which requires a knowledge of how the terms are defined.

Consider all of the objectives and try to anticipate any unknowns. Try to include such factors as signal and noise levels, desired accuracy, throughput rate, characteristics of the interfaces, environmental conditions, and size and cost limitations. Some of these may force performance compromises or the consideration of a different approach. Considering the different types of boards on the market and the complex manner in which some specifications may relate to a system application, selecting and properly utilizing the optimum board for an application is not always an easy task. Configuring the interface involves the following issues:

1. The objectives of the system and how they relate to the specifications.
2. How the system may be configured to relax the performance requirements.
3. How the system components limit and degrade performance.

Some general considerations for a control and measurement system are as follows:

1. Input and output signal ranges.
2. Data throughput rate.
3. Error allowed for each functional block.
4. Environmental conditions.
5. Supply voltage, recalibration interval, and other operating requirements.
6. Special environmental conditions such as radio frequency fields, shock, and vibration.

Multiplexer considerations include the following:

1. The number and type of input channels needed, single-ended or differential, high- or low-level, dynamic range.
2. Settling time when switching from one channel to another and maximum switching rate.
3. Allowable crosstalk error between channels and the frequencies involved.
4. Errors due to leakage.
5. Multiplexer transfer errors due to the voltage divider formed by the on-resistance of the multiplexer and the input resistance of the sample hold.
6. Type of hierarchy used for a large number of channels and the addressing scheme.
7. Channel-switching rate (fixed or flexible, continuous or interruptible, capable of stopping on one channel during testing).

Sample hold considerations include the following:

1. Input signal range.
2. Slewing rate of the signal.
3. Acquisition time.
4. Accuracy, gain, linearity, and offset errors.
5. Aperture delay and jitter.
6. Amount of droop allowable in hold.
7. Effects of time, temperature, and power supply variations.
8. Offset errors due to the sample hold's input bias current through the multiplex switch and sources.

A/D CONVERTER CONSIDERATIONS

The following considerations are important to A/D conversion:

1. Analog input range.
2. Resolution required for the signal to be measured.
3. Requirements for linearity error, relative accuracy, and stability of calibration.
4. Changes in the various sources of error as temperature changes.
5. Conditions for missed codes, if allowable.
6. Time allowed for a complete conversion.
7. Stability of the system's power supply.
8. Errors due to power supply variations.
9. Character of the input signal (noisy, sampled, filtered, frequency).
10. Types of preprocessing needed or desired.

Other A/D conversion circuits may be more acceptable for the application than successive approximation, including the integration and counter-comparator types. The integrating types are generally better for converting noisy input signals at relatively slow rates. Successive approximation is best suited for converting sampled or filtered inputs at rates up to the megahertz range. Counter-comparator types allow low cost but can be both slow and susceptible to noise. They are useful for peak followers and sample holds for digital storage applications.

A DATA ACQUISITION DESIGN EXAMPLE

Suppose that a data acquisition system is to be used to process data from a number of strain gauges. Signal-conditioning hardware can be purchased with the gauges. Assume that these conditioned signals are ± 10 volts full scale, with a 10-ohm source impedance. It is also desired that the signal channels be sequentially scanned in no more than 50 microseconds per channel. The maximum allowable error of the system is to be 0.1% of full scale. System logic levels are to be transistor-transistor logic, and output is required in either binary or two's complement code with parallel data readout. The temperature range in the equipment cabinets, which includes equipment temperature rise, is +25° to +55°C.

A technique that usually provides satisfactory results is to choose each component to perform 10 times better than the overall desired performance. For a system that needs 0.1%-grade performance, use a 0.01% converter (12 bits) with a compatible multiplexer and sample hold.

An A/D converter that completes a conversion in 35–45 microseconds is acceptable. Since the sample hold will probably add about 5 microseconds of settling time, the combination should be capable of meeting the 50-microsecond/channel scanning requirement. Since the multiplexer will scan sequentially, its settling time is not important. The multiplexer can be switched to the next address as soon as the sample hold goes into hold on data from the current address. Thus it has 35–45 microseconds to settle before a measurement is called for.

A proper error analysis will consider the details of errors to determine if the worst-case situation is within the allowable 0.1% system error. In the multiplexer, if the switches are MOSFETs with variable-resistance channels, they will not be subject to voltage offset errors. Errors will tend to be due to two factors.

There is leakage of current into the on-channel from the off-channels. This will develop an offset voltage across the source impedance. Suppose that the leakage current at 25°C is 10 nA and the source impedance = 10 ohms. Then the error voltage = $10 \times 10 \times 10^{-9} = 10^{-7}$ volts (which is negligible).

There is also an error due to the voltage division between the MOSFET on resistance and input impedance of the sample hold. Suppose that the on-resistance = 1000 ohms and the sample hold $R_{in} = 10^{12}$ ohms. Then the divider ratio attenuation error = 10^{-9} (which is also negligible). In the sample hold, let the nonlinearity be 2 millivolts over a 20-volt range, or 0.01%.

Suppose there is an input bias current of 10 nA. This will cause an offset error voltage in the source resistance. Let the total source resistance = 10 ohms (due to the source) + 1K ohms (due to the multiplexer switch). Then the offset error = $10^3 \times 10^{-8}$ = 10 microvolts, (which is negligible). Now suppose that the offset change versus temperature = 25 microvolts/°C. Since the temperature inside the equipment housing can change by 30°C, the total change over the range = 25×30 = 750 microvolts or 75 ppm in a 10-volt range.

Let the offset versus power supply equal 100 microvolts/% change in the supply voltage. Suppose that the supply can vary by 100 microvolts or 1% of 10 volts. The error contribution is 100 microvolts, or 0.001% of full scale.

A comparable analysis can be used to prepare a system-timing diagram and assign settling time allowances. However, in this case, there is more than adequate settling time and a formal timing analysis is not needed.

For the A/D converter the linearity error, which is the relative accuracy, is $\frac{1}{2}$ LSB, or 0.0125%. This is also the quantizing uncertainty. It is a resolution limitation and is not usually considered as part of the error budget.

The temperature error for the A/D converter is calculated as follows: Let the gain temperature coefficient = 5 ppm/°C. Then for a 30°C change, the temperature-induced error is 5×30 = 150 ppm. If the power supply sensitivity error is 0.002%/% change in voltage, for a 1% shift, the error = 20 ppm. If the differential nonlinearity temperature coefficient is 3 ppm/°C, then for a 30°C temperature change, the error = 90 ppm. This is less than $\frac{1}{2}$ LSB, and the 12-bit monotonicity can be maintained with no missing codes.

In this data acquisition example, the worse-case arithmetic sum of these errors is 0.07%, and the rms sum is 0.03%. These values are less than 0.1 of the desired error of 0.1%, so the system is acceptable. The component specifications can be relaxed if hardware cost is an important factor. This will reduce costs by using a more marginal design.

5

User Input Device Technology

Many types of user input devices can be used with the PC besides the common keyboard. The application of these devices is explained in this chapter.

The list of useful devices includes touch panels, trackballs, light pens, joysticks, and mice. Some devices, like joysticks, use potentiometric technology, while others, like light pens, use the properties of light. Mouse technology includes the use of both techniques. We will examine all of the useful techniques, including optical data input technology like laser scanners and digital cameras.

Verbal input is a fast-growing technique with PCs. We will show the differences between isolated-word recognition and continuous-speech recognition. The various speech-recognition problems will be examined, as well as several speech-recognition products. We will also discuss the problem of voice systems interfacing. Optical character recognition is also explained. We will examine this technology as it relates to character isolation and recognition and discuss input device selection.

It is important to be able to interact with the controlled process conveniently and quickly. Several types of input devices can be used for this activity, including keyboards. The alphanumeric keyboard may be used to enter commands, text, and parameter values. Each key causes a 7-bit code to be stored in a character register in memory. The interpretation of the code is determined by the CPUs program. It is also possible to accumulate consecutive characters in a buffer until a termination character is typed, thereby producing a character-string input. In the programmed function keyboard (PFK), depression of each button generates a unique code which is stored in a register. Several technologies are used to detect key depression, including mechanical contact closures, change in capacitance, and change in magnetic coupling.

Many factors make one keyboard preferable to another, including the following:

1. The key spacing.
2. The slope of the keyboard.
3. The shape of the key caps.
4. The pressure needed to depress a key.
5. The feeling of contact when a key is depressed.

Other important considerations involve details of the layout, such as separating keys like line delete from other frequently used keys like those for the return function and making the frequently used correction key easily reachable without the need to depress simultaneously the control or shift key.

The PFK is sometimes provided as a separate unit, but often the key or buttons are integrated with the main keyboard. Button devices differ from text devices in that text devices are always prelabeled.

Button device keys usually have no predefined character meanings. Buttons are generally used to enter commands or menu options. Dedicated systems use buttons with permanent labels. To allow the changing of soft labels, some systems use coded overlays on which the command names are printed. Button devices may also report pressure releases as well as depressions. This makes it easy to start a control activity when a button is depressed and then to terminate the activity when the button is released.

Another type of button device is the chord keyboard, which has five keys shaped like piano keys. It is operated by depressing several keys at the same time, like the playing of a chord. The five keys allow 31 chords to be generated. Learning the chords takes some training, but skilled users can operate these devices rapidly, and they allow fast-touch typing. They are generally not suitable as substitutes for the standard alphanumeric keyboard.

TOUCH PANEL TECHNOLOGY

Touch panels allow the user to give full attention to the screen and to indicate items on it using a finger rather than by moving a screen cursor to the item. The touch panel is mounted across the face of the cathode ray tube (CRT), and when the user's finger touches the panel, this position is detected using one of several different technologies.

One type of high-resolution touch panel uses two layers of transparent material; one is coated with a thin conductor and the other is resistive. Finger pressure causes a voltage drop on the resistive substrate, which is measured to calculate the coordinates of the pressure point. Similar techniques are used in digitizing tablet technology. Graphic pad digitizers use a flat tablet and a stylus pencil. The tablet provides a flat surface over which the stylus pencil is moved. The position of the stylus is monitored by the computer.

The stylus usually incorporates a pressure-sensitive switch which closes when the user pushes down on the stylus. This is used to indicate that the stylus is at a position of interest on the screen, and a menu choice is made.

Most tablets use an electrical sensing system to measure the stylus's position. A grid of wires can be embedded in the tablet's surface. The electromagnetic coupling between the electrical signals in the grid and in a wire coil changes as the stylus is moved. This induces an electrical signal in the stylus, and the strength of the signal is used to determine the position of the stylus on the tablet.

The digitizer senses the position of a pen, cursor, or some other pointer and relays this information electronically to the computer as an $x - y$ coordinate on the pad. A drawing placed on the tablet can be digitized by positioning the stylus over each point or line to be recorded.

The pointing devices that contain push buttons can start predefined computer routines or other operations to simplify graphic data entry.

Other digitizer technologies are electrostatic and magnetostrictive. The electrostatic tablets radiate an electric field which is sensed by the pen. The system changes the frequency of the radiated field as a function of the pen's location on the surface. This frequency changes for each tablet position. The digitizer circuits translate the frequency changes into $x - y$ coordinates, which are then sent to the computer. Electrostatic digitizers can sense the pen's position through most media with a low dielectric constant, such as paper, plastic, or glass. Electrostatic systems cannot digitize accurately in close proximity to metal or partially conductive materials such as pencil lead or some felt-tip inks.

Magnetostrictive tablets use magnetostrictive wires laid beneath the table to keep track of the pen's position. A magnetic pulse on one end of the wires produces a small strain wave which propagates across the tablet. The wave is detected using a pickup coil in the pen or cursor. The time elapsed from the start of the pulse to the sensing of the probe can be related to the position of the pen. Magnetostrictive tablets are not sensitive to conductive materials. However, magnetic objects near the tablet's surface can disturb the magnetic properties and degrade the tablet's accuracy. These tablets must also be periodically remagnetized with large magnets.

In electromagnetic tablets, either the table or the pen transmits a small ac signal. This signal is detected by receiver circuits, which produce the digital signal defining the pen's location. Electromagnetic digitizers are not affected by conductive or magnetic materials on their surface and generally require no periodic recalibration.

Special functions are determined by the software. Programmed routines are often used to simplify and speed the process. Typical routines are as follows:

Length calculations—The line length and sums of line lengths are calculated, accumulated, and displayed.

Area calculations—The sums and differences of areas are calculated, accumulated, and displayed. Automatic closure is a part of some systems, eliminating the need to close an area boundary with the cursor.

Volume calculation—The space volume is computed with slices digitized in the $x - y$, $y - z$, or $z - x$ planes.

Label insertions—Labels are assigned to the data.

Event counting—The operator selects events, such as End-of-coordinate or End-of-data, and starts an event counter. The eventer counter's contents are displayed or transferred as part of the output data.

Orientation conditions—This refers to how the digitized data will be scaled, formatted, and processed. Rectangular or polar coordinates may be measured in an absolute mode, as an absolute displacement from a fixed point, or in a delta (δ) mode where each point is output as the difference, or delta, between it and the last digitized point.

Tablet surfaces include formica, epoxy, laminated glass, and plastic. Many digitizers use glass or epoxy surfaces for their mechanical stability. The less expensive systems use plastic surfaces. Some systems utilize back-lighted or rear-projected images, which are then projected on the tablet.

Some Computer-aided-drafting (CAD) systems are designed around a digitizing table and are generally called *digitizing systems*. This type of system is designed specifically to digitize and plot drawings. The system consists of a digitizing table and cursor, video display microcomputer, and mass memory in the form of magnetic tape, floppy disk, or hard disk.

These systems accept input either in the form of drawings digitized with the cursor or from points entered through a keyboard. The system stores drawing information in mass memory and generally produces a serial digital output that can be used directly by plotters or line printers or sent to automated machines for cutting and other processing steps.

KEYPAD OPERATIONS

The keypad type of cursor used on many systems, in addition to recording points for digitizing, may also enter commands. The cursor commands can invoke the following responses on the display:

1. Provide coordinates of the current digitized position.
2. Prompt the user with requests for information needed by the system.
3. Display the results of computations.
4. Identify errors in the entered information.

5. *Scaling* refers to the number of units represented by each unit on the digitizing table. Different axes may be scaled independently or with functional relations.
6. *Translation and rotation* refers to the orientation of the display with respect to the digitizing table.

Digitizers are also designed for several operating modes. In the point mode, single coordinate points are entered one at a time. In the run mode, the digitizer sends a stream of $x - y$ coordinates to the computer as the pen moves along the surface. This mode may be used to position a cursor on a CRT or a pen on a plotter. The track mode is similar, except that data are sent only when the pen is activated. The true increment mode also uses a stream of data, but the data stream stops when the pen stops moving. These digitizing modes often can be selected by software. The point mode enters a coordinate every time the user presses an output key with the cursor centered over a point to be digitized.

Modes are also available to enter data as the cursor moves continuously. In the rate-digitizing mode, the rate of entered points can be selected so that data are entered at a rate of 1 to 100 points per second as the cursor moves. This allows the user to enter curves accurately. The cursor's movement can be slowed or accelerated to adjust the number of points representing the curve.

A number of digitizer specifications are used for evaluation and selection. The resolution is the number of points that can be digitized per inch; 100 lines per inch resolution provides an identifiable $x - y$ coordinate every 0.010 inch. Many systems provide between 400 and 1000 lines per inch.

Repeatability is a measure of how consistently the digitizer can identify the same position on the tablet surface with the same set of $x - y$ coordinates. Factors that tend to reduce repeatability include thermal expansion of the tablet surface, humidity, and variations in electrical characteristics.

POTENTIOMETRIC DEVICE TECHNOLOGY

Several types of devices which provide scalar values use potentiometers. A set of rotary potentiometers may be mounted in a group. Rotary potentiometers are often used to control cursor movement or object rotation. Slide potentiometers, in which linear movement replaces rotation, are more suitable for controlling values with no angular movement. The setting of a slide potentiometer is often quicker and easier for the user to gauge than that of a rotary potentiometer.

Potentiometers are sampled by devices whose analog values are converted and stored in registers to be read by the CPU. The values read from the registers are then converted to equivalent numbers to be used by the program.

An A/D converter and a power supply can be used to determine the potentiometer's position by measuring the voltage. The voltage is proportional to the amount of shaft rotation about the axis. When the converter is correctly adjusted, a zero position will correspond to a digital reading of all 0's and a full-scale position will correspond to all 1's.

The CPU can be directed to poll the pointing device to see whether it has been changed by more than some desired amount. If it has, the CPU is directed to generate an event, which is placed in an event queue.

TRACKBALL TECHNOLOGY

Trackball controls are large plastic balls mounted with a fraction of the ball's surface protruding from the top of the enclosing unit. The ball rotates freely within its mount and is typically moved by drawing the palm of the hand over it. Rotating the ball can be interpreted by the computer to move the screen cursor. Trackballs can be used to select commands displayed. The ball positions the cursor on the proper area of a menu. Pushing a keyboard key (usually Return) makes the selection. The computer reads the cursor's position at the time when the key is pressed and executes the procedure listed in the menu.

Trackballs (sometimes called *crystal balls*) are used as shown in Figure 5-1. The ball's motion turns potentiometers, whose output is converted into digital data for reading by the CPU. Large and rapid position changes can be difficult to make with a trackball.

LIGHT PEN TECHNOLOGY

Light pens are used as element-indicating or -picking devices. The light pen does not emit light to create lines on the screen; it senses or detects the light from the picture elements on the screen. Some light pens do use a narrow, focused light source called a *finder beam*, which is directed at the screen to indicate what picture elements are in the pen's field of view.

The hand-held light pen consists of a pencil-sized plastic cylinder with a light sensor in one end. One light pen design is shown in Figure 5-2. A number of

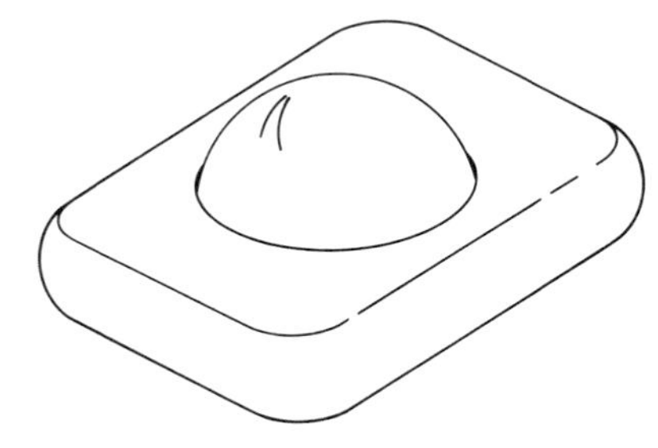

Figure 5-1 A trackball position control uses a large plastic ball which is moved with the palm of the hand.

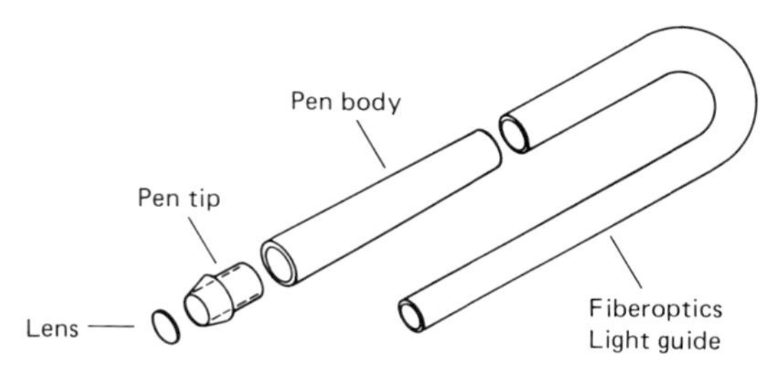

Figure 5-2 Light pen design concept.

different designs are used. The other end is connected to the computer by a cable. As the user positions the light-sensitive pen tip to select a point on the screen, the light pen sends a pulse when this screen area is bright. The pulse produced by the light pen is then used to calculate the former screen position of the pen. The pen senses the burst of fluorescent light emitted when the electron beam is bombarding the phosphor. This is the light emitted during the drawing of the picture element. The pen's output is usually connected to the system's control logic in such a way that the computer stops executing commands when the pen senses the light. This signal can be correlated with a menu list to detect the particular item being displayed at the time when the light from the pen was detected. Most light pens also produce a second signal that the user generates by pushing a button on the pen to signal the computer of the user's selection of a point, as shown in Figure 5-3.

Another control technique is to program a series of points as buttons in the lower portion of the screen. These can be treated as programmable push buttons. By selectively touching these buttons with the light pen, the user controls the program.

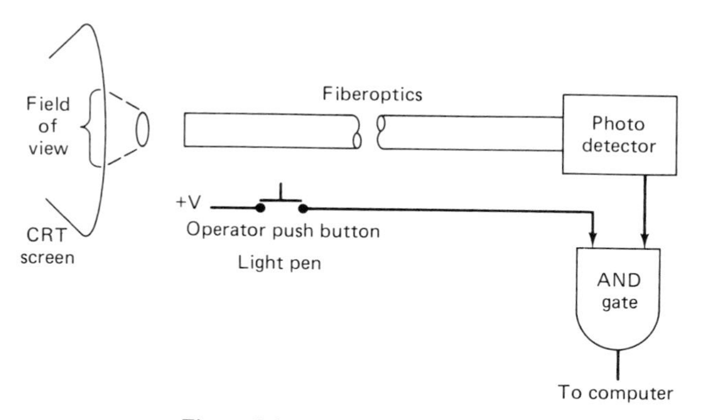

Figure 5-3 Light pen operation.

When the computer is interrupted by the light pen signal, its instruction counter contains the address of the next instruction. This address may be used to give the application program the location of the segment containing the detected element. Using this technique, the light pen acts as a locator for a raster display, stopping the raster scan at the detected pixels. The $x - y$ coordinates provide the locations.

Several factors can affect light pen performance. Many CRT screens use bonded implosion shields over the display tube, leaving a gap between the outer glass surface and the phosphor. This gap can cause accuracy problems when positioning the light pen.

Unless it is properly adjusted, a light pen can detect false points, such as adjacent characters, or fail to detect desired points. When used for several hours, the pen can be tiring to the user, who must pick it up, point it, and set it down for each use. It is often more convenient to use a tablet or mouse to simulate the picking function.

In those display systems that do not use a raster scan, light pens cannot detect blank portions on the screen. In these systems, some means of sensing the light pen's position over blank portions of the screen must be provided. One technique is to flash the entire screen for one frame to allow the system to give the pen's position. Some systems fill the entire screen with characters, line by line, until a character is generated under the light pen's position.

The light pen is one of the simplest and most flexible input devices, one must often supplement its function with one or more push buttons.

One of the disadvantages of light pens is that they cannot be used where no image is being displayed. In order to know where the light pen is located, the computer must find it. Other devices, such as the track ball, volunteer their position coordinates. The procedure required for correlating the position with the display list can become complicated, since the input is not synchronized with the display list.

JOYSTICK TECHNOLOGY

A joystick is like an aircraft control stick; the user controls the motion of a cursor on the screen by pushing it in the direction of the desired motion. The displacement-type joystick actually moves, but with the force-actuated joystick, the stick is fixed and the force from the user's hand causes the cursor's motion.

Force-actuated or isometric joysticks use strain gauges on the shaft to measure the deflections caused by the force. Joysticks may also be rotated or pushed on the end to control variables other than the basic position. The joystick can be moved left or right, or forward or backward. Potentiometers sense the movements. Springs are sometimes used to return the joystick to its center position. Some joysticks offer a third degree of freedom, since the stick can be twisted clockwise and counterclockwise.

With an absolute joystick, the travel of the joystick corresponds directly to the screen position. Moving the absolute joystick to its upper position places the cursor at the top of the display. Moving it to its lower left position places the cursor at the lower left corner of the display.

With the rate joystick, the motion of the joystick imparts a direction of motion to the cursor. If you move the rate joystick to the left, the cursor will move left from its present position. Push it to the lower left, and the cursor will move diagonally in that direction. When you return the joystick to the center, the cursor comes to a halt. In many designs, by varying the distance in any direction the rate joystick is moved, the user controls the speed at which the cursor will move.

The absolute/rate joystick is a hybrid design that utilizes both techniques. A switch on the joystick handle is used to select the mode. In the absolute mode, a motion to the upper left corner with the joystick will move the cursor to the upper left corner of the computer display. Once the general target area is reached, the switch is released and the cursor positioned using the rate mode.

It can be difficult to use a joystick to control the absolute position of a screen cursor accurately, since even a small hand movement is amplified in the position of the cursor. The cursor's movements can become erratic, since the joystick does not allow fine positioning.

A small dead zone is normally used to allow for drift in the joystick's center position. The relationship of cursor velocity to joystick displacement is generally not linear.

When the joystick does not have a spring return to zero, it can be used for absolute rotation rather than rate of rotation. The joystick can be used for three-dimensional orientation if the shaft can be rotated for a third degree of freedom. The joystick can then be used to control the rate of rotation on the screen about each of the three axes.

A variation of the joystick is the joyswitch. It can be moved in any of the following directions: up, down, left, right, and in four diagonal directions. A switch allows nine states. In each of the eight on states, the position of the screen cursor may be changed at a constant rate in that direction.

MOUSE TECHNOLOGY

Mice are hand-held units that use rollers or an LED and a ruled metal surface. Movement of the unit causes the rollers to turn internal potentiometers or the LED to sense the ruled lines. These voltage changes, in turn, are used to sense the relative movements of the mouse and track the relative $x - y$ position. The motion is converted to digital values and used by the CPU to calculate the direction and magnitude of the mouse's movement. The mouse is normally moved on a flat surface called a *mouse pad* (Figure 5-4). The computer maintains a current mouse position register which is incremented or decremented by

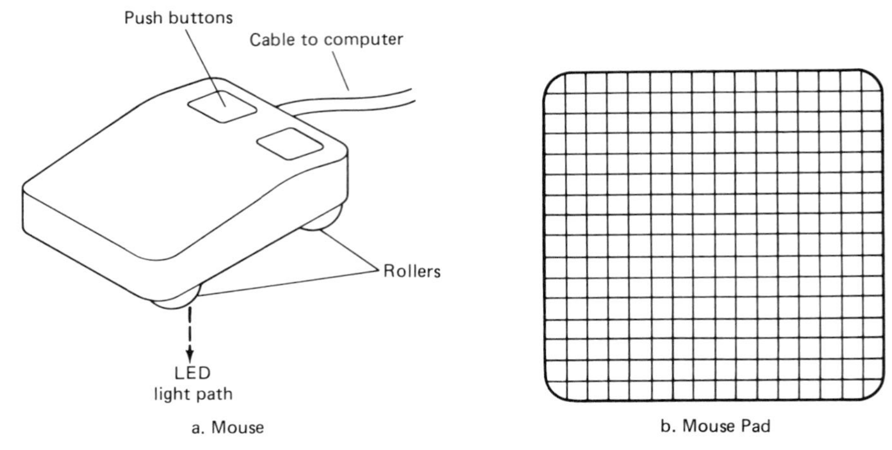

Figure 5-4 The mouse is a hand-held unit with rollers or an LED on the bottom.

the mouse's movements. The mouse usually has one or more push buttons which are used to input commands.

The accuracy of the mouse is less than that of a tablet, and with so many types of mice available today, choosing the right one can be a difficult task. There are optical, mechanical, opto-mechanical, optical-mechanical, acousto-mechanical, and analog tracking units with one to four buttons. Tracking may be dependent on or independent of the rotation of the mouse. Resolution ranges from 20 to 2000 counts/inch, and serial or parallel interfaces with relative or absolute position information are available in a variety of data formats and at different baud rates. Some are programmable, and the user can even buy a mouse with a built-in-speed transmission. With all these options, a mouse can cost from $30 to $1800.

Although mice were invented in the late 1960s, they have only recently become popular. Some studies indicate that mice make nearly ideal pointing devices for humans. In general, mice are designed so that the front of the device is low because that is the way the human hand naturally falls. Mice must also be lightweight because they are frequently lifted up and repositioned on the surface. The buttons are often parallel to the surface so that the mouse will not be moved accidentally when a button is pressed.

The shape of the mouse allows it to be gripped between the thumb and forefinger for drawing. Some mice are shaped to fit the hand. These may feel comfortable to some people, but not all hands are alike and the devices must be symmetrical to accommodate left-handed users. The cord should come out away from the user so that it does not interfere with hand motion.

Six different techniques that can be used to perform the tracking function of a mouse. However, only the optical, mechanical, and opto-mechanical methods are widely used.

Optical mice track position by counting the number of line (or dot) crossings on a special mouse pad, much as a barcode wand reads barcodes (Figure 5-4). Since there is no contact between the mouse and the surface, and since there are no moving parts, optical mice are reliable. The mean time between failures (MTBF) for a typical optical mouse is calculated per MIL-HDBK-217D to be over 25 years. It does not require maintenance, and tracking errors of less than 1 in 100 are typical throughout the life of the mouse. A special mouse pad is needed for optimum performance.

Some optical mice track motion relative to the orientation of the mouse pad. The mouse cannot be rotated more than 90° in either direction. This motion is much like the tracking characteristics of a digitizer. Other optical units track motion relative to the orientation of the mouse, which is similar to the tracking characteristics of mechanical mice.

Mechanical mice use a metal ball to drive two orthogonal mechanical shaft encoders. This device is low in cost and lets the user track motion on any surface, but these units have a limited life and require periodic cleaning.

Opto-mechanical mice use a ball to drive two orthogonal optical shaft encoders (Figure 5-5). This device is low in cost, has a long life, and lets the user track on any surface, but periodic cleaning is required. Although the same basic techniques are used for tracking, there are large differences in performance among products in this area.

Analog mice use either a ball or two orthogonal metal wheels mounted orthogonally to drive the shaft of a potentiometer. This device is low in cost and compatible with the interfacing schemes used by PCs. The disadvantages are that such mice have a limited life and a limited amount of travel in any one direction due to the use of the potentiometer.

Opto-mechanical mice use an optically sensed, spring-loaded plate to deter-

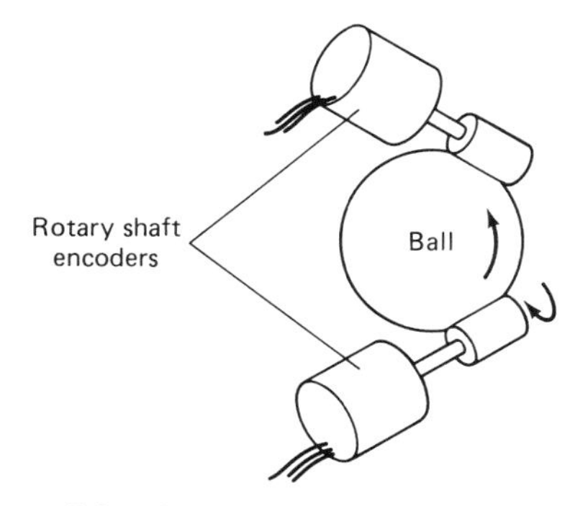

Figure 5-5 Opto-mechanical mouse construction.

mine the direction and to count optically the line crossings on a special mouse pad. These units have characteristics very similar to those of optical mice.

Acousto-mechanical mice use strain gauges to determine direction and a piezoelectric transducer to determine magnitude. There are no moving parts and no maintenance, and the mouse can track motion on almost any surface. However, the resolution (in counts/inch) depends on the type of surface on which the mouse is used and on the downward pressure on the mouse.

The main difference between optical mice and other types is that optical mice require a special surface. This is not as significant as it seems because, in practice, some kind of pad is often used with nonoptical mice to reduce slipping and audible noise and to protect the desktop. This pad is often rigid. The pad for optical mice meets all these conditions. Most nonoptical mice do not work well unless they are horizontal.

Mouse buttons

In many units, three buttons are used. A one-button mouse is simple, but advanced functions require double-clicking or the use of the mouse button in conjunction with keyboard keys. Many two-button devices use chording (holding both buttons down) to cancel a command.

A resolution of 20 counts/inch is adequate for text editing on a character-mapped screen, but most applications use at least 100 counts/inch. This resolution is adequate for high-resolution graphics.

The major differences among the mouse technologies can be summarized as follows:

1. *Maintenance:* Optical mice have no moving parts. Since mechanical, opto-mechanical, and analog mice require regular maintenance, some mice have access holes to facilitate maintenance.
2. *Accuracy:* Optical mice rarely miss a count, while mechanical mice may slip on the surface.
3. *Noise:* Some mechanical mice that use a metal ball make audible noise on some surfaces. It is possible to reduce this noise by running the mouse on a sheet of paper.
4. *Jitter:* When a mouse is positioned between two grid points, a type of jitter called *teasing* may occur. The mouse may jump ahead or back by one count. Many devices use hysteresis to eliminate this problem.

Interfacing

There are two basic type of interfaces for mice: parallel and serial. Parallel mice generally use quadrature encoding for motion and encode switch depressions as a TTL low voltage. Different 9-pin standards are used by the manufacturers.

Serial mice are interfaced to RS-232 ports. A wide variety of protocols can be accommodated because the interface is standard. Generally, these protocols send the change in the mouse state (buttons or position) whenever the mouse changes state. The Mouse Systems serial protocol is a 1200-baud, relative-position protocol that yields 48 updates/second with a maximum velocity of 50 inches/second. Information is transmitted only when there is a change in mouse state. Instead of polling the mouse, the processor must use interrupters to service the device. This procedure tends to be efficient, since the host services the mouse only when necessary, rather than at a fixed rate like 30 times/second every second.

There are transparent drivers for the IBM-PC that allow the mouse to be interfaced to existing application programs without modification. The motion of the mouse is translated into cursor keys and buttons that bring up pop-up menus that can be customized by a user. Programs such as Lotus 1-2-3 work well with this interface, even though these programs were not designed for use with a mouse.

OPTICAL DATA INPUT

Optical scanning technologies have been in use for more than 20 years. The first commercial application was the IBM card reader, which sensed light holes in the card. Early optical character recognition (OCR) techniques were aided by special tints and inks and were characterized by large-scale installations.

MANUAL DATA CAPTURE

In the past, all input was either by card input or by keyboard entry. Today a variety of techniques are available to allow manual graphic data input. Major technologies for hand entry include magnetostrictive tablets, electromagnetic tablets, and ultrasonic tablets. There is also the mouse. The advantages of hand methods include flexibility, low price, high accuracy (most manual digitizers are capable of 0.0005-inch repeatability in selecting the same point), and general ease of use. They are most useful when the user works with existing graphic data, interactively transforms the coordinate system, or modifies individual graphic elements. But for initial input, the major disadvantage is the large amount of time required.

VIDEO DIGITIZERS

For image input in the area of medium-to high-resolution color, video-based technologies are the most common technique. Examples include geographical, mining, remote-sensing applications, automated surveillance, and inspection where the final output is a color display or color area plot. Most video systems have certain characteristics in common. Resolutions of 256 × 256, 512 × 512, or 1024 × 1024 pixels are used. Color and gray scale capability ranges from 1 bit/pixel to 24 bits/pixel. Most high-resolution systems support 8 bits/pixel for display, giving 256 displayable colors. The use of 24 bits/pixel provides 256 levels of brightness for each of the three primary colors used, for a total palette of over 4 million color shades. The acquisition time is typically 1/30th of a second, the same as the standard frame rate for television production.

The early video systems were expensive because of the high cost of the A/D conversion components required and the memory required to store the image. Most video systems are used for pictorial imaging applications, since they have relatively low resolution capabilities compared to other graphic devices and because the video image is nonlinear and nonrepeatable.

There is a lack of hardware and software to convert the video image to graphic data. In the three-dimensional area, there has been some success in combining multiple simultaneous views and gray scale information to generate accurate three-dimensional shape recognition and position. This determination technique has been used in some robotics and flexible manufacturing systems.

LASER SCANNERS

In applications where high resolution is important, laser image digitizing can be used. These systems can have a repeatable input resolution of up to 2000 lines/inch (0.0127 mm) for document sizes of up to 25 × 36 inches. The digitizing and data transfer speeds can exceed 12 MHz (pixels/second), but most operate in the 3- to 5-MHz range. This allows an A-sized (8.5 × 11 inch) document to be scanned in about 2 minutes at 3 MHz, generating 46.5 Mb of data.

Laser scanners use a single laser whose beam is moved across the input document with rotating mirrors. As the beam travels across the document, reflected light is received by either a single- or multiple-element sensor.

MULTIELEMENT OPTICAL SCANNERS

Scanners of this type can provide large formats such as E-size drawings, which measure 34 × 44 inches. Some systems use a fixed scanning technique with a drum on which the paper moves. Others use a flatbed and move the sensor array. This group of systems includes microdensitometers, which are used for high-resolution applications in microscopy, with up to 100,000 pixels per scan line. Some systems use as many as 48 linear arrays of 2048 elements. Multielement scanners are designed for high-volume use, such as automated forms or invoice scanning. These systems are used for binary (black and white) imaging of high-contrast documents.

DIGITAL CAMERAS

Digital cameras are an application of linear or planar sensor chips to a camera-like fixture. The use of a lens allows remote viewing. One type of camera has a 1728-element linear charged coupled device (CCD) array that is scanned mechanically in the film plane behind a lens. Pixel data from the CCD array are then stored in main memory. These cameras have data rates varying from a few hundred to 4 million pixels/second. Gray scale rates of up to 16 bits/pixel, as well as color, are available.

Speed is the main advantage of digital cameras. Frame scan lines range from 0.33 second to 20 seconds. At the low end, a number of cameras use 256 × 128- or 256 × 256-pixel planar arrays. This is essentially the same chip used in solid-state home TV cameras. At the high end are systems using 4096-element linear arrays composed of two 2048-element CCDs optically combined using a beam splitter, scanned over a 4096-line field, for a total resolution of 4096 × 4096 elements.

Digital cameras can be used in much the same way as a 35mm still camera, with compatible lenses and other photo equipment. The higher-resolution systems compare drum and laser scanners for high-resolution applications.

RASTER-TO-VECTOR CONVERSION

Raster-to-Vector or raster-to-entity conversion is the ability of the system to read the binary image and extract the data necessary to duplicate the original image. The methods used may be thought of as analogous to viewing an image and writing down instructions to draw another one like it. The original binary image is stored and is analogous to a photograph in memory.

Several algorithms are used to interpret images for vectorizing purposes. A line-following system scans through the white space of the image until a black pixel is found to define an edge (transition) between white and black. The system then makes a series of explorations, looking for continuations of the edge found. This process continues until the algorithm returns to the original starting point of the edge. Since the algorithm interprets the image at the lowest possible level (pixel by pixel), it may look at the same pixel several times while following different edges around.

Another algorithm breaks the image into a hierarchy of hexagonal or square blocks. A table of statistics is maintained for each of these blocks which indicates to the system if anything of interest lies within a block. The statistical information also indicates if the object is a simple object such as a line, circle, or arc, plus its location in the block. If the statistics are not recognized, the entity is too complex to be described at this level in the hierarchy. The tile is then broken down into its next smaller components, and the same data are collected. After the first pass, the algorithm knows which regions are empty. It also has a tree structure of the objects from which it can build a top-down model of the image. This algorithm has been used in three-dimensional object recognition for robot vision applications and parts handling.

Another approach, employed in robotics vision applications, uses methods adapted from laser interferometry modeling and image processing. It includes Fourier analysis and fast digital methods for performing spatial filtering. This is similar to the high-pass, low-pass, and bandpass filtering used in audio and electronics, but it is performed in two or three dimensions and in space rather than time.

A grid of squares is interpreted as a two-dimensional square wave composed of a base frequency and the odd harmonics of the frequency. The image of the grid is scanned into the computer and passed through a two-dimensional filter that removes all signal components having a pure X component of a certain number of cycles/inch and related harmonics. The result is that the vertical lines are filtered out and the desired image and horizontal lines remain. In the same manner, the user can filter periodic diagonal lines and leave the horizontal and vertical lines.

OCR was an early component of many mainframe-based systems. A fixed- or adjustable-field scanner was used, with resolution of up to 200 dots/inch on an 8 1/2 × 11 inch input field (1728 × 2200 pixels). The display resolution was not usually sufficient to let the user view more than a small portion of the image at one time, and scan times were several minutes.

Digital images may be stored, retrieved, and transmitted the same way as any other files, although the files are larger than usual on small systems. Compressed data formats are available, and the basic scan-and-save capabilities of low-cost PC systems make them useful for many applications. With a 5:1 compression ratio, 1K byte will hold about 10,000 page-format images of 100K byte each. The same store will hold approximately 2000 pages uncompressed.

System enhancements include fast scanning, gray scale, and color image support, plus image processing hardware and software. Tasks such as convolution (filtering the image to improve quality or accentuate features) and correlation (used in character recognition) may be performed on these systems.

An example of a typical digitizing and recognition system is shown in Figure 5-6. Applications include automated inspection applications. The recent advances in vision systems and AI support fast image handling and processing. The camera uses a linear CCD array that moves across the image plane in steps to generate a digitized image. The camera has built-in compensation for any nonlinearities of the CCD sensing elements and is linked to the workstation by

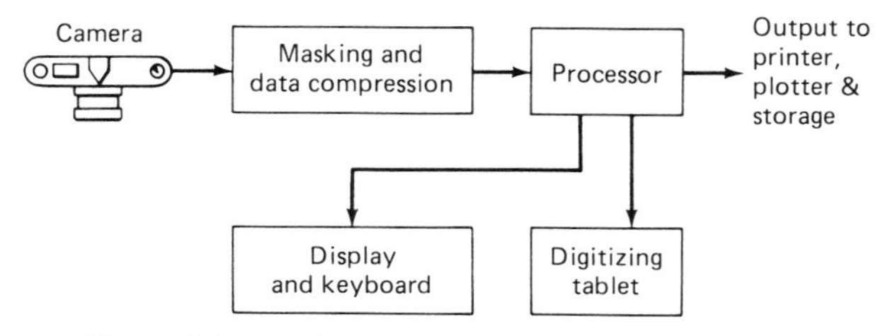

Figure 5-6 Typical digitizing and recognition system.

a dedicated direct-memory access (DMA) interface channel operating at 16 Mbytes/second.

The system maintains images as segmented rectangular arrays of any size and shape. Multiple images may be defined simultaneously. When a portion of an image is displayed, the displayed portion is copied to display memory and to an edit buffer, so changes to that portion do not affect the actual image until an update is selected.

A number of applications for these scanners exists, including the automated inspection of phosphor blemishes on high-resolution Vidicon tubes; the entry of images into and support of a parts control inventory system using a distributed network data base management system; the digitizing of topographical relief maps into a data base; and automatic visual fault checking of PC boards and hybrid circuit chips.

VERBAL DATA INPUT

Speech recognizers are available in relatively low-cost versions that recognize a few hundred words and in more expensive devices that recognize thousands of words. The words must be defined as discrete utterances, preceded and followed by fixed pauses in some units. Continuous-speed recognition is also available in commercial systems.

Most of these word recognizers must be trained to the unique characteristics of an individual's speech. The word is generally filtered into several frequency bands, and patterns based on the relative loudness or pitch of the sound are matched against a dictionary of recorded patterns. Training the recognized unit requires the establishment of a dictionary of patterns, during which the user must speak the word several times. Automatic speech recognition (ASR) is a more difficult task than automatic speech generation (ASG). It has become easier as advances have been made in algorithms and placed on VLSI chips. ASG offers the most natural form of computer interaction and the easiest one for humans to learn. There is the potential for lower data-entry cost and higher accuracy, since it eliminates mental encoding by the operator. The operator's hands and eyes are free so that other tasks can be performed at the same time.

Speech recognition generally requires the digitizing of waveforms and comparisons to stored-vocabulary reference patterns. There are two basic categories of speech recognition systems. In speaker-dependent recognition, individual speech patterns are used, and the system must be trained to recognize each operator. Speaker-independent systems compare a speaker's voice inputs with averaged voiceprint patterns.

Speech-recognition systems convert the acoustical information (waveform) into a written equivalent of the information contents of the spoken message. Speech recognition depends on the constraints placed upon the speaker. It must take into account the spectral voice qualities that differentiate speakers saying the same words. Less information is needed for voice recognition than for voice generation, since recognition is effectively performed by the use of voiceprint matching, where pitch and timber are less important.

Speech recognition is essentially a pattern-recognition problem. The system must digitize an analog signal or voice waveform and compare it to a stored reference pattern or vocabulary. Word templates can be formed by a variety of techniques, including bandpass filtering, A/D conversion, zero-crossing detection, and fast-Fourier transform (FFT) analysis. Speech recognition involves an extraction pattern-similarity measurement, time registration, and a decision strategy.

The input word template is processed and then compared to a series of templates or patterns stored in memory. Comparison algorithms are used, along with decision logic, to choose the proper templates. Speaker-dependent systems are based on the characteristics of the specific operator, and can handle both isolated words and connected speech.

Speaker-independent systems do not store individual voice patterns and do not have to be trained to the individual operator. The speaker's voice inputs are compared with averaged voiceprint patterns. The wide variations in human speech pitch, intensity, and duration allow recognition of only isolated words or connected speech from a limited vocabulary. These systems are not as accurate or flexible as speaker-dependent systems.

ISOLATED-WORD RECOGNITION

Isolated-word recognition (IWR) systems do not have the problems that continuous-word recognition systems have, since all words are separated by pauses. IWR systems can be 99% accurate in recognizing a word spoken in isolation. Typical vocabularies are 50 to 300 words.

An IWR system divides the acoustic signal into separate phonetic and spectral signals. Each of these is then detected and converted separately into a bit pattern. Since some bit patterns are longer than others, a time normalizer is used. A correlation is used to match these patterns with vocabulary words stored in memory.

The typical system includes microphone input, visual or acoustic feedback of word entries, acoustic feature detector, reference-pattern and memory-correlation processors, and recognition software. The operator trains the system by repeating the desired vocabulary words into a microphone so that templates can be made and stored. The key parts of IWR are the algorithms for reference-pattern representation and the search/decision strategy. With these algorithms for discrete or isolated word recognition, utterances exist as predefined, known patterns. Most variations among pattern occurrences, other than those of extreme duration or overall energy, do not have to be considered. A known, finite number of patterns needs to be recognized, and these can be used to train the IWR system.

CONTINUOUS-SPEECH RECOGNITION

This refers to the recognition of certain sets of words uttered without pauses. Continuous-speech recognition (CSR) systems use a hierarchical-symbol space where subparts are recognized and grouped into larger units. CSR techniques are based upon information theory.

With CSR, it is difficult to tell where one word ends and another begins, since the characteristic acoustic patterns of words vary, depending on the context. As the vocabulary size increases, matching becomes more difficult.

All the words in a sentence may need to be recognized, but only half of them may need to be interpreted correctly for the sentence to be understood. Recognition systems can keep their program structure simple by using only some task-specific knowledge and by requiring that the speaker speak clearly and use a quiet environment.

Since the system does not know where words begin and end, the beginning must be specified prior to a match. Error and uncertainty in segmentation, labeling, and matching require the use of such special algorithms as tree searching. An exact match cannot be found until the next word in the sentence or the ending is found.

In order to represent the sentence completed so far, alternate word sequences can be arranged in descending order of likelihood. Given the word sequence with the highest likelihood, task-specific knowledge, such as phonological rules, lexicon, and syntax, can be used to generate the words that can follow the sequence. Each of these words is then matched against the unmatched symbol string to estimate the conditional likelihood of occurrence. This process is repeated until the whole utterance is analyzed and an acceptable word sequence is determined.

SPEECH-RECOGNITION PROBLEMS

Problems are encountered with speech recognition, since a speaker rarely repeats words in the same manner and a listener does not always hear in the same way. Pronunciation pauses, background noise, speed, and personal stress levels vary in their individual contributions for both speaker and listener. A high-level mental function allows a person to weigh the words, appreciate the concept, or provide understanding of what a speaker is trying to convey. The speaker uses his or her knowledge of both the language and the context in recognizing or understanding a sentence.

This knowledge comes from the characteristics of speech sounds (phonetics), variability in pronunciation (phonology), stress and intonation patterns in speech (prosodics), the sound patterns of words (lexicons), the grammatical structure of language (syntax), the meaning of words and sentence (semantics), and the context of conversation (progmatics).

The advantages of direct voice input to computers are its speed and the freedom it provides for other tasks. Not only is it much faster than keyboard entry, it frees the user's hands and eyes for other tasks. Direct speech input is also potentially more reliable than keyboard devices, since it eliminates any intermediate data-preparation and data-entry steps.

Most speech-input systems have limited vocabularies. The inexpensive systems for PCs are speaker dependent and are only capable of recognizing discrete-word speech from relatively small vocabularies, but they do have relatively high acceptance and rejection accuracy rates. Vocabularies can usually be increased by adding more memory, though operators have to pause between words on most systems.

Dynamic programming requires more processing power, but it can be used for time warping to reduce pauses in CSR that result from unconstrained endpoints. Dynamic programming techniques can be used to account for variations in the rates at which people speak and for changing voice inflection. They can also recognize words accurately 99.5% of the time under adverse conditions such as the high-noise levels found in some environments.

SPEECH-RECOGNITION PRODUCTS

Commercial continuous-speech, speaker-dependent voice data-entry systems can operate at more than 99% accuracy with background noise as high as 85 dB. This noise may be conversation between workers, the sound of equipment, or day-to-day variations in the user's tone of voice. These systems can recognize the voice patterns and vocabulary of selected users, using a standard vocabulary of 100 words, with options of up to 360 words at accuracies of 99% or greater. They can be used for commercial and industrial applications, and include 1 Mb of floppy disk and 10 Mb of Winchester disk storage.

In addition to board-level products, there are also speech-recognition chips, generally n-channel metal-oxide semiconductor (NMOS) or complementary metal oxide semiconductors (CMOS) that can recognize a certain vocabulary. Speech-recognition chips are manufactured by General Instruments, Intel, IBM, and NEC. These chip sets include a controller, feature extractor, and RAM. Intel has based its speech algorithms on its existing digital signal processor and microprocessor chips, and offers chip sets, boards, and speech-transaction development systems.

Other manufacturers, such as Hitachi, offer systems with dynamic programming and word recognition in any language, with accuracy of over 99%. It can interface with PCs and numerical-control machinery. The word recognition accuracy is approximately 99%.

Other products include an integrated voice and Ethernet-compatible data network designed to work with the IBM-PC. Called Comnet, the system allows a user to phone the computer to leave or retrieve messages and request specific information. It uses voice digitizing and storage so that the PC can recognize words for each application. Vocabulary files can be transferred with Ethernet, providing a user with voice recognition on any PC in the network. Most user application programs can be controlled by use of the voice-recognition unit in conjunction with keyboard input without modification.

Products are available for such applications as materials handling, quality control, machine-tool programming, inventory control, and data entry. Some of these reduce the pause time between spoken words and allow voice input into computers at a rate approaching 200 words/minute, which is faster than normal conversation.

Two-way voice interaction systems are available which use an A/D converter for incoming signals, a digital voice processor, RAM, an external control microprocessor, and a D/A converter for analog audio output. A dynamic technique is used to match programming patterns effectively and allows for normal variations in the way people talk. These products are offered as board-level multibus units, as stand-alone systems, and as voice-application development systems that utilize the IBM-PC as a host computer.

VOICE SYSTEMS INTERFACING

Speech technology requires an integrated approach. The physics of sound must be understood. The computer-based system must suit the application's needs and cost limitations. Relevant factors include the cost of the hardware and the availability of key hardware components such as common bus architecture. The most appropriate way to approach voice I/O is in terms of traditional feasibility or requirements-and-practicality analysis. Potential problems include operator resistance to change, user acceptance, acceptable error rates, lack of appropriate speech models, methods of presenting feedback, intelligibility of voice messages, procedures to correct recognition errors, voice activation and exclusion procedures, and the use of speech-processing technology in noisy operational environments. Human factors are a critical part of successful speech applications because perfect synthesis and recognition systems do not exist.

The major synthesis techniques in use include waveform modulation, phoneme stringing, and linear predictive coding (LPC). Waveform encoding can provide high-quality speech, but the mass-storage age requirements are high for heavily sampled speech. Phoneme stringing allows a large vocabulary to be stored in a small memory space. The disadvantage is a robot-like speech quality. LPC requires fewer bits to produce each word because of the data compression used. The speech quality is better than that of most text-to-speech synthesizers, and emphasis and inflection can be added to the LPC modeling algorithms by concatenation of words and phrases.

Although most manufacturers claim that their speech-recognition units achieve accuracies between 97% and 99%, in many cases the test procedures are nonstandard and the input data base is carefully chosen. The interpretation and extrapolation of performance claims to field applications is not easy. Environmental factors such as noise, stress-related factors like fatigue, and applications issues tend to degrade recognition accuracy further. The incorporation of digital signal-processing techniques, retraining, syntactical constraints, and the use of more effective human-machine dialogues can enhance system performance.

Suitability and trade-off analysis can be done that uses a decision matrix to evaluate techniques and environmental parameters, applications, subsystems, and operational phases. The analysis should include both suitability criteria, such as decreased workload, faster response time, greater safety, reduced stress, and ease of training, and feasibility criteria, such as reliability, availability, maintainability, cost, and utility.

The result of this analysis will be applications recommendations, guidelines, and system specification. Voice-interaction systems will include informational requirements, vocabulary content and format, I/O modes, and priorities for subsystem failures. Application concerns will identify restrictions, limitations, and additional requirements based on specific tasks.

Applications of speech I/O technology can be viewed as a hierarchy of generic tasks, such as "enter data," that can be divided into levels of suitability. The first level is where a distinct advantage is gained. At the second level, the use of a speech interface is neutral. The third level is where the speech interface would be useful as a temporary solution. The fourth and fifth levels include situations where speech I/O would be disadvantageous and unknown.

FUTURE CONSIDERATIONS

Emerging technology in cybernetics and AI, and research into the conceptual structures of human processes, will eventually have their effect upon speech-processing technologies. In addition, more forgiving, mistake-tolerant, and informal ways to interact with machines will become available.

The intelligent-interface machine of the future will integrate many technologies, including adaptive input devices such as speech processing, interactive graphics, natural language, and parallel data-flow computer architectures. AI will be a major player in the success of such systems. Structured knowledge, learning, and problem solving are tasks that must be understood for an adaptive system to accept, access, manipulate, and present information. The intelligent-interface machine will be able to carry on an intelligent conversation with a person and make judgments. It will also be able to program itself, listen to and obey spoken commands, and treat images and graphics the same as words. It will also be able to see and interpret an image.

SONIC TABLETS

Another less common sound technique involves the sonic tablet. This method uses a stylus with two strip microphones on the sides of the tablet. A small spark is generated at the tip of the stylus, and the interval between the time when the spark occurs and when its sound arrives at each microphone is used to calculate the distance of the stylus from each microphone.

Where the position can be extended to three dimensions, the location is determined by the time of the sound waves. The intersection of three sonic cylinders, be used to determine the position of the stylus.

Mechanically coupled three-dimensional devices can also be used. These take the form of a box with slides that are moved anywhere in a cube. Slide potentiometers are used to record the position along each axis. These devices operate in a volume of several cubic feet. They may be used to sense the motion of the operator's hands and fingers and these may be used to define positions and shapes in 3D for robotic programs. This may be done using optical or electromagnetic coupling.

Optical methods require active light sources which are moved by the user. The sensors are located in the corners of a small enclosure. The sensors are organized to provide the parameters of the plane in which the source is located. Three sensors can be used to define three planes to determine the position of the light source. A fourth sensor can be used for error-checking and for use when one of the other light sources becomes obscured from a sensor.

OCR TECHNOLOGY

The continued demand for fast, reliable data-entry devices has supported the development of OCR technology. OCR devices recognize human-readable characters, convert them into electronic signals, and process the signals into usable data, using recognition and validation programs.

OCR systems fall into two general types: desktop document readers and portable hand-held readers. In the desktop units, the medium moves past the optical reader head. In the hand-held units, the user passes the reader head over the characters to be read. High-speed readers have been used by the U.S. Postal Service for a number of years. Low-cost document readers are now used in many word processing applications. These low-cost readers, are designed to interface with a host microcomputer.

Hand-held OCR systems were originally designed for retail applications and are now used in banks, libraries, hospitals, and factories. In these applications, it is not necessary to enter primary data but rather to recapture data or update a document.

OCR OPERATION

An OCR system must perform five basic functions:

1. Image detection.
2. Character isolation.
3. Character recognition.
4. Data validation.
5. Data transfer.

Image detection requires optical-to-electrical conversion of the printed data. Characters must then be isolated from extraneous information, such as borders, nonreadable data, and background images. The isolated characters are then classified and matched to the proper character in a stored character set. Validation takes the form of a repetitive loop for reading, validating, and confirming a successful data acquisition or indicating a failure so that the data can be rescanned or entered another way. Then the data can be transferred to the host computer.

Some systems contain the OCR hardware, except for the scanner, on a single plug-in board. Other systems have their own CPU, memory, and host interface and can be attached to a host computer via an RS-232 port, a printer-like parallel interface, or the keyboard interface. They may also provide a connector that can be plugged into a slot on the PC bus.

OCR text must be within user-defined fonts and printed properly. Characters must meet standards such as those established by the American National Standards Institute (ANSI) or the International Standards Organization (ISO). OCR-A is an ANSI standard that establishes a machine- and human-readable standard character set. The characters include letters and special symbols in different sizes.

The ink that is used must be nonreflective to provide the proper contrast.

Photo sensors have high sensitivity in the near infrared spectrum where human eye sensitivity is weak. Ink visible to the human eye but invisible to the optical character reader can be used to mark positions and descriptive material. This is called *nonread* or *blind ink*. Paper also affects reflectance.

Most data validation is done by application programs in the host computer. As unvalidated data are entered, the application program judges their validity and sends a signal to the operator to indicate whether the data are valid. In many systems, part or all of this task is performed by the OCR reader hardware in order not to overburden the host.

Nothing prevents the user from partially wanding the data, stopping after only some characters have been scanned. In most cases, the wand can be used to enter formatting instructions to be used to obtain a complete data field. Character tables are used to define character usage and field length. These define a field-identification character, which lets the system know that the information that follows is to be read in as valid. For example, C can be defined as a field ID with the field length as six. With the field defined this way, the entry C123456 is valid.

CHARACTER ISOLATION AND RECOGNITION

In older OCR systems, character isolation and recognition was performed by template matching. In this process, a diode array in the reader produced a two-dimensional image that was transmitted to the processor, where a match is made with a known image. A disadvantage of template matching is the large amount of memory required to hold all possible character templates for matching. Another problem with template matching is its relatively poor accuracy, since the images are sensitive to print quality and distortion from wanding performed at an angle.

Instead of two-dimensional arrays to capture the image of a character all at once, some OCR system use a linear array of diodes in the wand. As the wand is moved across the medium, a series of slices through a character are stored in RAM. This device reconstructs the complete image by assembling successive slices. Since fewer pixel signals need to be analyzed, both communication bandwidth and processing requirements are lowered. Accuracy is quite high, and an error rate as low as 1 in 10,000 characters is not difficult to achieve.

Microcomputer-based OCR systems are versatile reading systems, but as the text to be read becomes more complex, higher optical resolution is required and more data need to be processed.

We have seen that many different interaction devices may be employed in a control factory or system, these include the more common as well as innovative methods.

When computers control a process in real time, real time is the same as the computer time. The real-time interval between any two consecutive events can be under computer control to either compress or expand time. Unless the process is very simple or the PC is very fast, control may not be possible.

The events are usually organized in the computer as a number of separate parts. When they are run, it is important to interact with those that appear to need attention. Using this procedure of human-computer interaction, the desired effects can be produced by the computer system.

INPUT DEVICE SELECTION

The input device must be compatible with the requirements of both humans and machines, including the joints and muscles which have to operate the device. Not only the required positioning precision for the specific task (Figure 5-7), but also the duration of the task for the muscle group and its particular fatigue factor is important.

The two main muscle groups involved in input devices are those of the forearm/wrist and the thumb/finger/wrist. The relationships between these two groups and the various input devices should be considered in design.

The movement of the forearm/wrist group generally requires more energy and is more fatiguing than the movement of the thumb/finger/wrist group. The tablet uses forearm and wrist motions and closely approximates the activity of writing and drawing. It has high lineal resolution, since the thumb/index finger motions are used. The mouse requires forearm/wrist motion except for switch activation, while the joystick uses only the motion of the thumb/finger/wrist group. The operation of a keyboard or keypad requires the repeated striking motion of the finger and wrist.

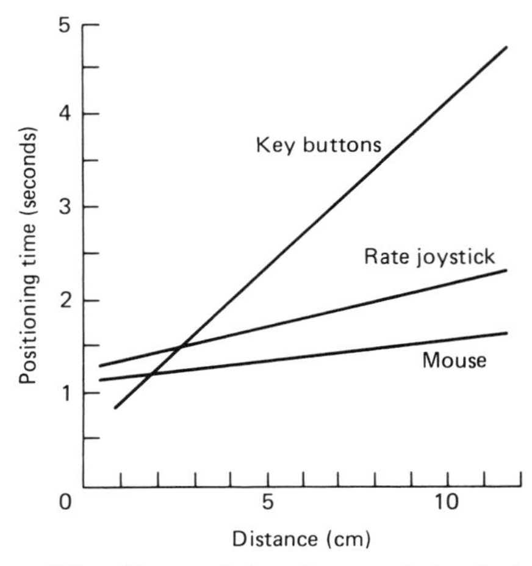

Figure 5-7 Characteristics of some pointing devices.

For high-resolution applications where cost is not a primary consideration, the digitizing tablet is usually the most convenient way for the user to enter x and y positions, since the tablet stylus is held like a pencil and moved over the tablet face. With most tablets, the position can be determined whenever the stylus is within a half inch or so of the tablet surface.

Picking devices may be used for text or data input and editing. A keyboard-mounted picking device leaves the wrist and forearm in a known relationship to the keyboard. The return hand motion can then be minimal and direct. In contrast, the return hand motion from either a tablet or a mouse may vary and is not automatically repeatable. Thus the input or editing process may be interrupted and delayed. For text editing, a joystick may be able to position faster and with fewer errors than a mouse.

After the joystick has been used to place the cursor in the region of a data row or column, the rate mode can be used to move it in a horizontal or vertical direction. This allows the selection of a character or line within a cell.

The major considerations of tablets and other locator devices are their resolution, repeatability, and size or range. These parameters are important in mapping and detailed drawing, but they are of less concern when the device is used to position a screen cursor, since the user has the feedback of the cursor to guide hand movements.

High resolution can be defined by the ability to resolve over one part in 2000 (usually 2048 × 2048 pixels). Medium resolution refers to 512 × 512-pixel displays and low resolution to 256 × 256-pixel displays. The cost of a picking device usually depends on the degree of resolution required; for comparable applications, a joystick is 20–30% less expensive than a mouse.

For medium-resolution control applications where cost is a consideration, either a mouse or an absolute joystick can be used. A joystick with a rate mode can always be adjusted to operate more coarsely. For lower-resolution control applications, either a mouse or an absolute joystick can be used. The absolute joystick offers high speed and tends to provide natural mapping with a low fatigue factor.

The tablet allows natural, absolute mapping of the user's motions to the display space, with resolution as high as 500 to 1000 points/inch and a low fatigue factor. For applications where cost may be an issue, either a mouse or a joystick can be used. The mouse offer good speed and a resolution of 100–300 points/inch, but it does not allow natural, absolute mapping and the fatigue factor is higher. The joystick allows a higher speed and a resolution of 100–400 points/inch in the absolute mode, and 1000–10,000/points/inch in the rate mode.

6

The PC Bus and Bus Standards

The electrical bus of the PC/AT computer is explained in this chapter. We also investigate the PS/2 micro channel bus and the micro channel option setup technique.

The parallel interfacing technique will be explained using the RS-232 and RS-449 standards. Current loop techniques are investigated using the RS-422, 423, and 449 standards.

Asynchronous and synchronous communication methods are compared. The IEEE-488 standard is explained and the characteristics of IEEE-488 interface boards are investigated, as well as those of RS-232/422 boards. We also show how to program these types of interface boards. The characteristics of RS-485 interface boards and RS-232/422 PS/2 interface boards are investigated.

The subject of component failures is examined, as well as software and noise related problems. Troubleshooting techniques and tools are examined. Logic probes are investigated, and logic state measurement is discussed.

The microcomputer bus is very important in interfacing. The bus moves information into and out of the microprocessor. It is also used to move information to and from the peripheral interfaces.

The bus has three sections: address, data, and control. Information is moved back and forth over the data bus. The 8088 processor uses an 8-bit data bus, and the 80286 uses a 16-bit bus.

Where the data go depends on the address bus. The microprocessor specifies what external hardware it wishes to communicate with by an address. Each memory location has a specific address.

The control bus sequences the flow of information over the data bus. When the microcomputer wishes to output data, it tells the external hardware when information is valid on the data bus. When the microprocessor needs to input information, the external hardware tells the microprocessor when those data are available.

The signals that pass over busses have a precise timing relationship. To show these relationships, a timing diagram is often used. The horizontal axis of the timing diagram is the time axis. Sometimes cross-hatching is used in the diagram to signify that the signals are crosshatched in a don't-care region. Also, if a signal like $\overline{\text{Strobe}}$ is shown on the diagram, the overbar indicates that the signal is asserted or true when it is at a low voltage level.

A typical diagram might show the behavior of the address and data buses with respect to the $\overline{\text{Strobe}}$ line (which acts as a control line). The $\overline{\text{Strobe}}$ line indicates that the buses are valid. Whatever device is receiving these signals will then look at the address bus, determine if it is being addressed, and if is, make use of the data. When the $\overline{\text{Strobe}}$ line is negated, the current cycle ends. The $\overline{\text{Strobe}}$ line indicates that the address and data buses are no longer valid.

Microcomputers always operate with some sort of bus. This bus carries a collection of signals that the microcomputer needs for the particular processor used.

Computer buses cannot be extended for great distances because it takes time for signals to travel over conductors. If the distance is great, the time increases to the point where signal timings on the bus cannot be maintained. For this reason, most buses operate at distances of a few feet in length.

Most peripheral devices are located at some distance from the processor. In addition, many peripherals are built by other manufacturers. As a result, the peripherals do not use the same collection of signals to communicate that microprocessors use.

To solve the problems of distance, signal, and timing incompatibility between the processors and peripherals, we use a specialized form of circuitry between them called the *interface*. An interface is a place where independent systems meet and act on or communicate with each other. Computer interface circuits allow a diverse group of peripheral devices to interact with the internal microprocessors.

The interface is a circuit board which plugs into the microprocessor bus. The connector that plugs into the bus allows the interface access to the microprocessor signals. At the other end of the board is another connector. A cable is connected between this connector and the peripheral device. The circuitry in the interface can perform four tasks:

1. Transform processor signals into signals compatible with the peripheral device.
2. Adjust the timing on the processor bus to the speed used by the peripheral device.
3. Adjust the signal levels so that long cables may be driven.

4. Adjust the data format from the processor so that it is compatible with the peripheral.

The three major types of electrical interface are parallel, serial, and analog. Parallel interfaces are like those used by microprocessor buses. Data are transferred over a group of wires called *data lines*. Variations occur in the data lines used and in the signals used for control or handshaking. *Handshaking* refers to the technique of controlling the flow of information from one device to another.

Serial interfaces use a single path to transmit data. Information is transferred one bit at a time. The two major types of serial interface are asynchronous and synchronous. The asynchronous serial interface is more common in microcomputers.

Analog interfaces are different from serial and parallel interfaces in that they do not use digital signals. Serial and parallel interfaces use digital signals to communicate with peripherals. Analog interfaces use signals that vary continuously.

Typical of values that vary continuously are temperature, pressure, and other physical quantities. Analog interfaces are needed to allow computers to interact with the real world.

PC/AT BUS STANDARDS

The original PC bus used a clock of 4.77 MHz with the Intel 8088 microprocessor. Besides the 8-bit bidirectional data bus and the 20 address lines, there are 6 interrupt lines, 3 sets of direct-memory-access control lines, and a number of data control and status lines. The 62 pins which make up the IBM-PC bus are divided into two rows (A1–A31 and B1–B31) in the edge connectors.

The IBM-PC is a single board computer. The processor, some of the I/O circuits, and memory reside on this board along, with the five edge connectors for expansion boards. Data are transferred on the PC bus over data line pins A2–A9. Addresses for bus transfers are specified on the 20 address pins A12–A31. The 8-bit version of the Intel 8086 16-bit processor was used in the PC and XT. This simplifies the bus by reducing the number of lines required and avoids the task of byte transfers in 16-bit systems.

The PC used the 8088 processor in maximum mode, which required an Intel 8288 bus controller. The 8288 control signals are brought out to the bus, so the two signals ALE (Address Latch Enable) and AEN (Address Enable) are on the bus. (pins B28 and A11.) ALE indicates that a valid address is on the bus address lines, and AEN signals if the processor or the direct memory access (DMA) controller is driving the bus during a DMA transaction. Other 8288 signals that can also be found on the bus include the following:

I/O Read (IOR) B14 I/O Write (IOW) B13
MEMory Read ($\overline{\text{MEMR}}$) B12 MEMory Write ($\overline{\text{MENW}}$) B11

A bus handshake line, I/O CH RDY (A10), can be used to increase the current bus cycle. This line can only be asserted for a few microseconds so that the dynamic memory is always refreshed. RAM refresh in the PC is handled by one channel of the system DMA controller, which requires the bus.

The six interrupt pins B21–B25 and B4 are connected to an interrupt controller on the main board which automatically generates vectors for interrupt servicing. There is no interrupt acknowledge signal on the PC bus.

The three pairs of DMA handshake lines include the DRQ1-3 lines (B18, B6, and B16 pins), which are used for DMA requests, and $\overline{\text{DACK1-3}}$ (B17, B26 and B15 pins), which are used as the acknowledge lines. $\overline{\text{DACKO}}$ (pin B19) is used to refresh dynamic RAM boards which may be plugged into the bus. T/C (B27) is used to indicate when the correct number of DMA bus cycles has occurred during a DMA transfer. OSC (B30) is used for a 14.31818-MHz clock, and CLK(B20) is the 4.77-MHz clock which runs the processor. RESET DRV (B2) is a reset signal for all cards on the bus. Power supplies available to the bus cards includes +5 volts (B3), −5 volts (B5), +12 volts (B9), and −12 volts (B7). There are three ground pins (B1, B10, and B31).

Bus cycles take four clock cycles, or 840 nanoseconds, while DNA cycles take five clock cycles, or 1.05 microseconds. The cycles are controlled by an 8288 bus controller running at 4.77 MHz.

The bus card has a metal plate attached to one end. This plate is used both as a card guide for the back of the card and as a support for I/O connectors which may be attached to the card.

Although it never evolved into an industry standard, the IBM-PC bus became the most popular microprocessor backplane bus ever introduced. Thousands of boards from hundreds of companies have been designed to plug into the PC bus.

THE PC/AT BUS

The AT is a full 16-bit computer based on the Intel 80286 microprocessor. The 80286 uses a 16-bit data bus and the AT expansion bus has a wider data path, more interrupt lines, and more DMA signals. Like the PC bus, the AT bus signals resemble the microprocessor signals on which the computer is based. In order to maintain compatibility with expansion boards designed for the PC, the original 62-pin connector and pin definitions remained the same. A second connector with 36 pins was added to carry the additional signals. The second connector is in front of the 62-pin connector; some older boards designed for the IBM-PC will interfere physically with the AT expansion connector.

Since there was never a standard board size and shape for the PC expansion boards, some manufacturers made use of the available space by dropping the bottom edge of the board just in front of the PC's edge connector.

Most of the signals on the AT's 62-pin connector retain the same names they had on the PC's bus; a few signals have new names but similar functions. For example, DACKO on the PC was renamed REFRESH for the AT bus. B8 on the PC bus was not used. It became OWS (zero wait state), which allows an expansion board to signal that it does not require the main board to insert wait states into the bus cycle. The MEM CS16 (D-1) and I/O CS16 (D-2) signals allow an expansion board to indicate that it can accept a 16-bit, one-wait-state transfer. These are added signals on the 36-pin connector.

Some of the address lines (C-2 through C-8) on the 36-pin connector partially replicate the addresses on the 62-pin connector. Unlike the PC's original address lines, the AT's 36-pin address lines are not latched on the computer's motherboard. Expansion boards that use these lines must latch the address values on the falling edge of the ALE signal.

Since the introduction of the original AT, the clock speeds of compatible computers have increased to over 20 MHz as Intel and other microprocessor vendors have improved their manufacturing processes. The manufacturers of compatible computers have adopted the AT bus for their 80286-based machines. Chip manufacturers such as Chips and Technologies and Western Digital developed ICs that reduce the size of the motherboard. In addition to machines based on the 80286 microprocessor, several 80386-based machines also use the AT bus, adding extensions for 32-bit memory boards. However, vendors of an 80386-based computer use different techniques for extending the AT bus to 32 bits, so memory boards for these computers are not compatible.

Some vendors offer extensions to the AT bus. AST Research offers an AT bus extension called Smartslot which adds an additional 8 pins to the bus. These pins allow multiple processors on several expansion cards to share the bus using an arbitration scheme. A central arbiter for all Smartslot cards grants the bus to one of the requesting cards.

The IEEE created the P996 Bus Committee to study the feasibility of standardizing the AT bus, but before the committee could produce a standard, IBM introduced its PS/2 line of PCs, which use a completely different and incompatible bus.

THE PS/2 MICRO CHANNEL BUS

In 1987, IBM introduced the PS/2 computer line and the Micro Channel bus. IBM patented several aspects of the PS/2's bus, so manufacturers need a license to use the Micro Channel. There are similarities and differences between the old and new PC buses.

The Micro Channel bus supports three types of cards: 16-bit, 16-bit with a video extension, and 32-bit. The 16-bit version uses 116 pins, and the extension adds another 20 pins. The Micro Channel bus is designed to support multiple bus masters. The processor or DMA controller on the PS/2 motherboard usually controls the bus, but the Micro Channel has a set of signals to allow expansion cards to take over as bus masters. The Micro Channel allows up to 15 masters to share the bus with the mainboard.

The Micro Channel assigns more power and ground pins than the earlier PC and AT buses. These additional power and ground pins allow expansion cards to draw more power from the computer, since the extra pins provide a lower-impedance path to the computer's power supply. The ground and power pins also provide a lower-impedance path to ac ground for radio frequency interference (RFI). This tends to reduce potential interference emissions and improve the general data integrity by reducing noise.

The Micro Channel bus also uses a set of AUDIO (B-2) and AUDIO GND (B-1) signals. This allows expansion cards to use the PS/2's audio amplifier and speaker. In a similar fashion, the video expansion connector allows an expansion card to override the video circuits on the PS/2 motherboard.

Normally, a bus transfer on the Micro Channel requires 250 nanoseconds, representing four clock cycles for a 16-MHz microprocessor. Matched memory cycles occur in three clock cycles and last for 187.5 nanoseconds.

MICRO CHANNEL OPTION SETUP

The Micro Channel design eliminates addressing and option-configuration switches from the expansion cards. The scheme that allows this is called the *Programmable Option Select (POS).* During power-up, the PS/2 mainboard addresses each expansion card individually with its -CD SETUP line. This line is not common across the PS/2 backplane. Each expansion slot has its own -CD SETUP signal. When a card is signaled by its -CD SETUP line, it issues a code. The processor reads this code from the expansion card. The following codes are used:

Code (Hexadecimal)	Definition
000	Device not ready
0001-OFFF	Bus master
5000-5FFF	DMA devices
6000-6FFF	Direct program control and memory-mapped I/O
7000-7FFF	Memory storage
8000-8FFF	Video adapters
9000-FFFE	Reserved
FFFF	No device present

The processor matches the code with the configuration data stored in the computer's nonvolatile memory and loads these data into the expansion card. Configuration data includes the card's bus-master arbitration level, the address range of the card's on-board I/O ROM (if any), and the I/O address range assignment for the card. Since each type of card from different manufacturers must have a unique POS code, the Micro Channel architecture allows IBM to control the types of cards available for PS/2 computers.

A Micro Channel card is 11.5 inches long and 3.475 inches high, including the edge connector. Additional dimensions are needed to locate the edge connector precisely on the board. This is because of the type of edge connector used by the Micro Channel. The PC and AT buses use edge connectors with fingers placed 0.100 inch apart. The Micro Channel uses fingers that are 0.050 inch apart, so a precise fit is needed to prevent shorted fingers. Plastic positioning keys which mate with notches in the expansion-card edge connectors are used to position the cards.

A full-sized Micro Channel expansion card is about 40% smaller than a full-sized PC/AT card. The reduced card size requires more highly integrated circuits and surface-mount components, which need less room.

SIMPLE PARALLEL INTERFACING

Microcomputer interfaces are designed to link microprocessor buses with peripheral devices. They often take the form of a board plugged into the microprocessor bus. A cable links the interface board with the peripheral. The interface is concerned with the signals which are passed through this cable and the circuits on the interface board that are used to generate these signals.

A simple parallel interface can be built with a single TTL IC. Other parallel interfacing techniques, such as the IEEE-488, require complex circuitry. Parallel interfaces are distinguished by two features. First, there is the number of bits transferred in parallel by the interface, called the *data path*. Then there is the type of handshake used to communicate the movement of these bits between the computer and the peripheral.

Data paths range from a single bit to 128 bits or wider. The most common size for 8-bit microprocessors is an 8-bit data path. This allows the microprocessor to transfer a full data word over the interface during each transfer. The 8-bit parallel interface is also popular for 16-bit microprocessors. One reason is the number of 8-bit peripherals available. These were originally designed for 8-bit microcomputers. Another reason is that ASCII, the most common character code, requires at least a 7-bit interface.

The type of handshake used to move information over the data line includes zero-wire handshakes, one-wire handshakes, two-wire handshakes, and three-wire handshakes. In any of these handshakes, there are variations in how the wires are actually used, including pulsed and interlocked handshake methods.

The zero-wire handshake is a simple interface that uses an 8-bit latch to store the state of the processor's data bus. On the rising edge of a write signal, the latch takes the states of the data bus lines and stores them in the latch. The states are reproduced on the output lines of the latch after a read signal occurs. A 16-bit interface can be built by adding a second 8-bit latch.

This type of zero-wire handshake, parallel-output interface can be used to drive simple outputs for lights or relays. Devices such as these do not have handshaking requirements. Each of the output lines from the interface can be used to drive a light or relay.

The single-wire handshake requires adding another wire to indicate when data are valid on the data lines. This signal has the effect of stretching the write pulse from the microprocessor. This allows slower devices to respond to the write pulse and provide some settling time.

A two-wire handshake adds another line so that the receiving device can indicate when it is ready for data. This provides a true handshake using this acknowledge line, and full interlocking is possible. The two-wire handshake is adequate for interfacing a single peripheral, but some interfaces use a third wire to create a protocol that allows several peripherals to use the interface. An example of this is the IEEE-488 bus, which is discussed later. But first, we will consider one of the most widely used serial interface standards: RS-232.

RS-232

A widely used serial interface standard is the RS-232C interface. *RS* stands for *recommended standard*, and *C* indicates that this is the third version of the standard. The type of signal provided is shown in Table 6-1.

The main RS-232 signal lines are those used to transmit and receive data (BA and BB). These lines are used to send serial information between the two systems. The following bit rates are available:

19,200	1,200	110
9,600	600	75
4,800	300	50
2,400	150	

Occasionally, other rates are also used.

When serial data are transmitted over telephone voice-grade lines, the data must first be modulated so that they can be transmitted. For bit rates of less than 300, the method of modulation used is frequency-shift keying (KSK). The mark or logic 1 condition is represented by a tone of given frequency, and the space or logic 0 condition is represented by a second, different frequency. Bit rates above 300 use phase-modulation techniques due to the lack of available bandwidth. Voice-grade lines are too noisy for high-rate communications, and more expensive data-grade lines must be used.

The other signals are used to indicate the status of the modulator-demodulator (modem) communications link. Signals such as request-to-send, clear-

TABLE 6-1 RS-232C Signals

Symbol	Signal Name
AB	Signal ground
CE	Ring indicator (modem to DTE)
CD	Data terminal ready (DTE to modem)
CC	Data set ready (modem to DTE)
BA	Transmitted data (DTE to modem)
BB	Received data (modem to DTE)
DA	Transmitter signal
	Element timing (DTE to modem)
DB	Transmitter signal
	Element timing (modem to DTE)
DD	Receiver signal element
	Timing (modem to DTE)
CA	Request to send (DTE to modem)
CB	Clear to send (modem to DTE)
CF	Received line signal (modem to DTE)
	Detector
CG	Signal quality detector (modem to DTE)
CH	Data signal rate selector (DTE to modem)
CI	Data signal rate selector (modem to DTE)
	Secondary channel
SBA	Secondary transmitted data
SBB	Secondary received data
SCA	Secondary request to send
SCB	Secondary clear to send
SCF	Secondary received line
	Signal detector

to-send, data-set-ready, and data-terminal-ready are used to control the modem link. The signals between the modem (communications equipment) and the computer (or terminal) implement a handshake similar to that used in other buses like the IEEE-488. The difference, for RS-232, is that the handshake is used only at the beginning and end of a block of serial data. RS-232C is popular and is used for many PC peripheral communications.

CURRENT LOOP

Another bus standard is *current loop*. This was used in the mechanical teletypewriters. A common interface converter that is very useful will convert loop devices to RS-232. This is sometimes called a *loop-to-EIA* converter. A loop-to-EIA converter circuit is shown in Figure 6-1a.

Also useful circuit is *auto loop back*, shown in Figure 6-1b. This occurs where the computer, terminal, or modem does not have the full RS-232 standard

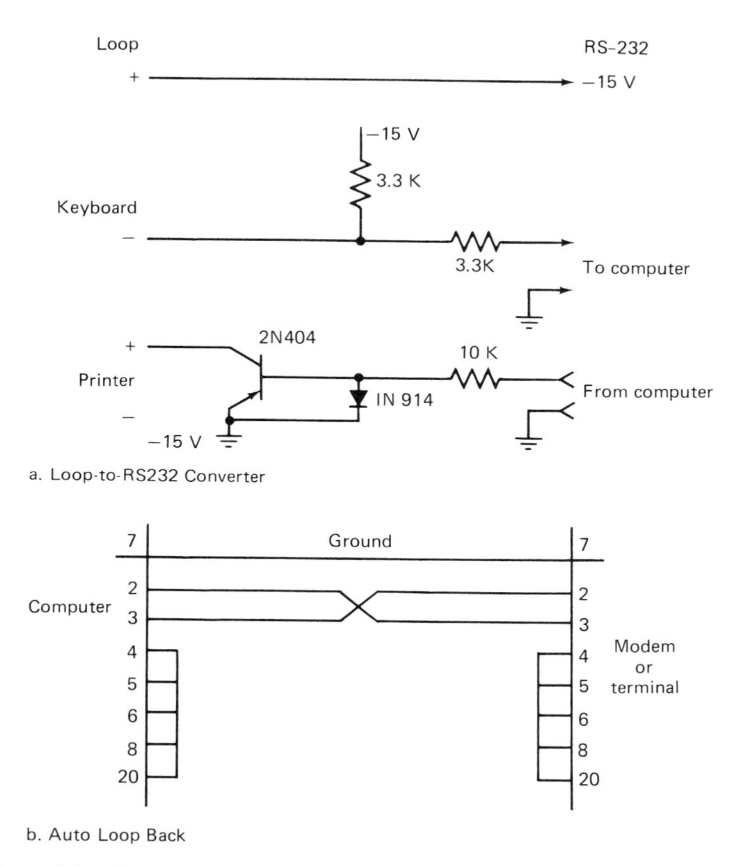

Figure 6-1 Current loop circuits. (a) Loop-to-R-232 converter; (b) Auto loop back.

implemented. The jumpers trick the devices into believing that all conditions are in place for data to pass.

RS-232 transmits signals as single-ended voltages. The mark or space condition is represented by the voltage between two wires. One wire is the common ground. The transmit path will use the transmit wire and this ground. The receive path will use the receive wire and this ground. If, instead of a single-ended connection, a differential connection is used, the path may be physically longer between devices due to the noise immunity of a differential channel. The data rate can also be higher due to the reduced noise effects. This is shown in Table 6-2.

Terminals that use the RS-232C standard use a 25-pin male connector (DB-25) which mates with a corresponding female connector. The RS-232 bus

TABLE 6-2 A Comparison of RS-232C, RS-422, and RS-423

Characteristic	RS-232	RS-422	RS-423
Maximum line length	100 ft	5000 ft	5000 ft
Maximum bits/sec	2×10^4	10^6	10^5
Data "1" = mark	−1.5 to −36 V	$V_a > V_b$	$V_a = -$
Data "0" = space	+1.5 to +36 V	$V_a < V_b$	$V_b = +$
Receiver input, minimum	1.5 V single-ended	100 mV differential	100 mV differential

standard was originally designed to interface data terminal equipment (DTE) and modems. Other equipment may not use all the signals, and not all devices will operate together without some modifications.

The simplest interface for a send/receive terminal requires at least three signals: signal ground (AB), transmitted data (BA), and received data (BB). The device at the other end of the transmission link might also require a clear-to-send (CB) signal before it can transmit the data.

Since this standard was originally developed to interconnect terminal equipment and modems, when the standard is used to interconnect other kinds of equipment, the modems used to interconnect the RS-232 devices may have the wrong connector types; in this situation, a null modem is required. This is a pair of identical connectors with the transmitted data pins wired to the received data pins (Figure 6-2).

Other control signals may also have to be wired to produce the proper operating conditions. A device that is required to recognize the clear-to-send (CB) signal may also produce a request-to-send (CA) signal. Wiring these two pins together ensures that the terminal will operate with equipment that may not produce a clear-to-send signal.

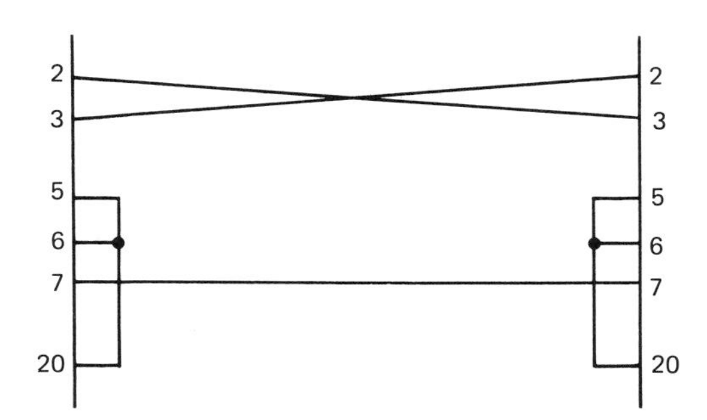

Figure 6-2 Null modem.

RS-422, -423, and -449

RS-232 has been one of the most popular standards in the industry. However, its electrical specifications are based on an earlier period of IC technology when logic levels were defined with two equal voltages of opposite polarity in the 5- to 25-volt range. Most contemporary logic uses a single positive 5-volt supply. Two newer standards, RS-422A and RS-423A, use positive 5-volt levels. These two standards differ from RS-232 only in the electrical characteristics of the interface signals. The functions remain essentially the same as those for RS-232C.

Other disadvantages of the RS-232 standard are limited speed and excessive RFI and crosstalk. The maximum rate of about 20,000 bits/second is too slow for many applications. A newer standard, RS-449, can be used for data rates of up to 2 Mbps, and cable lengths of up to 200 feet compared to 50 feet for RS-232. It also uses more signals compared to RS-232, as shown in Table 6-3.

Table 6-2 illustrates the difference between the three standards, Figure 6-3 shows the differences between the connections and the types of drivers and receivers used. RS-422 and -423 are not as common because they are newer and because of the widespread use of RS-232. They provide alternatives to RS-232 for higher data rates and longer line lengths.

The data sent over these standards can be formatted in many ways. The topics of asynchronous and synchronous data transmission and the common standards for information exchange are reviewed at this time.

TABLE 6-3 RS-449 Signals

Symbol	Signal Name	Symbol	Signal Name
SG	Signal ground	SQ	Signal Quality
SC	Send common	NS	New signal
RC	Received common	SF	Select frequency
IS	Terminal in service	SR	Signaling rate selector
IC	Incoming call	SI	Signaling rate indicator
TR	Terminal ready	SSD	Secondary send data
DM	Data mode	SRD	Secondary receive data
SD	Send data	SRS	Secondary request-to-send
RD	Receive data	SCS	Secondary clear-to-send
TT	Terminal timing	SRR	Secondary receiver ready
ST	Send timing	LL	Local loopback
RT	Receive timing	RL	Remote loopback
RS	Request-to-send	TM	Test mode
CS	Clear-to-send	SS	Select standby
RR	Receive ready	SB	Standby indicator

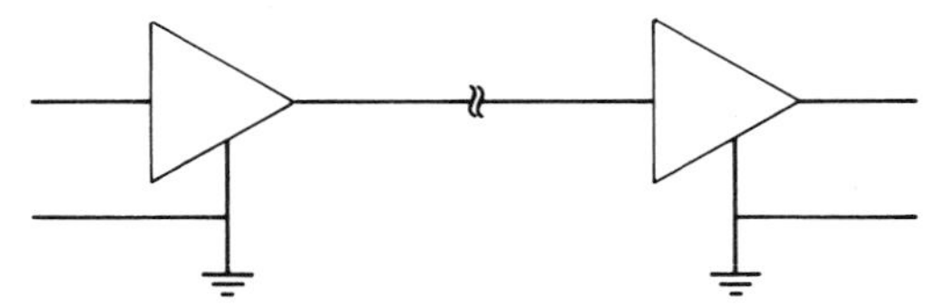

a. RS-232 Single-Ended Connection

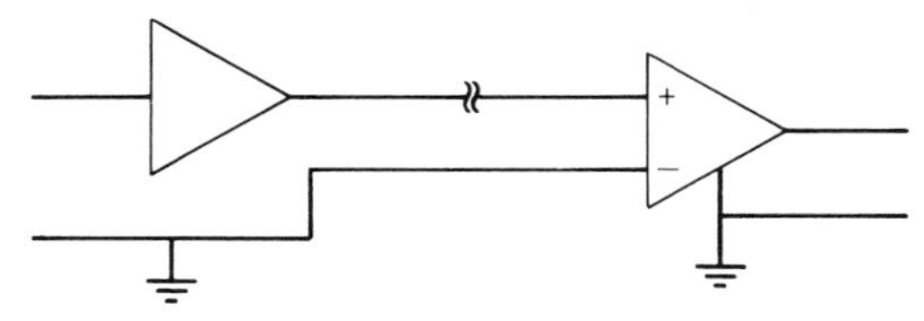

b. RS-422 Unbalanced Differential Connection

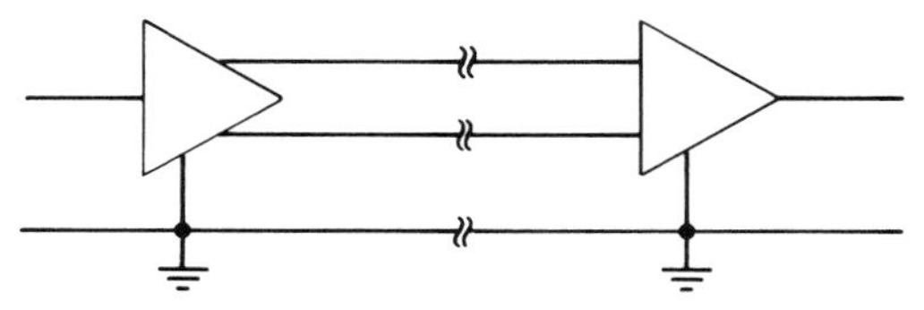

c. RS-423 Balanced Differential Connection

Figure 6-3 Serial standard connections. (a) RS-232 single-ended connection; (b) RS-422 unbalanced differential connection; (c) RS-423 balanced differential connection.

ASYNCHRONOUS COMMUNICATIONS

When data streams are sent in bursts of equal duration without the use of clock information, they are being sent asynchronously, and are not clocked. When data streams are sent with synchronizing bits, they are being sent synchronously and are clocked.

The most common asynchronous data structure consists of a 10- or 11-bit data stream. The start bit, 8 data bits, and 1 or 2 stop bits make up a character. The most popular standards for character coding are the ASCII and EBCDIC codes. ASCII uses 7 bits to encode 128 possible characters. An eighth bit may be used for parity. Typical asynchronous code functions are shown in Table 6-4.

The EBCDIC code is similar in function, except that the 128 codes are encoded differently. Code-conversion software or ROM can be used to convert ASCII to EBCDIC or EBCDIC to ASCII. ROM will need eight inputs: seven address lines for the data input and one address line to specify the conversion mode (ASCII to EBCDIC or EBCDIC to ASCII). It has seven outputs for the converted character. The size of the ROM needed is 256 bytes by 7 bits/byte.

TABLE 6-4 Typical Asynchronous Code Functions

Symbol	Code Function	Symbol	Code Function
ACK	Acknowledge	FF	Form feed
BEL	Bell	FS	Form separator
BS	Backspace	GS	Group separator
CAN	Cancel	HT	Horizontal tab
CR	Carriage return	LF	Line feed
DC1	Direct control 1	NAK	Negative acknowledge
DC2	Direct control 2	NUL	Null
DC3	Direct control 3	RS	Record separator
DC4	Direct control 4	SI	Shift in
DEL	Delete	SO	Shift out
DLE	Data link escape	SOH	Start of heading
EM	End of medium	SP	Space
ENQ	Enquiry	STX	Start text
EOT	End of transmission	SUB	Substitute
ESC	Escape	SYN	Synchronous idle
ETB	End transmission block	US	Unit separator
ETX	End text	VT	Vertical tab

SYNCHRONOUS COMMUNICATIONS

Asynchronous transmission requires at least 2 extra bits per character; these are the start and stop bits. When data are sent as a continuous stream of bits with no start or stop bits, the receiver can easily loose its timing, and the data will be in error. To prevent this from happening, synchronizing characters can be sent along with a group of bytes. The receiver will use these synchronizing characters to resynchronize the timing, and the transmitter and receiver will be in step. This method is known as *synchronous communication*.

In a synchronous data-link control (SDLC) scheme, the data are transmitted in blocks of characters called *frames*. Each frame contains a number of fields, and each field is made up of one or more bytes of data. In this example, the frame has seven fields. The frame is made up of bytes of data.

A start character will indicate to the receiver that a frame has begun. After a frame is received and checked for errors, a return frame will be sent back to the transmitter to indicate if the data were received intact, how many errors were present, and if retransmission is necessary.

An information frame will contain data, and a protocol frame will contain bytes concerning supervision and management of the transmissions. The data may be in ASCII or some other format.

During the transmission, resynchronization characters or bits may be inserted to maintain the system's timing. The start and stop characters may be special bytes of 1's and 0's that are easily detected by a hardware unit.

There are synchronous serial adapter ICs, such as Intel's SDLC controller, that will handle the protocol using a combination of software and hardware. The controller will recognize start characters, insert and delete sync bits automatically, and do some error checking. The software will assemble the frame from the fields and decode the frame from the fields.

Check characters are used to detect and correct single-bit errors and detect double-bit errors. The bits used contain enough redundant check information to do this.

Parity involves counting the number of 1's or 0's. In a byte of data, there is either an even number of 1's or an odd number of 1's. The eighth or sometimes ninth bit added to every byte to make the number of 1 bits even or odd is known as the *parity bit*. By recording or storing a parity bit with every byte, errors can be detected.

After the byte is received, the parity bit is generated from the 8 bits of the byte. If this parity bit does not match the original transmitted parity bit, there is an error of at least 1 bit. A change of 2 bits from 1 to 0 and from 0 to 1 will not be detected.

Checksums involve the use of check characters. A 1-byte check character is generated and added at the end of the block. When the block is read, a new check character is generated. The check character received is compared with the one generated. When it is different, there is an error in the block. The checksum may be generated by adding the bytes in the block or an Exclusive OR of all bytes in the block.

CONNECTOR TYPES

The RS-449 standard uses two connectors: a 37-pin unit (DB-37) designed for the modem-DTE interface and a 9-pin unit (DB-9) used as the secondary channel. In many applications, a single DB-37 connector is used. The RS-449 standard was designed to replace RS-232C equipment which conforms to the RS-449 standard. It requires an adapter to mate to RS-232C devices. Some interfaces may require cables with 36-, 9-, and 25-pin connectors to interconnect the various mixtures of equipment.

Many computers have an RS-232 compatible I/O port or board, and almost all data terminal equipment uses RS-232 interfaces. These interfaces are also common in tape drives and in other low-speed peripherals such as printers.

The serial interface trades efficiency for the general-purpose characteristics of an interface where any number of bits can be transmitted in any order in either the synchronous or the asynchronous mode. A major reason for the popularity of RS-232 is the wide range of equipment that can be connected using this interface.

Parallel transmission through the device interface overcomes the low throughput of serial transmission.

IEEE-488

The IEEE-488 standard was designed as an instrumentation bus standard with 24 conductors. Bit-parallel, byte-serial messages are used. The data I/O lines are either open collector or three-state. This allows all devices connected to the bus to listen to messages on the bus and any one device to send messages on the bus.

A three-wire handshake technique is used to ensure that each device on the system receives all the data sent to it. The handshake sequence also forces the talking device to operate at the speed of the slowest listener that is enabled on the bus.

The 488 bus can be used as the physical layer for equipment connections, and there are many bus-compatible devices, including computer peripherals. The bus can support data transfer rates of up to 1 Mb.

This standard was originally developed by Hewlett-Packard for interfacing some of the first programmable electronic test equipment. It was later adopted by both the IEEE and ANSI. It is also known as the general-purpose interface bus (GPIB), the Hewlett-Packard interface bus, and the ANSI bus.

The 24-pin cable has 8 parallel data lines (that use an asynchronous bit-parallel, byte-serial format) and 8 control lines, as shown as in Table 6-5. Devices that interface to the bus use a male plug. The cables use both male and female plugs at each end, which allow multiple devices to be connected in a daisy chain.

Every device on the bus is assigned a control address. Listeners may respond only to the messages addressed to it. Control devices can force a talker to send data to the desired listener by issuing the proper address. Up to 15 devices can share the bus. Devices are allowed to talk, transmit data, listen, receive data, and/or control the bus.

The 488 bus has a number of advantages over a serial interface such as RS-232. It is faster and more efficient, with the data being transferred 8 bits at once. The 8-bit bus is also compatible with many microprocessors, and the channel capacity is greater. A single 488 channel can accommodate as many as 14 devices.

TABLE 6-5 IEEE-488 Bus

Data Interface Management Bus	Data Byte Transfer Control	Data Bus
IFC—Interface clear	DAV—Data valid	8 Lines
ATN—Attention	NRFD—Note ready for data	EI01-DI08
SRQ—Service request	NDAC—Data not accepted	
REN—Remote enable		
EOI—End-or identify		

Devices can be separated by up to 20 meters, but for maximum speed, cable lengths should be 1 meter or less for each device connected to the bus.

Initially, many products did not adopt the 488 bus. This was partly due to the apparent complexity of the required control sequences. The sequential control patterns were costly to implement, and the availability and lower cost of the one-chip universal asynchronous receiver transmitters (UARTs) eased the adoption of the RS-232 type of serial interfaces in many products.

Since this time, many products that facilitate the use of the IEEE bus have appeared, including single-chip ICs like the 68488, which contain most of the circuits needed to interface to the 488 bus. The only additional circuits required are the bus transceivers. IEEE-488 bus controllers are also available in board form for PCs that provide the detailed functions required to transfer blocks of data using the 488 bus.

The 488 bus allows connection to devices that can have the following functions:

1. Control other units, acting as a controller.
2. Take information from the controlling unit, acting as a listener.
3. Provide information to the controlling unit, acting as a talker.

The bus consists of eight bidirectional data lines, three byte-transfer control lines, and five general control lines. The eight data lines carry device commands of 7 bits, as well as address and data words of 8 bits. The transfer-control lines are used to implement the handshaking required between the sending devices and the receiving devices.

The five management lines are concerned with the general conditions of the system. Their functions are as follows:

1. The *attention* line, when false, indicates that the data lines contain data ranging from 1 to 8 bits. When the attention line is true, the data bus holds a 7-bit command or 7-bit address.
2. The *interface clear* line places the system in a known state. It is like a system reset.
3. The *service request*, when set true, flags the controlling unit to indicate that a device needs attention.
4. The *remote enable* line sets the mode of each device to operate remotely or locally.
5. The *end-or-identify* is used to flag the controlling unit to indicate the end of a data transfer.

The handshaking function is used when devices must wait for information to become available. One line starts the dialogue and says "Ready for data."

Another line replies, ''Okay, I have something.'' The returning reply is: ''Send it to me; I am ready.'' This continues with ''Okay, here it comes.''

In the 488 standard, three lines are used for handshaking:

1. DAV, data valid on data lines.
2. NRFD, not-ready-for-data.
3. NDAC, not-data-accepted.

The timing of the handshake is complex, and all listening devices must accept the transfer of data before the next transfer is initiated. In order to initiate a talk, the controller sends the address and command-to-talk to the talker. Upon recognizing its address and the command, the talker sends information to a listener, via the data bus, using the handshake signals. When the transfer is finished, the end line can be used to indicate the end of the block.

A listen works in a similar way. The controller sends the address via the data bus. The next command directs the device to listen to a talker. The transfer of data is done, byte by byte, using the data bus and handshake signals. The end line then indicates that the transfer is complete.

The IEEE-488 bus provided an advance in intelligent data acquisition systems, and as more manufacturers produced compatible equipment, the standard became more widespread.

IEEE-488 INTERFACE BOARDS

GPIB/IEEE-488 interface boards* are available that plug into an I/O expansion slot inside a PC or compatible machine. A standard IEEE-488 connector is used at the rear of the PC to connect to a standard GPIB/IEEE-488 cable. The software driver/interpreter for the 488 board is usually on a floppy disk. The driver handles initialization and protocol conversions required for access to all functions covered in the IEEE-488 specification. The driver will allow the user to interface to the bus using high-level IEEE-488 commands such as REMOTE, LOCAL, ENTER, and OUTPUT.

The software can be designed as a DOS-resident driver, which simplifies interfacing to higher-level languages. The software may include routines which allow the software driver to be run from BASIC, FORTRAN, and Turbo Pascal.

The 488 bus can handle up to 14 other talker/listener devices. The controller may be the PC or any of the 14 devices on the bus. The hardware handles all of the system timing for talking, listening, and controlling the bus.

Control information is passed to the software driver in the form of an ASCII

*Specifications and programming examples courtesy of Keithley-MetraByte, Inc.

command string. Other parameters must also be passed to the driver. Typical parameter names and data transfer requirements will now be described.

COMMAND$ includes the device addresses or secondary commands and image terminators. This is usually a string, which is decoded by the command line interpreter.

COMMAND$ is separated from the operands (devices) by spaces; any other delimiters will cause a syntax error in the command line. The separator for devices is a comma ",'' and for the secondary address a period ".'' The image string is identified by brackets "[ ]''. The command line interpreter will check the syntax and send back an error code to isolate the error. This produces the following format:

```
CMD$="COMMAND dev1, dev2, . . . , devn[image]"
```

The image specifier permits the user to specify the variable field operations for the beginning and end of the data transfer variable. The variable can be a variable name, array identifier, numeric data value, or string.

VAR [$] (%) represents the data variable OUTPUT/INPUT to be transferred from/to. Data are transferred as specified by the image terminator/specifier. If the image specifier is not used, the data are treated as an integer. The data may be in the form of a string or an integer.

FLAG% is the transfer status of the call statement. If an error occurs, FLAG% will contain a hex number representing the error condition. A set of error and transfer message codes are generated at the completion of each call.

BASADR% is the address of the interface board being used. The user commands will now be discussed.

ABORT—This command terminates the current selected device and command. If no device is given, the bus is cleared and set to the state given in the last CONFIG command. The PC must be the active controller or an error message will be generated: "ABORT dev''; for example:

```
COMMAND$="ABORT"
```

CLEAR—This command clears or resets selected devices or all of the devices. If no device is given, the bus is cleared and set to the state given in the last CONFIG command. The PC must be the active controller or an error message will be generated; for example:

```
COMMAND$="CLEAR 10, 11, 12"
```

(Clears devices 10, 11 and 12)

CONFIG—This command configures the bus to the user's demands. The bus will remain in this state until reconfigured. The variable is not changed with this command. If the TALK-dev1 is omitted, the PC is assured the active controller and talker; for example:

```
COMMAND$="CONFIG TALK=5, LISTEN=11, 12, MLA"
```

(Device 5 is talker; devices 11, 12, and PC are listeners)

ENTER—This command inputs the bus data from a selected talker to a string array. The variable array must have been previously dimensioned. The FLAG% will contain error codes if an error occurs. The PC must have been previously programmed as a listener. If the PC is not the controller, then the ENTER command will wait until the talker sends a message to the PC; for example:

```
COMMAND$="ENTER 11 [$, 0, 20]"
```

(Enter from address 11, Array elements 0 to 20)

EOI—This end-of-input command sends a data byte on the selected device with EOI asserted. The bus must be programmed to talk before the command is executed. The variable contains the data to be transferred. The user must ensure the data and type match; for example:

```
COMMAND$="EOI 10 [$]"
```

(Issue an EOI with the last byte of the string to listener 10)

LOCAL—This sets the selected devices to the local state. If no devices are specified, then all devices on the bus are set to local. The PC must be the active controller or an error message will be issued; for example:

```
COMMAND$="LOCAL 12, 14"
```

(Sets devices 12 and 14 to the local state)

LOCKOUT—Performs a local lockout of the specified device. If no devices are specified, all devices on the bus will be set to local lockout. The PC must be the active controller or an error message will be generated; for example:

```
COMMAND$="LOCKOUT 5, 8"
```

(Lockout of devices 5 and 8)

OUTPUT—This command outputs a selected string to a selected listener on the bus. The variable will contain the data to be transferred. The image specifier will contain the data type and terminators. The FLAG% will contain the error codes if an error occurs. The PC must have been previously programmed as a talker. Devices are separated by commas and secondary commands by a period. Up to 14 devices may be accessed in the list; for example:

```
COMMAND$="OUTPUT 12, 14 [$E]"
```

(Outputs an even parity string to listener devices 12 and 14)

PASCTL—(Pass Control) The active control of the bus is passed to the specified device address, and the PC becomes the listener/talker. The PC must be the active controller or an error will occur. This command can be reissued to allow the PC to be the controller again. If listen or talk is not specified, the PC is set to the listen mode. The PC is not allowed to talk until programmed by the controller. A request to talk must be specified for the PC to transfer data to a listener when using the OUTPUT command; for example:

```
COMMAND$="PASCTL 8"
```

(Device 8 has control of the bus)

PPCONF—(Parallel Poll Configure) This command sets up the desired parallel poll bus configuration for the user. The PC must be the active controller or an error will occur; for example:

```
COMMAND$="PPCONG 10"
```

(Parallel poll for device 10)

PPUNCF—(Parallel Poll UnConfigure) This command resets the parallel poll bus configuration of the selected device. The PC must be the active controller or an error will occur; for example:

```
COMMAND$="PPUNCF 10"
```

(Remove parallel poll from device 10)

PARPOL—(Parallel Poll) This command will read the 8 status bit messages for the devices on the bus which have been set for a parallel poll configuration. The PC must be the active controller or an error will occur; for example:

```
COMMAND$="PARPOL"
```

REMOTE—This command sets the selected device on the bus to the remote position. The PC must be the active controller or an error will occur. If an error occurs, the FLAG% will contain the error code; for example:

```
COMMAND$="REMOTE 11, 12, 14"
```

(Set devices 11, 12, and 14 to REMOTE)

REQUEST—This command requests service from the active controller on the bus. It is used when the PC is not the active controller. An error will occur if the PC is already the active controller; for example:

```
COMMAND$="REQUEST"
```

(PC is in control)

STATUS—This command allows a serial polled devices status byte to be read into the selected variable. The PC must be the active controller or an error will occur. Only one device is allowed with one secondary address. If no device is specified, an error will occur; for example:

```
COMMAND$="STATUS 12"
```

SYSCON—This command provides SYStem CONfiguration and initialization of the bus. This command must be used before using the bus. If this command is not run first, an error will be generated. If SYSCON.COM is used, the error messages will be displayed on the current screen. BAx is in HEX (&H) or DECIMAL: The format is as follows:

```
COMMAND$=SYSCON MAD=dev, CIC=(0/1/2/3), NOB=(1/2),
            BAO=&Hdddd, (BA1=&Hdddd)"
```

where

dev = the address of the PC 00 to 30 decimal
NOB = number of IEE488 boards (one or two)
BA1 = base address for board 2
CIC = controller in charge, 0 = none, 1 = brd#1, 2 = brd #2, 3 = brd#1 and brd#2 (separate buses).

For example:

```
COMMAND$="SYCON MAD=3, CIC=1, NOB=1, BAO-&H300"
```

RXCTL—(Receive Control) of the bus; for example:

```
COMMAND$="RXCTL"
```

TIMEOUT—This command sets the time and duration when transferring data to/from the device. The variable integer VAR% is set to a number from 0000 to 65000. The approximate time is the VAR% * 1.5 seconds for the standard AT and the VAR% * 3.5 for the standard XT. No error flag is returned; for example:

```
COMMAND$="TIMEOUT"
```

TRIGGER—This command sends a trigger message to the selected device or a group of devices. The PC must be the active controller or an error will occur; for example:

```
COMMAND$="TRIGGER 10, 11"
```

(Devices 10 and 11 are triggered at the same time)

Program Example

The following is a list of BASIC statements for programming a 488 interface board:

```
100 PRINT "****** system initialization ******
110 DEF SEG=&H2000
120 BLOAD "488 BIN", 0
130 GPIB=0
140 FLG%=0
150 BRD%=&H300
170 CMD$= "SYSCON MAD=3, CIC=1, NOB=1, BAO-768"
180 CALL GPIB (CMD$, AS, FLG%, BRD%)
190 PRINT "FLAG RETURN CODE FOR SYSTEM INITIALIZATION'
    HEX$ (FLG%)
200 PRINT "INITIALIZATION COMPLETE"
210 PRINT "** NOW ENTER DATA FROM DEVICE 12 and PRINT IT
    ON THE SCREEN**"
220 B$=SPACES (18)
230 CMD$="REMOTE 12"
240 CALL GPIB (CMD$, B$, FLG%, BRD%")
250 PRINT "RETURN FLAG= "EX$ (FLG%)
260 CMD$= "ENTER 12[$, 0, 17]"
270 CALL GPIB (CMD$, b$, FLG%, BRD%)
280 PRINT "FLAG FOR ENTER COMMAND'"HEX$ (FLG%)
290 PRINT "DATA RECEIVED FROM DEVICE 12= ";B$
300 END
```

The mating cable required is CGIB-01, with a total bus length of about 20 meters, which cannot exceed twice the number of instruments. The typical data transfer rate is 2 kbps, with a DMA data transfer rate of 450 kbps. There can be only one talker at any one time for the 15 listeners for one board. The hardware handles all of the system timing for talking, listening, and controlling the 488 bus.

Power input is usually +5 volts at about .5 amp. The typical operating temperature range is 0° to 50°C, and the storage temperature ranges from −40° to +100°C.

Some 488 boards have a built-in ROM interpreter to handle initialization and protocol functions. Disk files with driver routines are not needed. The interpreter allows the use of commands in high-level IEEE-488 command syntax such as REMOTE or ENTER.

The interpreter can be in the form of a relocatable 16-kbyte block of code which is entered using a BASIC CALL statement or DOS interrupt commands using assembly language. Some interpreters include a group of subroutines which can be used to condition the data before data transfer when using assembly language programs.

Programming Examples

The following two examples in BASIC illustrate how the IEEE-488 interface board can be used to communicate with IEEE-488 instruments:

```
100 PRINT "******PASS CONTROL TO ANOTHER CONTROLLER******"
110 DEF SEG=&HC000
120 BRD%=&H300
130 CMD$="PASCTL 6"
140 CALL GPIB%(CMD$, X%, FLG%, BRD%)
150 PRINT "FLAG RETURN CODE FOR PASCTL IS = "EX$ (FLG%)
160 END
```

```
100 PRINT "******ENTER DATA FROM DEVICE 12 AND PRINT TO
    SCREEN******"
110 CLS
120 DEF SEG=&HC000
130 'DEFINE GPIB CALL VARIABLES
140 GPIB = 0:FLG% = 0:BRD% = &H300
150 B$ = SPACE$(18)
170 CMD$="REMOTE 12"
180 CALL GPIB(CMD$, B$, FLG%, BRD%)
190 PRINT "FLAG RETURN CODE FOR REMOTE= ":HEX$ (FLG%)
```

```
195 CMD$ = "ENTER 12($, 0, 17)"
200 CALL GPIB(CMD$, B$, FLG%, BRD%)
210 PRINT "FLAG RETURN CODE FOR ENTER = "HEX$ (FLG%)
220 PRINT "DATA RECEIVED FROM DVM IS -"; LEFT$(B$, 14)
230 PRINT "DO YOU WANT TO SCAN THE DATA AGAIN (Y/N)?"
240 YN$ = INKEY$:IF YNS = " " THEN 240
250 IF YN$<> "Y" THEN END ELSE PRINT: GOTO 200
260 END
```

ADDRESS SPACE

An IEEE-488 board will take one slot in the PC and 16 consecutive address locations in the I/O space. Some of the I/O address locations will be used by internal I/O or other peripheral cards. In order to avoid any conflicts with other devices, a DIP switch is usually provided to set the I/O addresses to be on any 16-bit boundary in the PC-decoded I/O space. This also allows the use of a second 488 interface board in the same computer. The user may transfer data to or from the two groups of devices connected to the boards. The maximum number of devices for the two boards is 30.

The interpreter will require a free 16 kbyte block of memory for its on-board 12 kbyte ROM and 4 kbyte RAM. The addressing DIP switch can be used to select the 16-kbyte block of memory on an even 16-kbyte boundary. The setting of the switches corresponds to the absolute 20-bit address location in 16-kbyte increments.

A 16-kbyte interpreter with 12 kbytes of ROM and 4 kbytes of RAM can be distributed as shown in the following map:

```
              IEEE-488 16 K BYTE INTERPRETER MAP
            -------------------------------------
0000
                    (12 K ROM INTERPRETER)
OADB          ASSEMBLY LANGUAGE LINK ROUTINE ADDRESS
OBB8            ASSEMBLY LANGUAGE UTILITY ROUTINES
3000          ------------RAM BUFFER BEGINS------------
              INTERNAL RAM BUFFERS FOR INTERPRETER
                           2 Kbytes
3800          ----------------------------------------
                  USER RAM AREA FOR SCRATCH PAD
                           2048 bytes
                     NOT USED BY INTERPRETER
3FFF        ------------END OF RAM BUFFERS------------
```

The following IEEE-488 function classifications are allowed:

Basic talker, serial poll, no extended talker.
Basic listener, no extended listener function.
Source and acceptor handshake capability.
Service request capability.
Parallel poll remote configuration capability.
Remote/local capability.
Device clear and device trigger capability.

RS-232/-422 EXPANSION BOARDS

An RS-232/422 I/O expansion board* for the PC and compatible machines is designed to plug into one I/O slot inside the PC. The board interfaces to the RS-422, RS-232, or current loop serial interface bus through a 25-pin (RS-232 and current loop) or 9-pin (RS-422) D connector.

The interface includes the standard bus control protocols such as data-set-ready, clear-to-send, data-terminal-ready, and others. These communication control signals can be disabled if desired.

The RS-422 interface will use clear-to-send and request-to-send control lines, or it can be set to only send-and-receive data. The selection of RS-232 or RS-422 and the protocol used are made by setting DIP switches on the board.

A typical board is able to communicate at speeds of up to 56 kilobaud at distances of 1 kilometer using the RS-422 interface. The data transfer rate can be selected by software. The board can be set up as a standard COM1 or COM2 interface, or it can be set at any other I/O address or interrupt level required.

A typical board is designed with the 8250 asynchronous communications chip. Even, off, or no parity, and 5, 6, 7, or data bits with 1 or 1 $\frac{1}{2}$ stop bits are usually software selectable.

The base address can be placed anywhere in the I/O address space. In most cases, the base address required will be for communication ports COM1 (which is at address hex 3F8) or COM2 (which is at hex 2F8).

The different user-selectable communication protocols available are typical of the following:

1. RS-232 with standard PC/XT/AT-compatible bus control signals, such as request-to-send, data-terminal-ready, or clear-to-send.
2. RS-232 without bus control signals, only data inputs and outputs.
3. Current loop.
4. RS-422 with request-to-send and clear-to-send control signals.
5. RS-422 without RTS and CTS control signals.

---

*Specifications and programming examples courtesy of Keithley-MetraByte, Inc.

These boards are usually designed to allow access to all of the available PC interrupt levels. When the board is to be installed as a COM1 or COM2 port, it is set to interrupt level 4 or 3. The level is usually selected in changing a jumper.

Power supply requirements are typically +5 volts at 500 milliamps and +5 volts and −15 volts at 30 milliamps each. Operating temperature is 0° to 50°C, with a storage temperature range of −55° to +125°C.

RS-422 INTERFACE BOARDS

The RS-422 interface boards for PC/XT/AT and compatible computers plug into a slot in the computer, and connection to the external serial buses are made through two 9-pin D connectors. Standard control signals such as request-to-send and clear-to-send can be used.

These boards allow communications at speeds of up to 57.6 kilobaud at distances of up to 4000 feet. Although most standard PC-based communication routines only allow the board to be set at up to 9600 baud, this limitation can be overridden in most cases.

The ports operate independently, and each has its own base address and interrupt selection controls. A channel can be set at COM1 or COM2 or at another available base address/interrupt level combination.

Some boards use the INS16450 UART, which is compatible with the 8250. It allows a number of communications parameters to be selected, as well as 5, 6, 7, or 8 data bits. The baud rate may be selected in a range of values from 120 baud up to 57.6 kilobaud.

PROGRAMMING EXAMPLE

The following illustrative program sets up one port as COM1 and the other as COM2. An RS-422 cable is installed between the two ports. This program will not operate properly if more than one communication device is set at the same address or interrupt level. Lines 60 through 140 of this example show how (in BASICA or GWBASIC) the communications ports can be set to communicate faster than the 9600 baud limit found in most DOS and BASIC communication routines.

```
10 OPEN "com1:4800" AS #1        'set up standard com1:
20 OPEN "com2:4800" AS #2        'set up standard COM2:
30 T1$="testing COM1 TX/         'define test transmission
   COM2 RX"                      data
40 OUT & H3FF,2                  'enable COM1:RS-485 driver
50 OUT &H2FF,1                   'enable COM2: RS-485
                                 receiver
```

```
60' ****start of baud rate
    override routine****
70 DUMMY=INP(&H3FB)              'read control register
80 OUT &H3FB,128:OUT &H2FB,      'select baud rate control
   128                           registers
90 OUT &H3F8,2OUT &H3F9,0        'set COM1: at 56 kilobaud
100 OUT &H2F8,2:OUT &H2F9,0      'set COM2: AT 56 kilobaud
110                              'note that if selecting
                                 38.4 Kbaud
120                              'write a 3 instead of 2 to
                                 &H318
130                              'and &H218
140 OUT &H3FB,DUMMY:OUT
    &H2FB,DUMMY                  'reset control register
150 PRINT #1, T1s                'transmit data from COM1:
160 INPUT #2,R2$                 'receive COM2: data, store
                                 in R2$
170 PRINT R2$                    'print received results
180'
190 IF T1$< >THEN PRINT          "error in transmission"
200'
210 CLOSE                        'close communications
                                 ports
220 END
```

The data transfer rate is 57.6 kilobaud for a maximum data transfer distance of 4000 feet. The required power is +5 volts at 850 milliamps. The operating temperature range is 0° to 50°C, with a storage temperature of −55° to 125°C.

The use of RS-422 is growing rapidly in the world of industrial serial communication. It is more robust than its RS-232 predecessor, and its ability to operate at distances of over 4000 feet brings communications capability to many industrial users.

RS-485 INTERFACE BOARDS

An RS-485 board allows IBM PC/XT/AT and compatible computers to be networked over the RS-485 bus. The RS-422 bus allows multiple receivers but only a single transmitter on the bus. RS-485 allows multiple transmitters and receivers to communicate on a two-wire bus. This provides a party-line type of network configuration.

An RS-485 board allows up to 32 different driver/receiver stations to communicate at 56 kilobaud. Although standard IBM communications software limits the speed to 19.2 kilobaud, this limitation can be overcome. The stations can be located up to 4000 feet away from each other. Typical applications include networking instruments, scanning and updating different input and output devices.

The board may be set up as a COM1 or COM2 serial interface port, or it can use any available base address/interrupt level combination. Some boards allow a single write to the BASE ADDRESS +7 (hex 3FF at COM1 or hex 2FF at COM2) to control the enabling/disabling of the RS-485 transmitter and receiver chip. Many boards use the 8250 peripheral interface adapter chip.

RS-232/422 PS/2 INTERFACE BOARDS

RS-232/422 I/O expansion boards* are also available for the IBM PS/2 models 50–80. These plug into a Micro Channel bus slot in the PC. The boards usually interface to the RS-232 or RS-422 port through separate 9-pin D connectors, with RS-422 connections using a female connector and RS-232 a male connector.

These boards can be configured to operate either as an RS-232 or as an RS-422 interface, but not both concurrently. The user can select the use of control protocols (data-set-ready, clear-to-send, data-terminal-ready for RS-232 or clear-to-send and request-to-send for RS-422) or can use only the data transmit and receive lines. Typical applications include instrument interfaces, industrial controller interfaces, printer/plotter interfaces, and interfaces to networks.

There are no DIP switches or user adjustments required for these boards. The selection of RS-232/RS-422, protocol/no protocol and the base address and interrupt level are usually set up by the operating system through software at power-up. A typical board allows the base address to be placed on any 8-bit boundary in the I/O space, and the interrupt level can be set at any available level.

Many of these boards use the 8250 UART chip. The baud rate can reach 56 kilobaud. Parity (even, odd, none), numbers of data bits (5, 6, 7, 8) and stop bits (1, 1 $\frac{1}{2}$, 2) are user programmable. These boards can communicate to distances of up to 1.2 kilometers with the RS-422 interface.

Programming Examples

The board can be configured as a standard COM port. The following BASIC program provides an example of programming. In this example, the computer operator is prompted to enter a command, which is subsequently transmitted. The 422 board waits for a response from the external device and prints the response on the CRT screen.

*Specifications and programming examples courtesy of Keithley-MetraByte, Inc.

```
10 INPUT "Enter Command" CMD$      'get command
20 OPEN "COM1:1200" AS # 1         'open communications
                                   at 1200 baud
30 OUTPUT #1, CMD$                 'send command to
                                   external device
40 INPUTS #1,RTN$                  'input response
50 PRINT "RESPONSE is ",RTRNS      'print the response on
                                   the computer screen
```

The following program is a communications routine whereby data are transferred between COM1: and COM2: ports:

```
10 OPEN "coml:4800" as #1          'set up standard
                                   COM1:
20 OPEN "com2:4800" as #2          'set up standard
                                   COM2:
30 TI$="testing COM1 TX/COM2 RX" 'define test
                                   transmission data
40 REM
50 REM
60'****start of baud rate override routine****
61 OPEN "COM1:1200" AS #1          'open the communi-
                                   cations port
62 D=INP(&H3F8)                    'read the UART
                                   control register
63 OUT &H3F8, 128                  'select UART baud
                                   rate register
64 OUT &H3F8,2:OUT &H3F9,0         'set the baud rate
                                   to 57.6 Kbaud,
                                   note that by
                                   writing a 3 to
                                   address &H3F8 instead
                                   of a 2, the baud rate
                                   can be set at 38.4
65 OUT &H3F8,D                     'reset UART control
                                   register with original
                                   contents
70 DUMMY=INP(&H3F8)                'read control
                                   register
80 OUT &H3F8, 128:OUT &H2FB,       'select baud rate control
   128                             registers

90 OUT &H3F8,2:OUT &H3F9,0         'set COM1: at 56 kilobaud
```

```
100 OUT &H3F8,2:OUT &H2F9,0    'set COM2: at 56 kilobaud
110                            'note that if selecting
                               38.4 kbaud
120                            'write a 3 instead of 2
                               to &H318
130                            'and &H218
140 OUT &H3F8, DUMMY:OUT
    &H2FB,DUMMY                'reset control register
150 PRINT #1, T1$              'transmit data from COM1:
160 INPUT #2, R2$              'receive COM2: data, store
                               in R2$
170 PRINT R2$                  'print received results
180 IF T1$ < > R2$ THEN PRINT "error in transmission"
200 CLOSE                      'close communications ports
210 END
```

The maximum data transfer rate is 57.6 kilobaud, with a maximum data transfer range of 4000 feet for RS-422. Power requirements are typically +5 volts and 500 milliamps and ±15 volts at 30 milliamps.

TESTING INTERFACES

An important part of interfacing concerns what to do when it does not work. It is important to know what went wrong and why. This testing or troubleshooting process is an integral part of the system design. A number of techniques are available to the designer for identifying and correcting problems. Problems related to component failure and methods for identifying them will be discussed.

The tools necessary in order to identify and locate these problems will also be described: voltmeter, logic probe, signature analyzer, oscilloscope, digital analyzer, in-circuit emulator, emulator, and simulator.

The most common problems that can occur in a system interface include the following:

1. Wiring faults, incorrect connections, open or short circuits.
2. Component failures or the wrong value component installed.
3. Software errors.
4. Noise or interference.

Wiring faults can be detected by a resistance check from point to point in the system. Each wire must go to the right pin and no other. It is a good idea to check circuit pinouts twice.

Wiring faults are usually the most common and troublesome problem, but they are easily solved, although this takes time. Most circuit paths can be buzz-tested using a simple continuity checker that emits a tone for a connection and is quiet for an open. Such a tester leaves both hands and eyes free to track the wiring.

COMPONENT FAILURES

Components such as resistors, capacitors, inductors, transformers, transistors, diodes, ICs, and connectors can all experience failures. Resistors open, and capacitors can leak electrolyte.

Each component can be given a figure of merit, known as its *mean time between failure* (*MTBF*). This is a statistical prediction, in hours, of how long the part will last under specific environmental conditions. Table 6-6 shows some typical values of failure rates for high-quality parts.

Some parts will last longer, on the average, than others. These values assume that all parts are being used properly. The figures are based on failures on a large sample for each part.

Since failure rates are defined as 1/MTBF, knowing the failure rate of each component in a system will yield the failure rate for the entire system. If the failure rates of all the components in the system are summed, this provides the system failure rate. The inverse of this number is the system MTBF.

Four ICs	.06
Crystal	.05
Ten resistors	.02
Ten capacitors	.50
PC board	.60 (assume 10 connectors, 500 soldered points)
Transformer	.50
Diodes	.052
TOTAL	1.82%/1000 hours

This means that if 1000 of these systems are used in the specified environment for 1000 hours, it is probable that 18 will fail. This does not tell when

TABLE 6-6 Failure Rates for Typical High-Quality Parts

Component	Failure Rate (%/1000 hr)	Component	Failure Rate (%/1000 hr)
Capacitor	0.02	Soldered joint	0.0002
Connector contact	0.005	Transformer	0.5
Diode	0.013	Transistor	0.04
Integrated circuit	0.015	Variable resistor	0.01
Quartz crystal	0.05	Wire-wrapped joint	0.00002
Resistor	0.002		

these 18 circuits will fail. This can be better understood by knowing the distribution of failures. Most failures occur when the parts are new, some when the parts are old, and fewer failures occur in between. An initial burn-in test will often produce most of the initial failures. This reduces the infant-mortality part of the curve.

SOFTWARE PROBLEMS

Software can be at fault. For example, there may be a routine in a program for handling power failures. Suppose that a mistake was made when copying this program, which restores conditions when power returns. If this routine is not actually tested when the power fails, the unit does not meet specifications.

Consider another simple example: An arithmetic calculation causes an overflow-and-halt condition when a measured input value exceeds a register's capability. The system may work well until this condition is exceeded. Then the system may stop mysteriously. These types of software problems, or bugs, can be difficult to identify.

NOISE PROBLEMS

Noise was discussed earlier with respect to its effects on linear circuits. It is reviewed here as a more general problem. When there is current in a wire, there is an electromagnetic field. There are fields from power transformers, motors, and electrical wiring. When these fields become large enough to generate errors, they become a problem. Any length of wire can act as an antenna.

When ICs switch, they can cause current changes in their power requirements. When many ICs switch at once and the system logic changes are not balanced, the power-supply voltage changes may affect other parts of the circuit. Bypass capacitors may be installed near each IC to prevent this type of noise.

A noise spike can result from turning on a piece of equipment on the same power line. In a system without noise filters, if that spike occurs at a critical time, data can be lost.

If two wires are close together, a pulse traveling along one induces a pulse in the other due to transformer action between the two wires. This induced pulse may cause reflections in the wiring and toggle a flipflop or cause read/write data errors. In order to prevent these problems, twisted and shielded-pair transmission lines can be used, as discussed earlier.

The power supply must be properly designed. There is always a small amount of 60-cycle ripple at the output of any filter. This can affect the contents of memory and cause an improper read or write. The power-supply design also affects the droop in voltage under heavy load before regulation occurs.

A number of tools can be used to find faults and identify them. These are discussed next.

TROUBLESHOOTING TOOLS

Most of the troubleshooting tools available are designed for certain kinds of problems. Thus, it is important to know the limitations of these tools.

Shorted and open conductors and incorrect voltages are the most common problems. They are also the easiest to detect. An ohm meter can be used to check for opens and shorts, and a digital voltmeter (DVM) or volt-ohm-milliampmeter (VOM) can be used to check voltages and currents. If you have the proper schematics and the time, it is possible to make sure that every component is where it belongs, draws the right currents, and receives the proper voltages.

To measure a voltage, a VOM is placed in parallel with the circuit element. Consider the measurement of a power-supply output voltage. The VOM can be used to measure such voltages, but it will not detect excessive ripple or noise riding on the power supply outputs. An oscilloscope must be used to measure this.

In order to measure a current, the VOM must be placed in series with the component. This means that the circuit must be broken. It is always possible that some connections can be made without cutting wires or traces. In the power-supply example discussed above, the VOM can measure the voltage across the load; then, by disconnecting the load and reconnecting it in series through the meter, the current can be measured. These measurements should be within the required tolerances. Incorrect values may indicate later problems.

COMPONENT SUBSTITUTION

Resistors, capacitors, diodes, and transistors can all be checked against known good devices. They can be measured with the DVM or VOM to determine if they are basically functional. Special test equipment is needed for diodes and transistors to measure the actual device characteristics. ICs are difficult to test without expensive equipment. It is much easier to have several of each device available in order to replace a device with a possible malfunction.

Intermittent problems are often due to connector or solder-joint failures. These can be checked first, before going on to other components. Intermittent problems will generally require an oscilloscope (preferably with some storage capabilities), or a logic analyzer which stores logic states can be used.

Static problems are easier to solve. This is usually the first step. These problems must be completely solved before continuing.

DESIGN PROBLEMS

Design errors are divided into two general types: improper use and improper specification. An example of improper use is passing too much current through a resistor. This can cause it to change value, damage other parts, or open. Applying too much voltage to a capacitor can cause arcing and eventual breakdown. Every device has its limits. Another common problem is too many loads on a single output line. This can cause the system to read or write improper data values on an intermittent basis, which may depend on temperature.

An example of improper specification is discussed next. Suppose we believe that a part can drive 20 bus loads when it can only drive 10. This is improper application of the specifications that may not have been noticed in the data sheet. More subtle errors include, for example, misreading the timing diagram of a particular part. Suppose that the address for a particular memory part must be stable for 30 nanoseconds. If the address gated to this memory part must be stable for 20 nanoseconds before the data and write pulses and this is overlooked, the system timing will be incorrect.

Some problems require specialized equipment for quick completion of the troubleshooting effort, but a VOM and an oscilloscope can be used if time is not as critical.

Intermittent problems require that all I/O loading and device specifications be checked. The system can also be operated at different temperatures to localize sensitive components. A can of freeze spray and a heat lamp can be used to locate temperature-sensitive problems. This is done by heating and cooling the suspected parts.

LOGIC PROBES

Logic probes are a tool for checking logic levels on ICs. They are helpful in isolating static conditions. The probes will indicate if a signal is a 0, a 1, or undetermined by lighting an LED or a small light bulb. Undetermined states usually indicate problems unless the output is a tri-state bus floating output.

The VOM and logic probe do not indicate time, so they are of little use for dynamic problems. What is needed are devices which can indicate that the timing is correct.

USE OF THE OSCILLOSCOPE

In order to obtain timing information, an oscilloscope can be used. By using one or more traces, critical events can be measured as a function of time. In microcomputer systems, events as short as 10 nanoseconds may need to be observed. A 10-nanosecond square wave will be filtered and will look like a sine wave on a 10-MHz oscilloscope. A unit with a 50- or 100-MHz bandwidth is needed to see these faster events clearly.

Most microcomputer logic levels are +5 volts standard TTL. The logic 0 signal is from −0.6 to +0.8 volt. The logic 1 signal is from +2.0 to +5.5 volts. The region from +0.8 to +2.0 volts is undefined.

Transitions from one level to another generally occur in less than 1 microsecond to avoid noise problems. An oscilloscope can be used to indicate if a logic level error is present. Such a measurement, along with the knowledge of the correct logic timing, will indicate to the troubleshooter that a line is at fault.

By observing chip-select, control, and bus lines with an oscilloscope loading, timing and noise problems can be found. The logic levels should be well defined. This means that for +5 volt TTL logic, the 0 level should be from −0.6 to +0.8 volt and the 1 level should be from 2.0 to 5.5 volts. In-between states indicate problems.

LOGIC STATE MEASUREMENT

Individual system timing and system logic levels may appear to be correct when observing a single bit, line, or pair of lines, but often a number of lines may need to be observed at the same instant in time. This is where logic analyzers (also called *digital-domain analyzers*) are used. These instruments allow the observation of up to 32 nodes simultaneously. They can generally display these bits in binary, octal, or hexadecimal or in the form of conventional oscilloscope traces. The display can be triggered when a given combination of bits occurs.

State analyzers can store a number of clock cycles, and they are able to display several sets of signals before and after the trigger set. Each set of signals in time is known as a *state*.

The available analyzers are designed to provide two types of information. Some provide mostly timing information, and others provide mostly logic state information. The timing information analyzers are configured like multichannel oscilloscopes. These devices are most useful where logic spikes, noise, or logic-level problems are to be investigated.

SIGNATURE ANALYZERS

These instrumentation units are based on storing repetitive sequences of signal values in a recirculating shift register. The values are clocked into a display each time around. The sequence of values will generally have a unique signature unless similar points or lines are being measured.

Each node in a system will have its own signature when it is working properly. It will also have a special signature for each possible problem. By using fault-tree methods, developed by using the signature analyzer, faulty equipment can be debugged quickly down to the faulty component. It will not find software problems or the cause of intermittent failures in a system. A typical unit is the HP 5004A Signature Analyzer.

COMPARISON TESTS

In this method, a device or board under test is compared to a known good device or good board. They share the same common input, and outputs are compared.

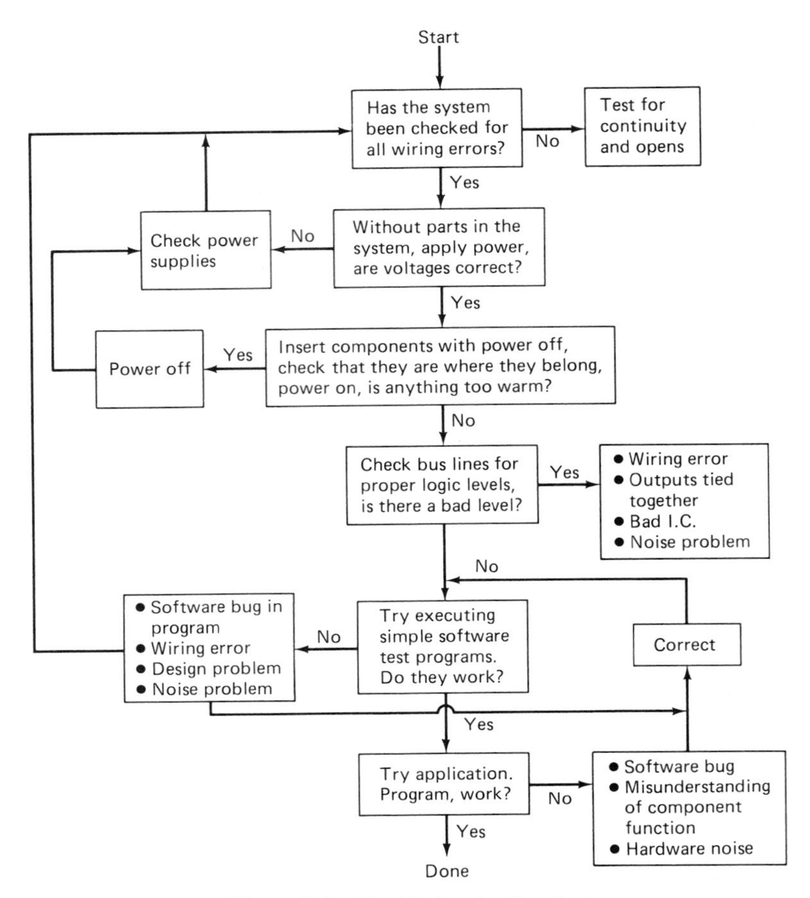

Figure 6-4 Troubleshooting flowchart.

This is a hardware method, and the required tools have been discussed. The underlying principle of this testing technique is to compare an existing board, component, or system with a properly functioning unit. The problem is to know how a proper unit functions and how to implement a reasonable procedure for performing the comparison in a systematic manner. Other problems include making the measurements themselves and recording a time history of these measurements for the comparison. The test instruments and techniques needed to perform such comparisons have already been described.

Components, software, and noise are some of the factors that can cause problems to occur. The flowchart shown in Figure 6-4 describes some basic methods of approaching typical interface-related problems. The equipment needed for debugging has been discussed, and several examples have been given.

Future hardware debugging tools will be oriented toward the state type of analyzer previously discussed. A large number of state, trace, and trigger capabilities, as well as the ability to format the display of the states in any machine's mnemonics, will be features of the new machines. Their use on minicomputers and large computers will also become widespread, with some systems including an analyzer in the unit for self-diagnosis.

FINAL CONSIDERATIONS

Don't handle an IC unless you are properly grounded or the ambient humidity is high. A static charge, such as the one generated by walking on carpeting on a dry day, can destroy many MOS chips. Don't install a board in the computer unless the following has been done:

1. Power is off.
2. You have waited at least 15 seconds for all charges to be dissipated.

Finally, don't trust specified voltages and currents; always measure them.

7

Interfacing Pressure Transducers

Pressure is one of the parameters most often measured. This chapter explains some of the most common techniques used for pressure measurement. It shows how to interface these transducers to the PC, using counters and programmable timers.

We will describe interfacing schemes and circuits that can be used to connect the various types of these transducers to the PC. Counters and programmable interval timers will be discussed. We will show how to set up the various counter modes available with counter/timer boards.

We will illustrate how to use counter/timer I/O boards, as well as A/D converter/timer counter boards. Interrupts and direct memory access are also explained, and the use of multiplexer boards is discussed.

PRESSURE MEASUREMENT

Pressure transducers are used in many control and automation applications. Many electrical output pressure transducers detect pressure using a mechanical sensing element. These elements may consist of a thin-walled elastic member such as a plate or tube, which provides a surface area for the pressure to act upon.

When the pressure is not balanced by an equal pressure acting on the opposite side of this surface, the element moves as a result of the pressure. This deflection is then used to produce an electrical output. When another separate pressure is allowed on the other side of the surface, the transducer measures the differential pressure.

If the other side of the surface is evacuated and sealed, then absolute pressure is measured. The transducer measures gauge pressure when ambient pressure is present on the reference side.

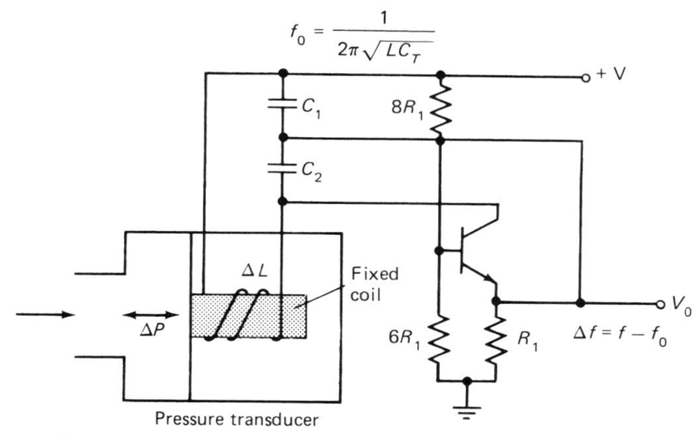

Figure 7-1 Single-coil inductive pressure transducer interface.

Inductive Pressure Transducers

Inductive pressure transducers use the pressure to move a mechanical member, which is then used to change the inductance of an electromagnetic coil. The inductance is a function of the relative motion of the movable core and the inductive coil, as shown in Figure 7-1 (lower left). This figure also shows how inductive single-coil transducers can be used in oscillator circuits to control the oscillation frequency.

Single-coil transducers are prone to suffer from problems in compensating for temperature effects. Reducing these effects requires matching of the core and winding materials for temperature versus permeability changes.

Another type of inductive transducer uses the ratio of the reluctance of two coils. The measurement of force due to external pressure is accomplished by a change in the inductance ratio of the coils. The force being measured changes the magnetic coupling path of the transducer due to the displacement of the core (Figure 7-2). These reluctance transducers are less sensitive to temperature effects than the single-coil types. A motion of about 0.003 inch provides an ac output voltage of about 100 millivolts. This can be rectified and filtered, as shown in Figure 7-4b.

In the diaphragm type of variable reluctance pressure transducer, as shown in the illustration, a thin diaphragm of magnetic material is supported between two symmetrical inductance assemblies. The diaphragm is deflected when there is a difference in pressure between the two input ports.

The gap in the magnetic flux path of one core will increase while the gap in the other core decreases. The reluctance changes with the pressure, and the net effect is a change in the inductance of the two coils of the transducer.

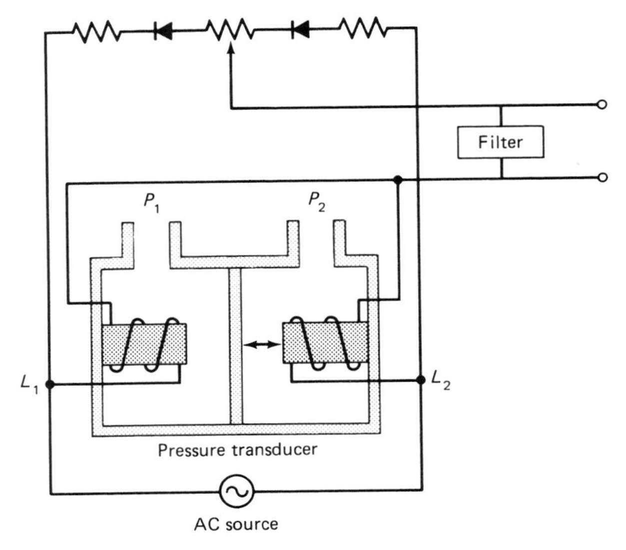

a. Circuit

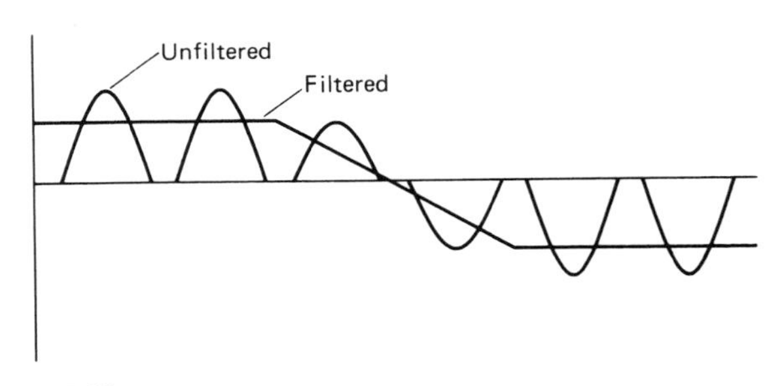

b. Bridge output

Figure 7-2 Differential reluctive pressure transducer interface. (a) Circuit; (b) Bridge output.

The inductance ratio L^1/L^2 can also be measured in a bridge circuit to detect the voltage proportional to the pressure difference.

Hysteresis errors are limited to the mechanical components. An E-shaped core is often used to maintain good balance and low phase shift. The diaphragm can also be used as part of the inductive loop. The basic characteristics of inductive pressure transducers are shown in Table 7-1.

Some transducers contain built-in dc to ac to dc conversion circuitry. Dc excitations of 28 and 5 volts are available for absolute, gauge, and differential pressure measurements. The range is typically 1 inch of water up to 12,000 psi. The ac transducers normally use a carrier frequency between 60 Hz and 30 kHz.

TABLE 7-1 Characteristics of Inductive Pressure Transducers

High output
Can be used for both static and dynamic measurements
High signal/noise ratio
Frequency response normally limited by the mechanical construction
Both reactive and resistive balances are needed at null
Magnetic objects and fields can cause transient errors
Volumetric displacement tends to be large
Mechanical friction can cause wear and errors

When dc conversion circuitry is used, the internally generated carrier frequency is usually much higher. This allows smaller coils and produces a more compact package.

The frequency response range is 50 to 1000 Hz, depending on the mechanical design. Some constructions, like the diaphragm types, have a reasonable tolerance to shock and vibration.

The static error is typically $\pm 0.5\%$, most of which is due to nonlinearity. Errors due to hysteresis and nonrepeatability are less than 0.2%. Proof pressure or overrange ratings of greater than six times the normal range are available.

Temperature effects can be minimized by using similar sensing element and coil materials. These errors range from 1% to 2% for temperatures up to 100°F. Transducers without dc conversion circuits can operate up to temperatures of 350°F.

The linear variable differential transformer (LVDT) type of construction, as shown in Figure 7-3, uses a sliding core connected to the pressure-sensing element. The output is rectified and filtered as shown. LVDT transducers are also used for displacement velocity measurement, as discussed in Chapter 9.

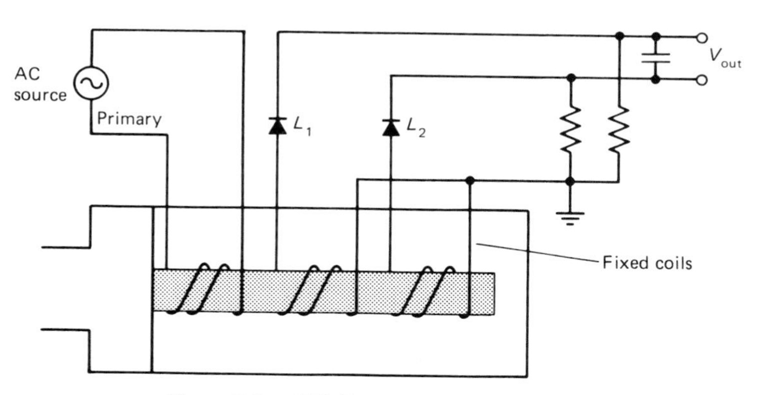

Figure 7-3 LVDT pressure transducer interface.

Piezoelectric Pressure Transducers

Piezoelectric pressure transducers depend on a crystal to generate a charge or voltage when it is mechanically stressed. A diaphragm is normally used to react with the crystal to produce the stress on the crystal. Piezoelectric pressure transducers can operate over a wide temperature range, with relatively small errors due to temperature changes.

A number of crystal materials are used. Some quartz units use crystals, including those found in the natural state. Grown crystals include ammonium dihydrogen phosphate (ADP). Various ceramic materials are also used. The piezoelectric elements are cut from the crystal along the existing crystallographic axes.

Ceramic elements are pressed from powdered materials into the required shape and then fired at high temperatures. The piezoelectric characteristics occur when the ceramic is polarized by an electric field during the cooling process.

Ceramic elements, when heated, can reach the Curie point. This is the temperature at which the crystalline structure changes and polarization is lost. The element will then cease functioning as a piezoelectric device. It can then be polarized. Curie points range from 300°F to over 1000°F. Quartz elements can be used in the temperature range of −400° to 500°F.

The output of piezoelectric elements can be influenced by the pyroelectric effect, which causes changes in the output proportional to the rate of change of temperature experienced by the crystal.

Some quartz crystal transducers have been used with amplifiers which permit static measurements, but most piezoelectric pressure transducers are used for the dynamic measurement of rapidly varying pressures. Typical applications include sound pressure levels of up to 180 dB, which can be measured with an accuracy of 3% of full scale. Units have been mounted as spark plugs to measure pressures in internal combustion engines. Piezoelectric transducers are also used on turbines, pumps, and hydraulic equipment to measure dynamic stresses.

Piezoelectric sensors respond to high-shock levels, which can be a problem in some high-level applications. A high-level shock can produce an overvoltage, which can saturate the amplifiers for a period of 50 or more times the shock duration. There is a loss of data during this time.

Frequency response is in the range of 10 to 50,000 Hz, and the pressure range is 5 to 10,000 psi. Quartz and ADP crystal units generally have a higher natural frequency than ceramic units; however, ceramic crystals provide higher output levels.

The normal output of a piezoelectric crystal is low and the impedance is high, so a high-impedance amplifier is normally used for signal conditioning. An operational amplifier can be used with capacitive feedback to compensate for the capacitance due to cabling.

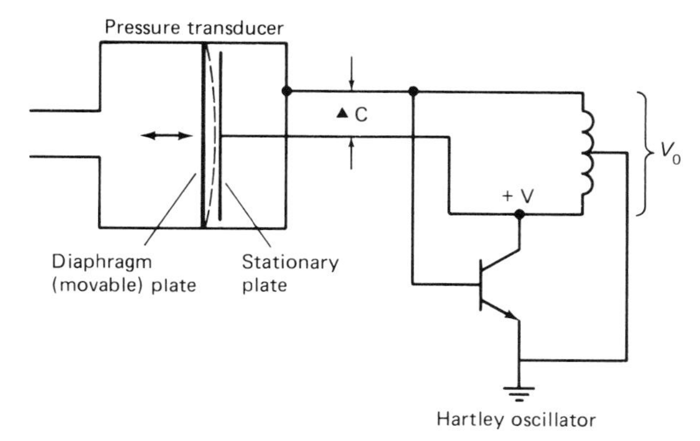

Figure 7-4 Capacitive pressure transducer interface.

Capacitive Pressure Transducers

Capacitive pressure transducers use a metal diaphragm for one capacitor plate, while the other plate is positioned next to the diaphragm, as shown in Figure 7-4. The pressure causes a movement of the diaphragm, which changes the capacitance between the two plates.

An ac signal is placed across the plates to sense the change in capacitance. The capacitive sensor can be used as part of a resistor-capacitor (RC) or inductor-capacitor (LC) network in an oscillator (as shown), or it may be used as a reactive element in an ac bridge. When it is used in an oscillator circuit, the final output may be ac, dc digital, or in the form of a phase shift.

Capacitance transducers must be reactively as well as resistively matched. Long lead lengths or loose leads can cause a variation in capacitance due to capacitive coupling. It is sometimes necessary to use a preamplifier close to the transducer, as well as matched cables, to reduce these effects.

Capacitive transducers are usually small in size, with a high frequency response. They can be operated at high temperatures and allow the measurement of both static and dynamic quantities. The range is .01 to 10,000 psi, with a typical error of .25%. Higher-priced units with accuracies of up to .05% are also available. Table 7-2 shows the basic characteristics of capacitive sensors.

TABLE 7-2 Characteristics of Capacitive Sensors

High-impedance output must be reactively and resistively balanced
Most units sensitive to temperature variations
Higher frequency response
Inexpensive and has a small volume
Has a low shock response
Can be used for either static or dynamic measurements

Some capacitive sensor designs use materials such as quartz for the capacitor plates. Two thin quartz disks are plated with platinum electrodes on the inner surfaces. The disks form a small capsule, and the electrodes are separated by a 0.0002-inch gap.

Strain Gauge Pressure Transducers

Strain gauge pressure transducers use the force from a pressure change to cause a resistance change due to mechanical strain. The pressure-sensing element may be a diaphragm or even a straight tube, since the deflection required is small. When a tube is used, the strain gauges can be mounted on the tube, which is sealed at one end. The pressure difference will cause a small expansion or contraction of the tube's diameter.

Some designs use a secondary sensing element or auxiliary member in the form of a beam or armature. Four or two arms of a Wheatstone bridge may be used for temperature compensation. The force being measured displaces and changes the length of the member to which the strain gauge is attached.

Strain gauge transducers may be unbonded or bonded. An unbonded gauge has one end, fixed while the other end is movable and attached to the force member. The bonded gauge is completely attached by an adhesive to the member whose strain is to be measured. Strain gauges may be made from metal and metal alloys, semiconductor materials, and thin film materials. The strain gauge property known as the *gauge factor* produces a change in resistance proportional to the change in length. Strain gauges may be arranged in the form of a Wheatstone bridge circuit, with one to four of the bridge legs active.

The gauge factor is defined as the unit change in resistance per unit change in length. All electrical conductors have a gauge factor, but only a few have the necessary properties to be useful as strain gauges.

The more common metal strain gauge materials have gauge factors that range from 2.0 to 5.0. High-gauge-materials tend to be more sensitive to temperature and less stable than low-gauge-factor materials.

Strain-sensing filaments must have stable elastic properties, high tensile strength, and corrosion resistance. The alloys used must have the proper combination of gauge factor and thermal coefficient of resistivity for optimum performance.

When the strain-gauge filaments function as the four elements of the bridge circuit, the circuit is divided into four parts of equal resistance value. A resistance of 350 ohms is typical, but bridge resistances ranging from 50 to several thousand ohms are used.

In an unbonded unit, a stationary frame is used with a moving armature. The filaments of strain-sensitive resistance wire are wound between rigid insulators mounted on the frame and armature.

The resistance of the four strain elements shown can be trimmed during assembly, so no signal appears in the bridge output circuit when there is no external force. If the bridge is balanced in this way for zero output, the unbalanced electrical output of the bridge will tend to be linear as a force is applied.

The mechanical arrangement can limit the travel of the armature to ± 0.0003 inch. This is achieved in many miniature transducers. These units provide excellent frequency response due to the low displacement and small mass used in the armature.

In bonded strain gauges, electrical insulation is provided by an adhesive or insulating material on the strain gauge. The force needed to produce the displacement is larger than that required with an unbonded strain gauge because of this additional stiffness. The bonded strain gauge member can be used in a tension, compression, or bending mode.

Semiconductor materials such as silicon can also be used for strain gauges. By changing the amount and type of dopant, the strain gauge properties are modified for specific applications. Silicon strain gauges have a higher gauge factor than metal strain gauges, as well as a higher temperature coefficient. Gauge factors of 50 to 200 are typical and may be either positive or negative.

Since the unbonded wire elements are stretched and unsupported between a fixed and a moving end, these sensors can have high sensitivity; they are also sensitive to vibration. Unbonded strain gauges can be used for pressures of less than 5 psi. Nichrome and platinum wire are commonly used.

Bonded elements are attached permanently with adhesives to the active strain element. Foil, thin film, and semiconductor strain elements are usually bonded or deposited such that the semiconductor, thin film, or foil and a pressure diaphragm appear as a single part. This makes them much less sensitive to vibration. Cut or etched foil types allow a strong bond, as well as automated trimming.

Thin-film strain gauges use a molecular bond instead of adhesives. Manufacturing techniques such as sputter deposition, as well as others, such as vacuum deposition, are used for electronic microcircuits. Control of the materials and the deposition process allows the strain gauge properties to be modified to produce the desired characteristics.

The metal substrate provides the desired mechanical base. Then a ceramic film is vacuum-deposited on the metal to provide the required electrical insulation. Four strain gauges can then be vacuum-deposited on the insulator and interconnected into a bridge configuration by vacuum deposition.

The multiple evaporations can be made during a single vacuum pumpdown period with the use of multiple sources and substrate masks. The leads are attached to the film by either microwelding or thermal compression bonding of the noble metal wire. Lead wire attachment is made directly to the film. The strain gauge pattern can be designed to optimize the characteristics of a specific

TABLE 7-3 Strain Gauge Elements Used in Pressure Transducers

Unbonded metal wire gauges
Unbonded metal foil gauges
Bonded metal wire gauges
Bonded metal foil gauges
Thin-film deposited gauges
Bonded semiconductor gauges
Integrally diffused semiconductor gauges

sensing element. The sensing elements can take on many configurations, since the strain gauges can be deposited on diaphragms, beams, columns, and other mechanical elements.

A common pressure element is the flat plate diaphragm with four strain gauges arranged in a bridge configuration. When a force is applied normal to the plate, the plate is deformed, causing the strain-sensitive elements to elongate and increase in resistance. This change in resistance is proportional to the change in the length of the strain gauge and alters the balance of the bridge.

Diffused semiconductor strain gauges are made using the same technology used for ICs. The sensing element is a four-arm strain gauge bridge which is diffused on the surface of a single crystal silicon diaphragm. The types of elements used in pressure transducers are listed in Table 7-3.

The silicon diaphragm is elastic and hysteresis free. These sensors have high gauge factors and provide relatively high output at low strain levels. IC manufacturing techniques such as electrostatic or thermal compression bonding or electron beam welding are used to encapsulate the sensing element.

The shock, vibration, and overload ratings of the silicon diaphragm are similar to those of other high-quality microcircuit devices. The combined linearity and hysteresis effects are less than 0.06%. The silicon diaphragm can be small enough to give a fast response with minimum sensitivity to accelerations from shock and vibration. The specification range of strain gauge pressure transducers is shown in Table 7-4. Stainless steel, hastelloy, or other materials are

TABLE 7-4 Strain Gauge Pressure Transducer Specification Range

Design pressure	To 200,000 psig (1400 MPa)
Design temperature	Typically to 250°F (120°C) Special designs to 600°F (316°C)
Materials	Stainless steel and other corrosion-resistant metals
Error	±.1 to +1% of total span
Range	3 in. H_2O to 200,000 psig (.08 kPa to 1400 MPa)

used for isolating the diaphragms for those applications where the media are not compatible.

Output Considerations

The full-scale output of a four-element strain gauge bridge for a pressure transducer with metal wire or foil gauges is 50 to 60 millivolts for bonded gauges and 60 to 80 millivolts for unbonded gauges, with 10 volt excitation. Compensating and adjusting resistors can reduce the output to about 20 to 30 millivolts for bonded gauges and 30 to 40 millivolts for unbonded gauges. These resistors are used for zero and balance adjustments, full-scale adjustment, thermal zero shift compensation, thermal sensitivity shift compensation, and shunt calibration (Figure 7-5). Semiconductor strain gauge transducers provide an output of 200 to 400 millivolts for 5 to 7 milliamp excitation.

Many transducers have an upper limit on the excitation voltage to prevent heating. Voltage limiters may be used to provide the proper level required for bridge operation. A constant current source can be used for thermal compensation. In order to provide TTL output levels, a signal conditioning amplifier is required for all metal and some semiconductor strain gauge transducers.

The zero setting can shift with temperature. This occurs when there is an expansion of the sensor components. The calibration factor or sensitivity can also change with the ambient temperature due to a change in the elasticity or spring constant of the sensor parts. Many metals have a temperature coefficient for Young's modulus of elasticity of about −0.0007°C.

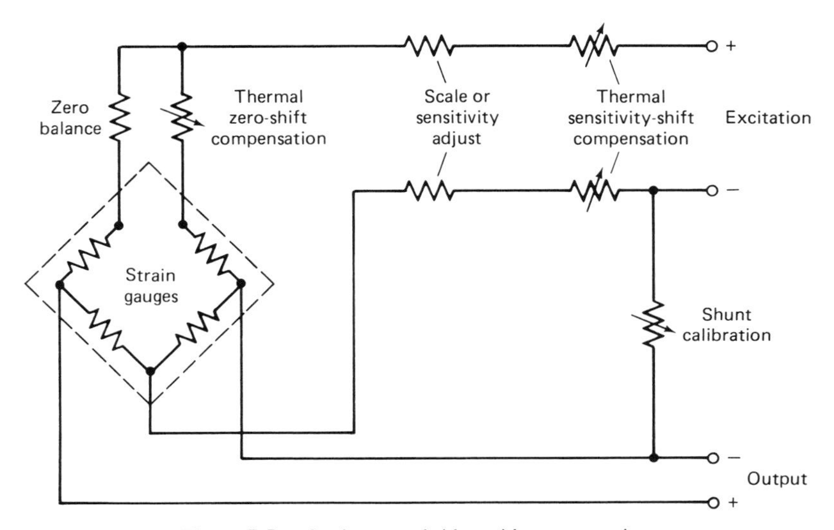

Figure 7-5 Strain gauge bridge with compensation.

In order to minimize the zero error with temperature, the differential expansion of the mechanical components must nearly balance; otherwise, the armature of the transducer can be pulled off center, and the range of span becomes incorrect even if the zero shift is corrected. This differential expansion must be reduced until the total change over the ambient temperature range is a small fraction of full scale. Then the remaining error can be compensated electrically using the resistors shown in Figure 7-5.

Ideally, the compensating resistors should be at the same temperature as the transducer. The compensating resistors should not dissipate any appreciable heat; if heat dissipation occurs, an allowance must be made for the resulting increased resistance. The temperature coefficients of the compensating resistors' material can be found by testing the actual resistors to be used. The basic characteristics of strain gauge transducers are summarized in Table 7-5.

Potentiometric Pressure Transducers

A potentiometric transducer uses a resistance element which changes due to the action of a movable slider. The motion of the slider can be caused by a pressure change to vary the resistance. Deposited carbon, film, and resistive wire elements are used. The electrical contact between the resistive element and the slider is a potential problem area due to noise and errors from inconsistent wiper pressure. Despite these limitations, potentiometer sensors are widely used. Their electrical efficiency is high, and they provide direct output in many operations without the need for additional amplification.

The output due to a pressure change is a function of the design of the potentiometer's resistance curve, which can be linear, sine, cosine, logarithmic, or exponential. The unit can be excited with ac or dc, and no amplification or impedance matching may be required. A high output can be obtained with a high input voltage.

TABLE 7-5 Characteristics of Strain Gauge Transducers

Fast response time
Good resolution
Minimal mechanical motion
Good accuracy
Predictable compensation methods for temperature
Low source impedance
Low acceleration effects for bonded type
Zero output at zero pressure due to bridge imbalances
High vibration errors for unbonded types (especially for ranges lower than 15 psi)
Low output levels result in noise problems
Isolation of grounds is required
Other signal-conditioning requirements

TABLE 7-6 Characteristics of Potentiometric Pressure Transducers

Major advantages
ac or dc excitation
Inexpensive
Amplification or impedance matching usually not required for transmission
Wide range of output functions available
Major disadvantages
High mechanical friction can cause limited life
The resolution can be finite in some devices
Noise can develop from contact wear
Sensitive to vibration
Accuracy is usually a function of the force required to overcome friction
Frequency response is low due to the mechanical contact
Large size
Large displacements may be required from pressure-sensing elements

Typical potentiometric transducers have an error of ±1%. The pressure range is 5 to 400 psi for typical units, and high-pressure devices of up to 10,000 psi are available. Devices are available with the following advanced specifications: resolution 0.2%, linearity ±.4%, hysteresis .5, and temperature error ±.8%.

Potentiometric pressure transducers were first reported in 1914. They are still used today due to low cost and connection simplicity. The more recent trends in potentiometric pressure transducers have taken two paths. There are miniaturized devices with relaxed tolerances. Other units use a control force to supplement the force of the diaphragm or capsule. These motor-or force-driven systems allow the use of low-resolution, multiturn potentiometers for improved accuracy. The advantages and disadvantages of potentiometric transducers are summarized in Table 7-6.

INTERFACING TO THE PC

If a single-coil inductive transducer is used, the transducer can be part of an oscillator circuit, as shown in Figure 7-3. A frequency counter is needed to measure the output of the oscillator. This facility is provided by many of the available I/O expansion boards. Typical of the counter/timer chips are the Intel 8253 and 8254 and the AMD 9513. Any of these can be used to count a frequency output.

COUNTERS

Most data acquisition boards incorporate counters. A precise frequency is generated in the system using a crystal-controlled oscillator. A counter can then be used to count the oscillations. If the counter is used to interrupt the microprocessor after a certain number of oscillations have occurred, this provides a relative timekeeping method. A signal is generated every time the counter reaches its maximum count. Then the count goes to zero and starts again. This type of signal can be used to generate an interrupt to the microprocessor.

Suppose, for example, that the oscillator is running at 8 MHz and the counter is set to count up to 1000. The interrupt will occur every .125 millisecond. The microprocessor can then be forced to do something every .125 millisecond. This periodic time can be used for scanning an input signal. The interrupt period needed depends on the application.

The counter can be under program control, which allows the time interval between interrupts to be controlled by the microprocessor. The processor and its program can be set to determine when and how often the interval timer will interrupt. This type of time interface is useful when differing intervals are needed by a system at different times.

The oscillator can also be used to drive a counter, which is then divided down to provide a real-time clock. Suppose that the counter takes the frequency output of the crystal oscillator and divides it down to one pulse/second. Then this signal is fed to a series of counters, which divide it down by the ratios required for keeping track of time. A pair of divide-by-10 and divide-by-6 counters will keep track of the seconds by counting the one-pulse/second signal. The output from this counter pair is a one-pulse/minute signal which can be sent to a similar pair for counting minutes. A divide-by-24 counter then can then be used to count the hours from the signal received from the minutes counter. Four interval signals are available: one pulse/second, one pulse/minute, one pulse/hour, and one pulse/day.

In most PCs, if these counters are preset to the time of day, they will keep the absolute time as long as power is not interrupted. The microprocessor can check the current time by reading the counters. This scheme is the basis for time-of-day ICs.

THE 8253 PROGRAMMABLE INTERVAL TIMER

The 8253 Programmable Interval Timer is typical of the counters used in I/O expansion boards. It has three 16-bit programmable counters in a 24-pin package. Eight pins are used for a bidirectional data bus, a chip select ($\overline{CS}$), two address lines, and the read and write control signals $\overline{RD}$ and $\overline{WR}$.

Each counter chip has three I/O pins associated with it. There are two inputs and an output. The inputs include a clock to drive the counter and a gate to enable the counting function. The output may be used to control an interrupt or provide a timing function.

The two address pins allow the chip to have a maximum of four read and four write registers. The 8253 actually has three read and four write registers. Each of these 16-bit counters has an 8-bit write register.

A write-only mode register allows the processor to configure each of the three counters. The mode register is used to set the mode of operation of each counter.

A set of select counter bits (SC1 and SC0) indicate to which counter this particular mode word applies. These bits indicate which of the three mode sub-registers is to be accessed. SC1 and SC0 form a 2-bit binary number indicating the affected counter. Bit patterns 00, 01, and 10 indicate counters 0, 1, and 2, respectively. Bit pattern 11 is not allowed.

Each counter can be independently programmed for a maximum count. The counters are 16 bits wide but use only an 8-bit path to the microprocessor. A set of read/load bits is used to indicate how the microprocessor will access the counter. These can be set to indicate that the microprocessor should read or write only the least significant byte of the counter (start at zero) and when the microprocessor does not need to read the counter in the lower 8 bits of the counter. This last function is used when the counter is to be treated only as an 8-bit counter.

The microprocessor can also be set to read and write the full 16-bit counter value. Writing to the counter must be done first to the least significant byte and then to the most significant byte. When a read of the counter contents is performed, the counter will supply the least significant 8 bits first, followed by the most significant 8 bits. This sequence must always be followed or the 8253 can lose synchronization with the program.

When a counter is counting, a read of a counter value may not always be accurate. Inaccuracy can be caused by ripple delays through the counter. The counter may be changing just as the read takes place. A latch function can be used to avoid this problem.

In the 8253 chip, this function allows the microprocessor to latch the current count in a special holding register. This register can then be read for the true count.

COUNTER MODES

Three mode bits are used to select one of five operating modes for each counter. In the interrupt-on-terminal count mode, the counter acts like an interval timer which operates once and then stops. The counter output goes low when the mode word is written to the 8253. When the initial count is written to the counter, it starts to count down. When it reaches zero, the counter output goes true and counting stops. The output goes low when the microprocessor writes a new mode byte or reinitializes the count. This mode is useful when it is necessary to have events occur after a certain number of time pulses. The processor can start the counter and then work on other tasks. When the counter reaches zero, it interrupts the processor, alerting it that time has run out.

The interrupt-on-terminal count mode can be used as a watchdog timer to monitor the operation of a process. A watchdog timer is used to ensure that certain events occur. If these events do not occur at a particular time, the computer assumes that something has gone wrong and shuts down the process.

A watchdog timer is part of most process control systems where the process system is monitoring and controlling an industrial process. Noise or some other factor can cause the processor to lose its place in the program and start executing the wrong sections of memory. If this happens, the processor will not be controlling the process and will stop communicating with the watchdog timer at the desired time out. This will cause a reset of the processor.

The interrupt-on-terminal count mode works the following way as a watchdog: When the processor writes the first byte of the count to a counter running in this mode, it stops counting. Then, when the second byte is written, the counter is reinitialized and starts counting down again.

When the 8253 is used as a watchdog timer, a software routine that tells the watchdog timer that everything is satisfactory is required. Periodically, the program will call this routine, which writes 2 bytes to the 8253 to prevent the watchdog timer from asserting a reset. This routine will be located in or called from the program's main loop. If the program loop is large, several watchdog-routine calls may be needed throughout the program.

There is also a programmable one-shot mode, which can be used to generate precise pulses. In this mode, the counter output is initially high and the counter is started from a low-to-high transition at the counter's gate input. The counter output remains low until the counter reaches zero; then it returns to a high level. If another low-to-high transition occurs on the gate input while the counter output is low, the count starts again. This allows the one-shot action to be retriggerable.

The rate generator mode can be used to generate pulses at precise intervals. The counter continually counts down, reaches zero, and then reinitializes. When the counter is at zero, the counter output is low; otherwise, it is high. This mode will produce a series of short, low-true pulses. Driving the gate input low causes the counter to reinitialize and stop counting. When the gate input goes high, counting resumes. This allows the count to be synchronized with the gate input.

The square-wave rate generator mode operates similarly to the rate generator mode, except that the counter output is high for the first half of the count and is driven low for the second half.

In the software-triggered strobe mode, a pulse is generated after a programmed number of counts occurs. Counting starts when the microprocessor loads the count into the counter. The pulse is produced for one clock pulse when the counter reaches zero. This mode is similar to the interrupt-on-terminal count mode, except that the output is low for only a single count pulse.

In the hardware-triggered strobe mode, the counter is initialized and a low-to-high transition at the gate input causes the counter to start counting. After the programmed number of counts, the counter reaches zero and the counter output pin is driven low for a single count. This sequence can be retriggered by the gate input. If a second trigger occurs before the counter reaches zero, the counter is reinitialized and the sequence is restarted. This mode can also be used as a watchdog timer.

A BCD/binary control bit in the mode register is used to operate the counter as either a binary or a BCD counter. The counter can operate as a 16-bit binary counter or a four-digit BCD counter. The maximum count in the binary mode is 65536; it is 10000 in the BCD mode.

A three-channel programmable interval timer (Intel 8254) can provide trigger pulses at any rate from 250 kHz to one pulse/hour. Two channels can be operated in fixed divider configurations from an internal 1- or 10-MHz crystal clock. The third channel may be uncommitted. It can provide a gated 16-bit binary counter that can be used for pulse counting, frequency or pulse generation, or delayed triggering, and in conjunction with the other channels for frequency and period measurement.

COUNTER/TIMER I/O BOARDS

A five-channel counter-timer interface board will plug into the IBM-PC/XT/AT and compatible machines (Figure 7-6). It offers the user five general-purpose 16-bit counters (AMD 9513). Various internal frequency sources and outputs can be selected as inputs for individual counters with software-selectable active-high or active-low input polarities. Each counter may be gated in hardware or by software. The counters can be programmed to count up or down in either binary or BCD. All five counters may be connected for an 80-bit counter.

Each counter is associated with a load register and a hold register. The load register is used to reload the counter to a predefined value and controls the effective count period. The hold register is used to save the count values without disturbing the count process, thus permitting the microprocessor to read intermediate counts. The hold register can also be used as a second load register to generate complex output waveforms.

Two of the counters are associated with additional alarm registers and comparators and with the logic for operating in a 24-hour time-of-day clock mode. For real-time control applications, the time-of-day logic will accept 50-, 60-, or 100-Hz input frequencies.

Each counter uses a single dedicated output. It can be inhibited when the output is not of interest. It is possible to configure both the input and the gating of individual counters with dynamic reassignment of inputs using software control. Thus, multiple counters can use a single input, and a single gate pin can be used to control more than one counter.

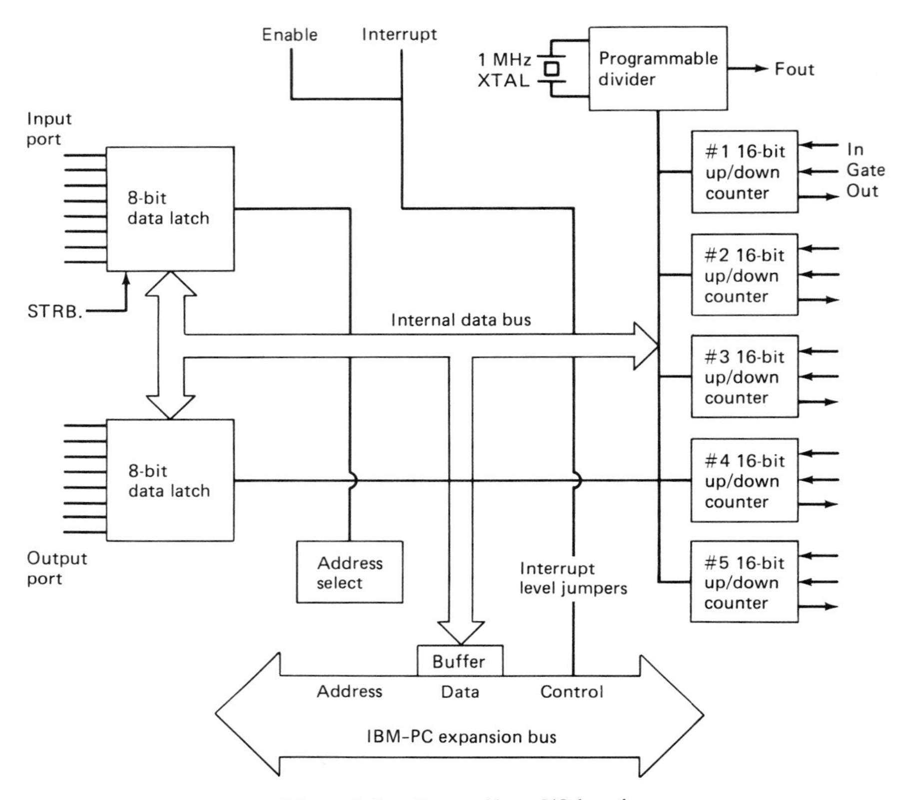

Figure 7-6 Counter/timer I/O board.

CONNECTOR INTERFACE

The counter and digital I/O signals are sent through a standard 37-pin D male connector that projects through the rear panel of the computer. The connector pin assignments are as follows:

Signal	Pin	Pin	Signal
Counter 2 input	19		
		37	Counter 1 gate
Counter 2 gate	18		
		36	Counter 1 input
Counter 3 input	17		
		35	Counter 1 output
Counter 3 gate	16		
		34	Counter 2 output

Group	Function	Pin	Pin	Function	Group
	Counter 4 input	15			
			33	Counter 3 output	
	Counter 4 gate	14			
			32	Counter 4 output	
	Counter 5 input	13			
			31	Counter 5 output	
	Counter 5 gate	12			
			30	Oscillator out (Fout)	
	Digital Common	11			
			29	IP0	
	OPO	10			
			28	IP1	
	OPO1	9			
			27	IP2	
	OP2	8			
			26	IP3	IP0-7 Digital inputs with latch
	OP3	7			
			25	IP4	
OP0-7 Digital outputs with latch	OP4	6			
			24	IP5	
	OP5	5			
			23	IP6	
	OP6	4			
			22	IP7	
	OP7	3			
			21	$\overline{\text{IP Strobe}}$	
	$\overline{\text{Interrupt enable}}$	2			
			20	+5 v power (from computer)	
	Interrupt input	1			

A/D CONVERTER WITH TIMER/COUNTER INTERFACE BOARDS

There are also interface boards with an A/D converter and timer/counters for the IBM-PC family that fit into a half slot. Connections are made through a standard 37-pin D male connector that projects through the rear of the computer. The following functions are typical:

1. An 8-channel, 12-bit successive approximation A/D converter with sample hold.

2. Full-scale input for each channel of +5 volts with a resolution of 2.44 millivolts.
3. Maximum A/D conversion time of 35 microseconds.
4. Throughputs of up to 4000 samples/second operating under BASIC.
5. An 8254 programmable timer/counter for periodic interrupts, event counting, pulse and waveform generation, frequency period, and pulse width measurements.
6. An external interrupt input that is jumper selectable for interrupt levels 2–7.

The half slot board includes status and control registers for interrupt handshaking. Figure 7-7 shows how a strain gauge can be connected to this type of board.

The interrupt input may be externally connected to the timer/counter or another trigger source. There are three 16-bit down counters in the 8254. One of these is connected to a submultiple of the system clock. All the functions of the other two counters are accessible to the user. Input frequencies as high as 2.5 MHz can be handled by the 8254.

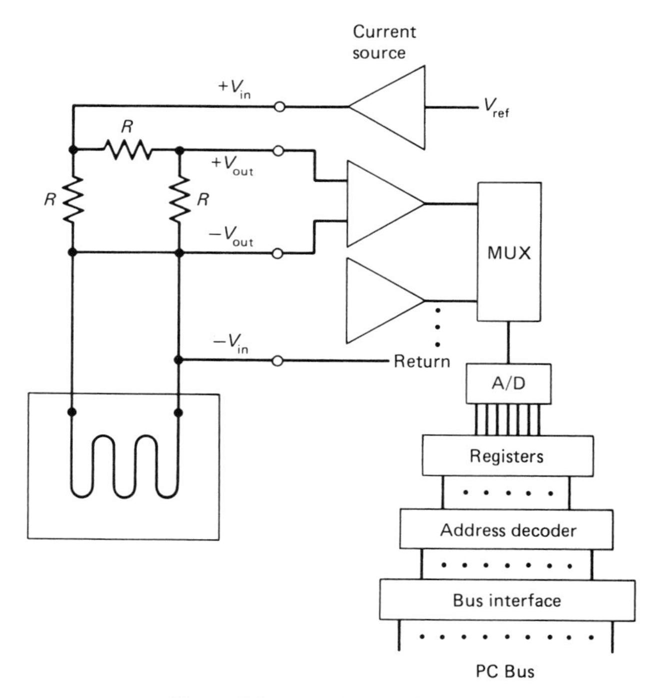

Figure 7-7 Interface strain gauge.

Programming

I/O is accessed using a call statement like the following:

```
CALL DA (MD%, DIO%, FLAG%)
```

The 18 modes of operation available to the programmer are represented by the integer variable MD%. Following is a list of the typical functions and modes:

Function	Operating Mode
0	Initialize, input base address
1	Set multiplexer channel low and high scan limits
2	Set multiplexer channel address
3	Read multiplexer channel address
4	Perform a single A/D conversion Return data and increment multiplexer address
5	Perform an N conversion scan after trigger Scan rate set by counter 2 or external strobe
6	Enable interrupt operation
7	Disable interrupt operation
8	Perform conversions on N interrupts and dump data in segment of memory
9	Unload data from memory segment and transfer to BASIC array variable
10	Set timer/counter configuration
11	Load timer/counter
12	Read timer/counter
13	Read digital inputs
14	Output to digital outputs
15	Measure frequency with timer/counter
16	Measure pulse width with timer/counter
17	Tag lower nibble of data with channel number

The integer variable DIO% represents input or output data. FLAG% contains the error codes. If FLAG% is returned nonzero, an error has occurred.

After initializing the unit, the following example program can be used to obtain 4000 samples/second:

```
xxx10 MD%=1                       set mux scan limits
xxx20 LT%(0)=2                    start scan on channel 2
xxx30 LT%(1)=5                    end scan on channel 5
xxx40 CALL DA (MD%,
      LT%(0), FLAG%)              setup
xxx50 IF FLAG% < > 0 THEN
      PRINT "ERROR"               checks for errors
```

```
xxx60 DIM A% (100)               dimension array of 100
xxx70 MD%=5                      mode 5
xxx80 TR%(0)-VARPTR
      (A%(1))                    start as element 1
xxx90 TR%(1)=100                 100 conversions
xx100 CALL DA (MD%,
      TR%(0), FLAG%)             do conversions, load array
xx110 IF FLAG% < > 0
      THEN PRINT "ERROR"         check for errors
```

In addition to an assembly-level driver utility, software includes linearization routines, graphics, and plotting routines for the display of data, data loggers, and a strip-chart recorder emulator.

The analog input channels can usually accept +5 or −5 volt signals with an overvoltage of +30 or −30 volts. The single-ended input current is usually less than 100 nA max at 25°C.

A differential reluctive pressure transducer can use a diode bridge which is filtered to provide a dc output, as shown in Figure 7-4. This signal can be sampled with a standard data acquisition facility, such as that provided by an I/O expansion board.

There are a wide variety of plug-in interface boards for data conversion and digital interfacing. They offer high data throughput, high accuracy, and low noise. Multilayer board construction with integral ground planes and the use of on-board dc/dc power supplies allows low-noise analog measurements and analog outputs at high data transfer rates.

PLUG-IN I/O BOARDS

These expansion boards use a slot in the PC's main board. Connections are made through a connector that extends out the rear of the computer. Field wiring may be connected to a mating connector, or a screw terminal board can be used to bring all connections out.

A typical configuration uses 16-channel single-ended or 8-channel differential inputs. The single-ended or differential mode is selected by an on-board switch. Software-selectable input ranges can include unipolar (0 to 10 volts) and bipolar (+10 volts) configurations. Gains can range from .5 to 100, and input resolutions can range from 3 millivolts down to 2 microvolts. A user-installed resistor allows the input range to be set to the requirements of the application.

A typical I/O board uses a successive approximation A/D converter with an 8.5-microsecond conversion rate. This conversion rate, combined with an 800-nanosecond sample hold time, allows conversions at a rate of over 100,000/second. In the DMA data transfer mode on a standard PC -or XT-compatible machine, these data can be written to memory at the full 100,000 sample/second analog sample rate. On the PC/AT and some of the slower AT-compatible machines, the data transfer rate is lower (60,000 samples/second), depending on the system clock speed and the number of wait states used by the computer and other software tasks being performed.

The A/D converter may be triggered using several methods: a software command, an on-board counter/pacer clock, or an external pulse. Transfer of the data can also be performed in different ways: software/program control, interrupt, or DMA.

Following are typical data throughput rates for different transfer modes:

Operating Mode	Throughput (Maximum) (Conversions/Second)
Program transfer to simple variable	200
Program transfer to array variable	4,000
Interrupt driven transfer	4,000
DMA transfer of scan channels	100,000
DMA transfer on single channel	100,000

On-board channel/gain queuing RAM will allow different channel gains to be stored. The board can be directed to follow the desired sequence. A typical sequence is as follows:

Sample No.	Channel No.	Gain
0	0	× 1
1	5	× 100
2	1	× 10
3	5	× 100
N	Return to Sample 0	
Where N can be in the range 2 to 2047		

This facility allows different channels to be sampled at different rates under hardware control. The timing of the system can be controlled by the clock and not by software, which can be interrupted by other system functions. The data points can be sampled at jitterless intervals, which is critical if frequency analysis is to be performed.

Since the RAM queuing list can be set to run under hardware control, the computer can perform other software tasks in the foreground while the board is acquiring data in the background.

Digital I/O of 16-TTL-compatible lines is usually available. These lines can be divided into one 8-bit output port and one 8-bit input port. They can be used for a wide variety of external triggering, switch sensing, and instrument control applications. The digital outputs can also be used as controls by accessory boards.

On-board five-channel counter/timers can be used to count the frequency output from Figure 7-3. Typical of the counter chips used are the AMD-9513 chips. Some of the counters may be connected to a crystal-controlled oscillator and are used to control A/D and D/A converter sample timing. Other counters can be connected to external signals and used as frequency or pulse generators; they can also measure frequency or pulse widths and count events.

Software and programming

Utility software includes machine language drivers to simplify programming. Programming is simplified by allowing functions to be accessed using simple calls. Drivers are available for interpreted and compiled BASIC, C, FORTRAN, Pascal, and Turbo Pascal.

Utility programs are available for the following procedures:

1. Installation and setup.
2. Calibration and testing.
3. Graphic displays.
4. Sensor linearization.

If a BASICA-compatible, machine-level driver (DAS) is loaded into memory at the start of the program, it can then be accessed using a BASICA "CALL" statement as shown below:

```
200 CALL DAS (MD%, DIO% (0), FLAG%)
HERE
```

MD% represents the mode of operation that will be used.
DIO% represents the data to be sent to the data acquisition system (DAS) or the data which will be returned.
FLAG% will return any errors found.

Typical software modes are as follows:

Mode	Function
0	Initialize DAS for base address, interrupt level, and DMA level.
1	Load channel/gain RAM.
2	View current channel/gain.
3	Perform a single A/D conversion, and load the data into a BASICA variable.
4	Perform an N conversion scan and store the data in a BASICA array. The sample rate can be set by the pacer clock or an external trigger, with a maximum sample rate of about 4000 samples/second.
5	Perform an N conversion scan, and store the data in memory under interrupt control.
6	Perform an N conversion scan under DMA control. The conversion rate can be set by an on-board pacer clock or an external trigger. The maximum conversion rate is about 100,000 kHz.

Mode	Function
7	Perform a single D/A conversion.
8	Perform N D/A conversions under program control. An array can be written consecutively on one of the D/As based on the pacer clock or an external trigger.
9	Perform N D/A conversions under interrupt control.
10	Perform N D/A conversions with DMA data transfers. Conversion timing is set by a pacer clock or an external trigger. Up to about 260,000 conversions/second are possible.
11	Cancel DMA- or interrupt-driven operations. Return control to program software.
12	Return current status of DMA- or interrupt-driven data transfers.
13	Transfer data from memory into BASICA arrays. This mode is needed, since interrupt- and DMA-driven conversions write to and read directly from memory locations without regard to BASICA variables.
14	Read the digital input bits.
15	Write to the digital output bits.
16	Set the analog trigger mode. This mode can cause any other mode to wait until a certain specified input condition is met before proceeding.
17	Initialize the counter/timer chip.
18	Set the 9513 counter's master mode register.
19	Set the counter N mode register.
20	Set the multiple counter control register.
21	Set the counter load register.
22	Read the counter N hold register.
23	Measure the frequency with a counter/timer.

A variety of accessory products are available. These will now be discussed. A screw terminal board allows each of the output connections to be brought out to screw terminals. A 16-channel expansion multiplexer allows expansion boards to be connected, providing up to 128 analog input channels. A cold junction temperature-sensing device allows accurate thermocouple measurements. Up to 4 channels of analog input data are allowed to be sampled simultaneously, within 30 nanoseconds. Boards can be added, allowing up to 16 channels of simultaneously sampled input.

A typical expansion board allows analog inputs using 8 differential or 16 single-ended channels. The mode is switch selectable with software-readable status outputs. A resolution of 12 bits is common, with an accuracy of 0.01% of reading or $\pm$ LSB. Input ranges include 0 to +10 volts, +10 volts, +5 volts, 0 to +1 volt, +0.5 volt, 0 to 100 millivolts, and +50 millivolts. Coding can include left-justified two's complement (bipolar) or left-justified true binary (unipolar). An overvoltage of +35 volts is allowable in a continuous mode per single channel.

The A/D converter is often the successive approximation type, with a resolution of 12 bits. A conversion time of 8 to 9 microseconds is typical. The A/D can be triggered by a software command, generated by an internal timer, or generated externally with a programmable edge.

The typical sample hold amplifier has an acquisition time of about 1 microsecond. An aperture of 0.3 nanosecond is typical.

Analog output channels are often of the 12-bit, nonmultiplying type with double buffering. Output ranges usually go up to +20 volts. Coding can be right justified true binary (unipolar) or right justified two's complement (bipolar). Output resistance is usually about 0.2 ohm. The maximum settling time for the 20-volt range is about 3 microseconds.

INTERRUPTS AND DMA

The following capabilities are typical:

Interrupt	
Level	IRQ2-IRQ7 software programmable
Enable	Software programmable
Source	ADC end of conversion
	End of ADC Queue
	Timer 2 terminal count
	External or DMA terminal count
DMA	
Level	1 or 3, software programmable
Enable	Software programmable
Termination	By interrupt on terminal count
	Autoinitialize
Transfer mode	ADC data to host
	From host to DACs paced by timer 2
Transfer rate	100,000 words/second (ADC)
	100,000 words/second (DACs)

A typical board may use up to five counter/timers. These are usually 16-bit, multimode programmable counters. Some counters are used to delay or pace the ADC start of conversion, while others are used to pace D/A converters during DMA or a real-time clock interrupt. Some of these counters are also available for external use, such as for frequency counting.

MULTIPLEXER BOARDS

A multiplexer offers an inexpensive way to expand the input capabilities of boards. Differential inputs can be multiplexed into a single A/D board input channel. Daisy chaining allows up to 256 analog input to be monitored. 256-channel operation is possible when the A/D board is set for 16-channel, single-ended operation. The channel to be enabled is usually determined by a 4-bit digital input (generated by the A/D board's digital output port). An on-board instrumentation amplifier can provide switch-selectable input gains.

A temperature sensor allows the monitoring of the cold junction temperature for measuring thermocouple inputs. Cold junction compensation (CJC) outputs

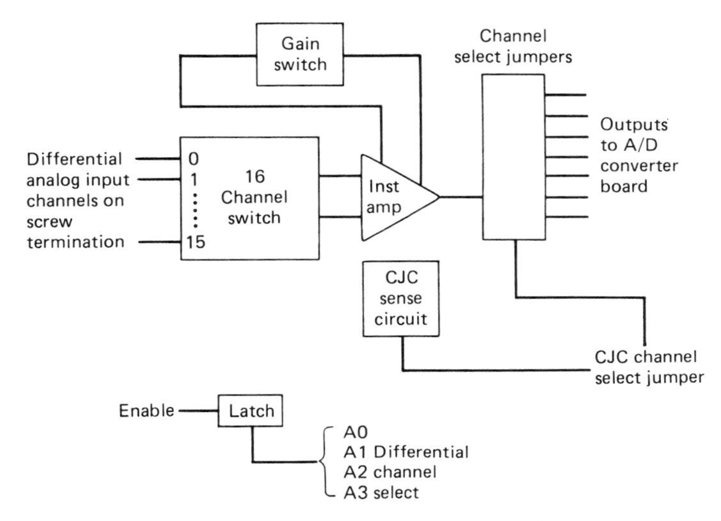

Figure 7-8 Expansion multiplexer block diagram.

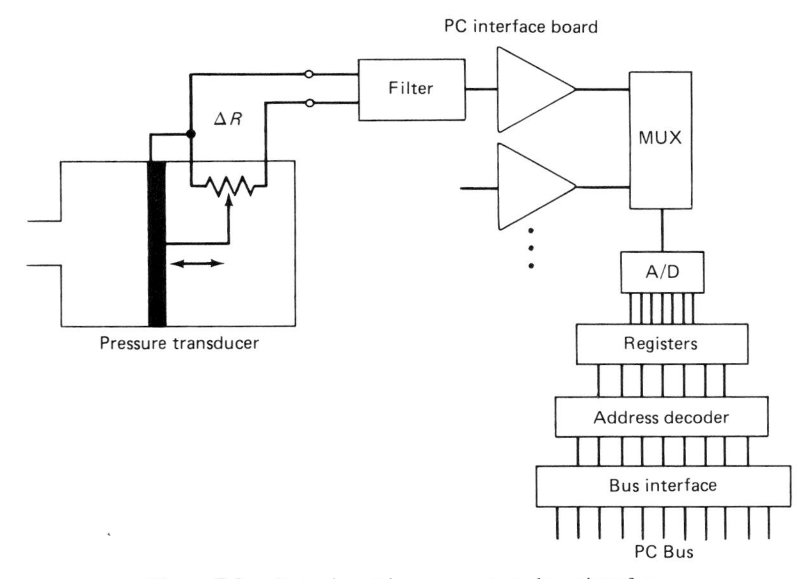

Figure 7-9 Potentiometric pressure transducer interface.

of 20 and 200 millivolts/°C are selected by a jumper. Jumpers are also available to connect an open thermocouple detection and a low-pass filter circuit to each input channel. A typical expansion multiplexer configuration is shown in Figure 7-8. Figure 7-9 shows how a potentiometric pressure transducer can be interfaced using an A/D expansion board.

8

Interfacing Temperature Measurement Sensors

This chapter will examine the most common types of temperature measurement, including resistance temperature detectors, platinum resistance sensors, semiconductors, and thermocouples.

The use of low-speed analog and digital I/O interface boards for interfacing these transducers is explained. Resistance sensor input boards and expansion multiplexers are also examined. We will investigate the more important specifications, such as sample rates, input range, and resolution.

INTRODUCTION

In the design and development of control systems, increased accuracy of measurement provides for a more successful and economical system operation. Whenever computer control or remote readouts are used, the long lines required can often cause problems due to the low output of some temperature sensors. Resistance thermometers can be used, since they have relatively high output and good linearity.

RESISTANCE TEMPERATURE DETECTOR CHARACTERISTICS

Resistance temperature detectors (RTDs) are based on the temperature dependence of the resistance of a material on electric currents. The resistance of metals increases with increasing temperature (positive temperature coefficient), while that of most semiconductor materials decreases (negative temperature coefficient).

Part of the total resistivity of metals is due to the impurities in the metal.

This is the residual resistivity, and it is lowest for the pure metals. The residual resistivity will change if the detector is used at too high a temperature or if the wire is contaminated by the environment or by materials in contact with the wire. The change is relatively independent of temperature. It is present when the resistance difference ($R_{T1} - R_{T2}$) between two temperatures remains constant while the ratio (R_{T1}/R_{T2}) decreases. These changes are irreversible.

Another part of a metal's resistivity is a function of deformation, depending on the physical state of the metal. In a well-annealed metal with low resistivity, when the crystal structure is stressed, the resistivity will increase. The resistance can be increased by mechanical shock vibration, thermal shock, or nuclear radiation at low temperatures.

The difference in thermal expansion coefficients between the wire and its supporting structure can cause stress on the wire. The RTD will then act as a strain gauge, as well as a temperature transducer. The effect is reversible, and the original resistance can be restored by reannealing the resistance element.

At very high temperatures, changes in the wire dimensions can occur due to evaporation. In this case, R_{T1}/R_{T2} remains constant but ($R_{TI} - R_{T2}$) increases. These effects should be considered when using RTDs.

When the elements are designed so that the wire remains annealed and strain-free for the operating temperature range, the resistance versus temperature characteristic will be similar to that of the wire alone. The repeatability can be 0.1°C for industrial sensors and 0.001°C for platinum wire standards.

Strain-free enclosure designs match the expansion characteristics and allow the wire to expand or contract freely as the temperature changes. In some applications, RTD devices tend to age in use until the stresses are equalized.

RTD MATERIALS

Copper is inexpensive for resistance temperature sensing and is one of the most linear metals for a wide temperature range. However, it oxidizes at moderate temperatures and has low stability and reproducibility compared to platinum. The low resistivity of copper is also a disadvantage.

Nickel is useful for temperatures from −100°C to +300°C. It is a low-cost material with a high temperature coefficient. Above 300°C, the resistance/temperature relation for nickel begins to change greatly. Nickel is also susceptible to contamination by certain materials, such as sulfur and phosphorus. Its resistance/temperature relation is not as well known or as reproducible as that of platinum.

Tungsten is a material whose characteristics are not as well known as those of platinum. Tungsten sensors tend to be less stable than platinum sensors. They do have good resistance to high nuclear radiation levels, as does platinum. The mechanical strength of tungsten allows it to be made into fine wires, and the sensing units can have high resistance values.

PLATINUM RESISTANCE SENSORS

Platinum resistance thermometers can be used with high accuracy from the triple point of hydrogen (13.81°K) to the freezing point of antimony (630.74°C). The resistance/temperature relationship of platinum is well documented, reproducible, and linear over this temperature range.

Platinum is chemically inert and not easily contaminated, since it does not readily oxidize. It can be used at temperatures of up to 1500°C with reduced accuracy. Platinum RTDs are more expensive than other resistance temperature sensors, although some industrial types are competitive in the same accuracy range.

Thin-film platinum temperature sensors consist of a film of platinum deposited on an insulating substrate. This provides high resistance in a small package with a relatively fast response time. The fabrication techniques are similar to those used to manufacture ICs. Platinum wire RTD elements can be delicate, and repeated thermal cycling will cause aging. The thin film deposition technique can be used with laser trimming of the resistance to provide a small, rugged sensor at a competitive cost.

EXCITATION TECHNIQUES

Resistance sensors do not produce voltages, so they must be energized from external power supplies. The resistance cannot have an arbitrary zero value at any temperature, since it must change value from R_{T1} to R_{T2}.

A zero-based output signal at the reference temperature is obtained when the sensor is used as an arm in a resistance bridge network. The Wheatstone bridge circuit (Figure 8-1a) can be used only when the sensor's resistance value is high

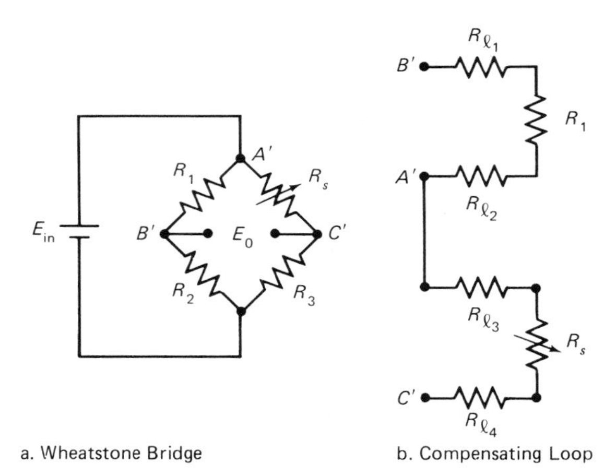

Figure 8-1 Resistance sensor bridge techniques. (a) Wheatstone bridge. (b) Compensating loop.

enough to mask the resistance due to the lead wires joining the bridge to the sensing element.

Cable runs in control systems may vary from a few feet to several hundred feet in order to connect the sensor to the bridge. If temperature-sensitive wire is used to connect the temperature-sensitive resistor, both of these sources must be measured.

Either ac or dc bridge excitation may be used, depending on how the output signal is to be used in the control system. Ac excitation eliminates thermal electromotive forces (emf's) in the sensor leads. However, reactive effects such as phase shift from the cable capacitance, as well as noise pickup and other interference due to cable coupling, can prove bothersome. Thermal emf's are normally insignificant.

Dc excitation allows unwanted noise to be filtered out. The accuracy and stability of the dc power source directly affect the accuracy and stability of the bridge output signal. Power supplies with stabilities of up to 0.01% are needed in some high-performance applications, although a stability of 0.1% is usually adequate for most control and automation applications.

A compensation loop can reduce the effects of lead wires. The basic technique is shown in Figure 8-2b. Three leads are used. One lead is connected in series with the power supply where any resistance changes have negligible effects on the voltage. The other two leads are placed in series with an opposite leg of the bridge where their effects tend to cancel. Another technique is to use separate connections for excitation and sensing leads, as shown in Figure 8-2 for two-, three-, and four-wire arrangements.

SEMICONDUCTORS

Transistors can also be used as temperature sensors. A common technique is to use the change in emitter-to-base voltage. When the device is properly characterized, a silicon transistor can provide a linear change in emitter-to-base voltage over the −40°C to 150°C range. The correlation between the temperature and the transistor base-to-emitter voltage is approximately 2 millivolts/°C.

Most transistor sensors of this type use a 400-millivolt total output change; the accuracy is ±2% or 5°C. The thermal time constant is typically 3 seconds in a flowing liquid and 8 seconds in moving air. A stable constant current source is normally used.

IC temperature transducers may contain the complete compensation circuit required, along with the sensor itself. Laser trimming can be used to produce a precalibrated temperature transducer for control and automation applications below 150°C. These sensors operate as a current source and provide an output of 1 microampere/°K.

Trimming of the internal calibration film resistor's manufacture allows a

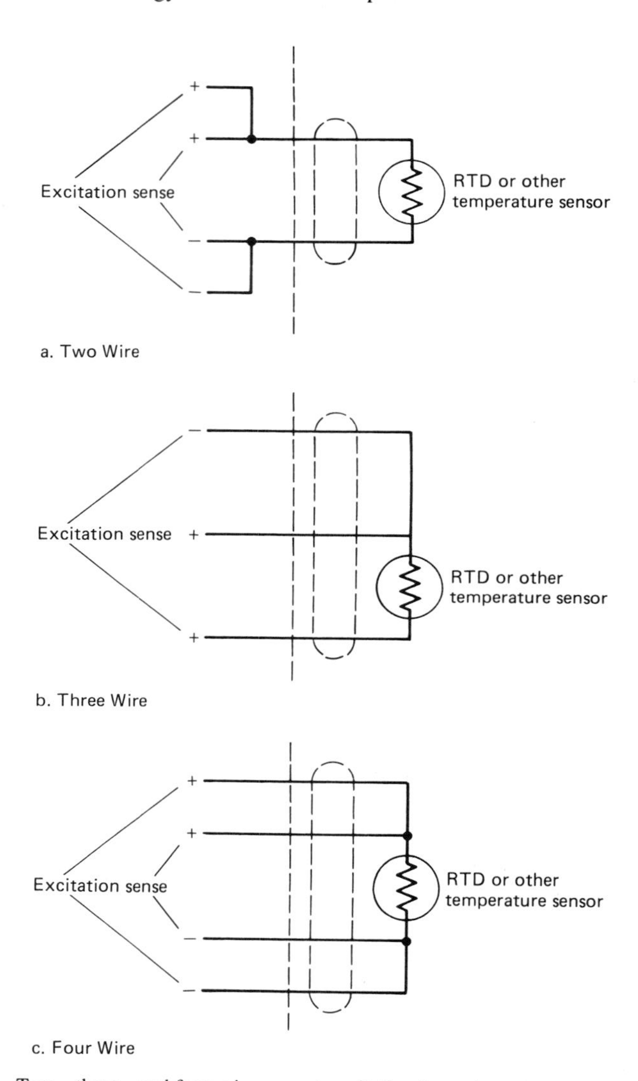

Figure 8-2 Two-, three-, and four-wire current excitation for resistance measurements. (a) Two-wire current excitation; (b) Three-wire current excitation; (c) Four-wire current excitation.

±0.5°C accuracy with a linearity of ±0.3°C over the temperature range. The output impedance is often greater than 10 megohms. This allows the rejection of much of the hum, ripple, and noise.

The high-output impedance allows these sensors to be used in remote sensing systems with lines that may be hundreds of feet long. These sensors can also be powered directly from 5-volt logic, which allows simple interfacing to microprocessor systems.

Some types of IC sensors output the temperature changes as a frequency change. A voltage-to-frequency converter is used to provide a digital pulse output. Celsius, Kelvin, or Fahrenheit scales can be obtained using external components. The pulse train is a function of the substrate temperature of the IC.

THERMOCOUPLES

The basic thermocouple circuit consists of a pair of wires of different metals joined or welded together at the sensing junction and terminated at the other end by a reference junction. This junction is maintained at a known temperature called the *reference temperature*. The load resistance of the signal-conditioning or readout equipment completes the circuit.

Suppose that a temperature difference exists between the sensing and reference junctions. Then a voltage will be produced which causes a current to flow through the thermocouple circuit. This voltage effect is due to the contact potentials at the junctions and is known as the *Seebeck effect*.

The wires between the sensing and reference junctions must be made of the same material. The connecting leads from the reference junction to the load resistance should be copper whenever the associated wiring in the signal-conditioning or output circuits is copper. This avoids error-producing junction potentials due to the different metals.

In addition to the Seebeck effect, there are the *Peltier* and *Thomson effects*. When current flows across a junction of two dissimilar conductors, heat that is not due to I^2R heating is absorbed or liberated from the wire. This is the Thomson effect.

The magnitude of the thermoelectric potential produced depends on the wire materials used and the temperature difference between the two junctions. Common thermocouple materials are Chromel-Alumel, Iron-Constantan, Copper-Constantan and Chromel-Constantan. Tables showing thermal emf versus temperature for most of the standard materials have been developed by the National Bureau of Standards, the Instrument Society of America, and many thermocouple manufacturers.

Most of thermocouples use a reference temperature of 0°C. In the laboratory an ice bath was used in the past. Modern techniques used in place of the laboratory ice bath include the following:

1. Automatic icepoint references.
2. Automatic oven references at temperatures other than 32°F (usually 150°F).
3. Electrical compensators.

Most electrical compensators use an ambient temperature thermocouple in a bridge circuit with ambient-sensitive elements. A typical type of the electrical compensator is shown in Figure 8-3. This is a useful method for single-channel measurement systems with accuracies of about ±1°F.

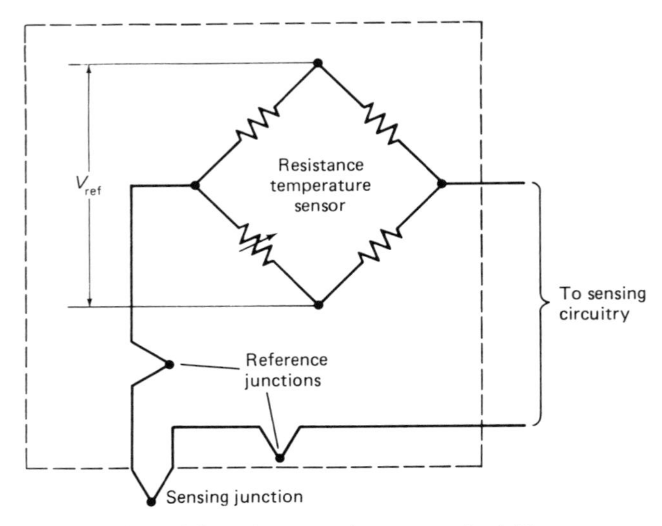

Figure 8-3 Thermocouple compensation bridge.

The oven type of reference is normally used as a reference for a number of thermocouple channels. Oven references, as well as ice-point references, are cost effective whenever a number of reference couples are required.

When the leads from the thermocouples are brought to a switch box, the contact resistances of the switches can cause errors. Even a small amount of nonconducting film on the contacts can cause a problem, since only a few millivolts are available to penetrate this film.

The following precautions should be taken when installing thermocouples:

1. The thermocouple should always be located in an average-temperature zone.
2. The thermocouple should always be completely immersed in the medium so that the true temperature can be measured.
3. The thermocouple should not be located near or in a direct flame path.
4. Whenever possible, the thermocouple should be located at a position where the hot junction can be seen from an inspection port in high-temperature-zone applications.
5. There should be no cable runs parallel to or closer than 1 foot to ac supply lines to avoid errors from induced currents.
6. Connections must be clean and tight in order to avoid errors due to contact resistance.

A self-heating error can occur if excessive power is dissipated in either the standard or the unknown. This can cause an error by heating the resistive element.

The use of a pocket or well tends to impede the transfer of heat to the sensing element and reduces the speed of response of the sensor. Agitation of the fluid around the sensing element tends to reduce this effect and improves the speed of response.

A potential problem is heat transfer. If there are two bodies and one is hotter than the other, there will be a net transfer of energy from the hotter body to the colder one.

When the bodies are at the same temperature and are enclosed in a space insulated from heat, each body is considered to radiate energy into the surrounding medium and to absorb energy continuously at the same time. Since there must be equilibrium, the processes balance one another and the temperature of each body remains constant. This theory is known as the *Prevost theory*.

If the walls of a chamber are lower in temperature than the internal hot gases, then a thermocouple placed in the hot gas stream will be hotter than the walls. The thermocouple radiates more heat than it receives and may measure a lower value than the true temperature of the gases.

Stem conduction errors occur when there is heat transfer along the stem of either the standard or the unknown couple. The thermocouple should always be immersed about 10 inches in a liquid for the larger thermocouple wires. Stem conduction error can be tested by measuring the temperature at a number of depths when the temperature is constant throughout.

A thermal lag error can exist if the standard couple and the unknown couple tend to respond to a changing temperature at different rates.

Most newer thermocouples react more rapidly than older units due to their smaller thermal mass. The typical response is in milliseconds, since wires 8/10,000ths of an inch in diameter are used to form the junction. These are then inserted into a quartz insulator, which is assembled into a 0.008-inch O.D. metal sheath or probe. The small-diameter junction (less than 0.002 inch) provides a faster response. Errors due to radiation and conduction are also reduced.

The thermopile is a combination of several thermocouples of the same materials connected in series. The output of a thermopile is equal to the total output of the number of thermocouples in the assembly. All reference junctions must be at the same temperature.

The differences between resistance thermometers and thermocouples include the following:

1. A reference junction temperature or compensating device is not required for the resistance temperature sensor.
2. A larger output voltage is obtained, with the resistance unit typically greater by a factor of 500 or more.
3. Recording, controlling, or signal conditioning equipment can be simpler, more accurate, and less expensive with the greater output signal of resistance units.

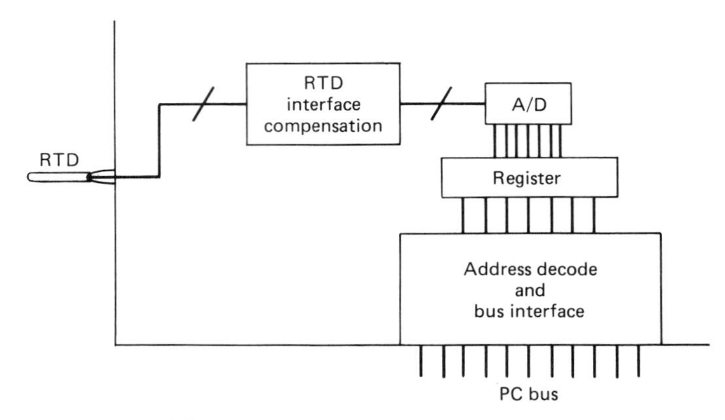

Figure 8-4 Using an I/O expansion board for RTDs.

4. Electrical noise is less of a problem with resistance sensors, allowing lead wires to be used.

The major problem in the use of thermocouples is the errors due to the spurious and parasitic emfs in the leads. This effect is responsible for the difference in precision between thermocouples and resistance thermometers. In moderate-temperature applications, the absolute accuracy of calibration, as well as the stability of calibration for resistance elements, can be higher by a factor of 10 to 100.

Variations in the state or composition of a wire in a thermocouple tend to produce Seebeck-generated emfs wherever a temperature gradient exists. This can become critical in applications where the temperature gradient on the leads is changing. Annealing reduces the causes of the effect but does not eliminate it. Periodic recalibration of the thermocouples is required to offset this effect. Sensitivity to small temperature changes is greater. RTDs also allow the output voltage per degree to be selected by adjusting the excitation current and/or the bridge design. The shape of the output curve versus temperature can also be modified by changing the resistance sensor bridge design.

An RTD can be interfaced using the analog input channel of an I/O interface expansion board. This type of board is discussed next. The essential functional elements are shown in Figure 8-4.

LOW-SPEED ANALOG AND DIGITAL I/O INTERFACE BOARDS

A typical low-speed, A/D I/O expansion board for the IBM-PC family is designed to allow use of the IBM-PC/XT/AT in low-speed, high-precision data acquisition and control applications. It combines in a single board most of the features needed for acquisition systems. These include plotting and storing graphs in real time and for later analysis, transducer and RTD linearization, and CRT-assisted calibration and setup procedures. A single CALL statement accesses all analog and digital I/O.

Other useful features include the following:

1. Analog input switchable filters.
2. Switch-selectable RTD interfaces.
3. Precision-adjustable voltage references.
4. Precision-constant current sources.
5. Two-, three-, and four-wire RTD bridge operation.

Digital I/O may be composed of one port of 8 bits and another of 4 bits. Each port can be independently programmed as an input or output port. Electromechanical relay boards and solid-state I/O module boards can use these ports to monitor and control ac and dc loads.

External interrupt control allows the user to select any of the IBM-PC interrupt levels (2–5) for programmed interrupt routines. This permits background data acquisition and interrupt control.

Some channels of the A/D converter can be equipped with instrumentation amplifiers by plugging them into sockets provided on the board and selecting the desired gains by switches. The instrumentation amplifiers can be used to provide gain scaling for thermocouples and resistance bridge transducers such as load cells and strain gauges. Other inputs are switch selectable for direct input for use with interfaces for two-, three-, or four-wire RTDs for temperature measurement.

Typical instrumentation amplifier gains and input voltage ranges are as follows:

Gain	Input Voltage Range
1	0.5 mV to 2 V
10	50 μV to 0.2 V
100	5 μV to 20 mV
1000	0.5 μV to 2 mV

The following specifications are typical: The analog input is ± 2 volts with an overvoltage of 120 volts (root-mean-squared), continuous for a single channel or 5 seconds for all channels. The common-mode range is the same as that of the analog input, with a common-mode rejection of 60 dB minimum. The input current is less than 1 nA at 25°C. The input filter is switchable with 30 dB attenuation at 60 Hz. It adds about a 1-second settling time for a full-scale step.

The following gains and related errors are typical for the instrumentation amplifiers:

Gain = 10, error 1.5% maximum, 0.6% typical
Gain = 100, error 0.5% maximum, 0.1% typical
Gain = 1000, error 1.5% maximum, 0.4% typical

The input current is usually less than 10 nA, with an average close to 2 nA at 25°C. The common-mode range is usually between −3 and +4 volts. The common mode rejection can be greater than 90 dB.

If the input signals are noisy, additional attenuation at 60 Hz may be needed; this can be switched into each channel filter on some boards. This mechanism engages a single-pole (resistor-capacitor) filter. The filter time constant can introduce a settling time penalty of up to 1 second for a full-scale step input.

If adjustable voltage reference outputs are available, they can be used for exciting strain gauge bridges or semiconductor voltage sensors (Figure 8-5). Constant current sources are useful for exciting RTDs, semiconductor temperature sensors' measuring resistances, or providing offsets.

RTD interfaces include built-in RTD (platinum resistance thermometer) interfaces. The temperature range is typically −200° to +650°C. A resolution of 0.2°C is possible using industry standard 100-ohm platinum RTD probes (alpha = 0.00385). Lead resistance compensation is included for most three- and four-wire probes. Excitation current is usually 1 milliamp.

If an integrating dual slope A/D converter is used, the resolution is usually 12 bits plus the sign bit, and the conversion rate is at least 30/second. Digital I/O is often via the PB and PC ports of an 8255-5 programmable peripheral interface.

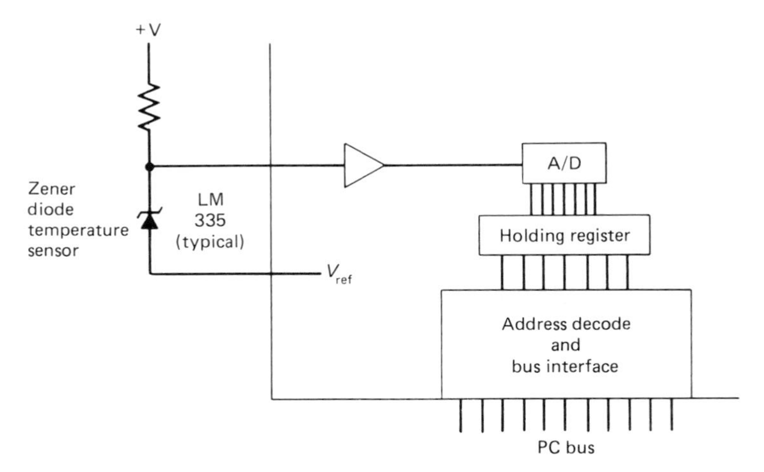

Figure 8-5 Using an I/O expansion board for semiconductor temperature measurement.

Some boards also contain a battery backed-up, real-time clock/calendar. This clock is used to update the PC's time and data functions automatically, thus eliminating DOS prompts for manual entry upon power-up. It is accurate to 2 seconds per month. Under normal use, the batteries will keep the module's time and date for 2 months without recharging on the system's power. The clock can also provide reference pulses of 1-second, 1-minute, or 1-hour intervals or output a frequency of 1024 Hz, which can be used as a source for interrupts or external timing applications.

Addressing

The board requires 16 consecutive address locations in I/O space. Some I/O address locations will be occupied by internal I/O and other peripheral cards. The I/O address can be set by the base address DIP switch to be anywhere in the PC-decoded I/O space. A good choice is to put base address at hex &H300, &H310 or &H320 (decimal 768, 784, 800). The procedure is as follows. First, type

```
A> BASICA BASWITCH
```

When the "Desired base address?" prompt appears, type in the address in decimal or IBM &H—format and press Return. The program will round the address to the nearest 16-bit boundary and check for possible conflicts with installed I/O devices.

Software

The I/O driver subroutine can be accessed by a single BASIC "CALL" statement. There are also utility programs for installation, graphics, a polynomial approximation that can be used to linearize transducers including RTDs, strain gauges, thermistors and thermocouples, and CRT-assisted calibration and setup procedures.

The single CALL statement can be used to access all analog and digital I/O on the board. The standard INP and OUT BASIC commands can also be used to access all I/O. The graphics package allows the user to plot predicted versus actual measured data from an experiment in real time and store the graph for retrieval at a later date.

Following is a typical CALL statement:

```
CALL (MD%, CH, DIO% (0), DIO% (1), BASADR%
```

The values in parentheses are integers.

In the CALL statement

MD% = Mode
CH% = Channel number
DIO%(8) = A Data I/O Integer Array
BASADR% = The baseaddress currently accessed

The following modes are typical, as shown in Table 8-1.

TABLE 8-1 Typical Low-Speed Analog and Digital I/O Interface Boards

Mode	Function	Channel	Data I/O Integer Array
0	Free scan of all analog inputs	Does not matter	DIO% (0–3) channel 0–3 data DIO% (4–7) channel 0–3 error flags DIO% (8) mode, channel error flags
1	Conversion on one analog input Data transferred when finished	0–3	DIO% (CH%) channel data DIO% (CH% + 4) channel error flag DIO% (8) mode, channel error flag
2	Conversion on all analog inputs Data transferred when finished	Does not matter	DIO% (0–3) channel 0–3 data DIO% (4–7) channel 0–3 error flags DIO% (8) mode, channel error flags
3	Conversion on all analog inputs Initiated by interrupt Data transferred when finished	Interrupt level Interrupt level (2–5)	Use mode 6 to obtain data after interrupt DIO% (8) mode, interrupt level error flag
4	Terminates interrupt processing Initiated by modes 3 and 5	Does not matter	DIO% (8) mode error flag
5	Free scan of all analog inputs Data collected on interrupt	Interrupt level (2–5)	Use mode 6 to obtain data after interrupt DIO% (8) mode, interrupt level error flag
6	Collect data after an interrupt using modes 3 or 5	Does not matter	DIO% (0–3) channel 0–3 data DIO% (4–7) channel 0–3 error flags DIO% (8) new-old data flag
7	Single-channel analog output	0 or 1	DIO% (0) channel 0 or 1 output DIO% (1) not used DIO% (8) mode, channel error flag

TABLE 8-1 (*Continued*)

Mode	Function	Channel	Data I/O Integer Array
8	Output data to both analog output channels	Does not matter	DIO% (0) channel 0 output data DIO% (1) channel 1 output data DIO% (8) mode error flag
9	Digital I/O on 8-bit port PB and 4-bit port PC	0-PB output, PC input 1-PB output, PC output 2-PB input, PC input 3-PB input, PC output 4-PB strobed output, PCO-2 handshake PC3 input 5-PB strobed input, PCO-2 handshake PC3 input	DIO% (0) If PB output, data (0–255) If PB input, data (0–255) DIO% (1) 1 PC output, data (0–15) IF PC input, data (0–15) DIO% (8) data range, mode, channel error flag
10	Enable/disable clock output pulse	Enabled, CH% = 1 Disabled, CH% = 0	DIO% (8) mode error flag

Following is a BASIC programming example for temperature control.

```
 10 CLEAR 32768!
 20 DEF SEG=0
 30 SG=256*PEEK (&H511) +PEEK (&H510)
 40 DASCON=0
 50 SG=(32768!/16)+SG
 60 DEF SEG=SG
 70 BLOAD "DASCON.BIN", 0
 80 DIM DIO% (8)
 90 DIM REL% (8)
100 BASADR%=&H300
110 MD%=0
120 RD%=9
130 RH%=0
140 CLS
150 FOR J=1 TO 100
160 FOR I-1 TO 1000
```

```
170 NEXT I
180 CALL DA (MD%, CH%, DIO% (0), DID% (1), BASADR%)
190 IF DIO% (2)/5>20 THEN RIO%(0)=1.GOTO 220 'TURN ON
    Air Conditioner
200 IF DIO% (2)/5<19 THEN RIO% (0)=2.GOTO 220 "Turn on
    Heater
210 RIO% (0)=0
220 CALL DASCON (RD%, RH%, RIO% (0), RIO% (1), BASADR%)
230 NEXT J
240 END
```

The above example shows how the board can be used to control temperature. Each bit is equivalent to 0.2°C; thus, 5 bits is equivalent to 1°C. Program line 190 examines the value of the temperature read from the RTD on channel 2's analog input. If the temperature is greater than 20°C (68°F), the air conditioner or chiller is switched on. On line 200 the temperature is less than 19°C (66.2°F), so the heater is switched on. If the temperature is between 19° and 20°C, a "zero" is written into the port connected to the relays controlling the environment deenergizing both units.

RESISTANCE SENSOR INPUT BOARDS

There are also analog input boards for most variable-resistance sensors. They can be used for the measurements of temperature sensors operated in a current excitation mode or in standard bridge configurations. Sensors in these categories include two-, three-, and four-wire RTDs, thermistors, strain gauges, and variable potentiometer devices. Some are designed with cold junction compensation for thermocouple measurements.

A typical board accepts several input sensors and multiplexes these into a single A/D board channel. The channel is usually selected by digital control lines provided as outputs from the host A/D board. Each input channel will have its own differential input amplifier with jumper-selectable input gains. Also included for each input channel is a precision current source for current-excited measurements and a jumper-selectable, precision-reference voltage for voltage-excited sensors.

These boards often provide thermocouple cold-junction compensation (CJC) circuitry that measures the actual input terminal temperature and allows the linearization software to subtract out the CJ error.

When used for bridge measurements, the board allows the bridge completion resistors to be mounted on or off the board. Two user-installed resistors can be set to form one-half of the bridge. A relay and calibration resistor allow shunt calibration.

Typical gains for instrumentation amplifiers are given below, along with the offset drift, common-mode rejection, and gain nonlinearity.

Gain	Input Offset Drift (μv/°C)	Common-Mode Rejection (dB)	Gain Non-Linearity (%)
250	5.1	100	0.075
100	5.1	100	0.075
25	6	100	0.045
10	6	100	0.045
2.5	15	94	0.045
1	15	94	0.045

Typical settling times along with the temperature coefficients are as follows:

Gain	Settling Time (μs)	Gain Temperature Coefficient (ppm max)
250	350	20
100	350	20
25	35	15
10	35	15
2.5	3.5	10
1	3.5	10

The following thermocouple types can be supported: J, K, T, E, S, R, and B with a CJC of +24.4 millivolts/°C (.1°C/bit).

The input bias current is typically 5 nA. The analog output voltage is usually less than 5 volts, and the current is less than 10 mA.

EXPANSION MULTIPLEXERS

These boards allow the expansion of any analog input to 4 channels. They can supply CJC for thermocouple inputs, and shunt terminals are provided for current measurements.

An expansion multiplexer allows isolated inputs to be connected to an analog input board (Figure 8-6). The board uses digital output bits from the A/D converter board to decide which board is to be monitored and which channel is to be enabled. Each input is isolated from all other inputs and from the A/D converter board. Each input includes an instrumentation amplifier that can be set for input gains with a user-installed resistor.

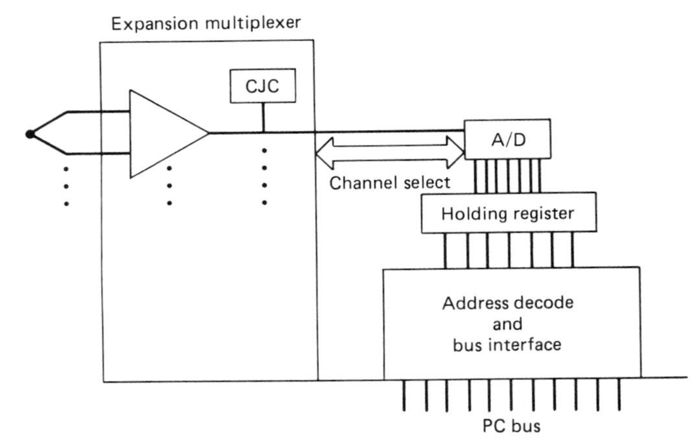

Figure 8-6 Using an I/O expansion board for thermocouple temperature measurement.

The multiplexer is controlled by 4 digital output bits from the master A/D board. Each multiplexer has an address selection switch. To allow expansion capability, the most significant 2 bits of the digital control lines are used as an enable control. This allows up to four boards to be connected to the same A/D input channel (one will be enabled and the other three will be disabled). The two LSBs are used to control which channel is enabled.

Thermocouple measurements can be handled with a CJC device which provides an input signal of 24.4 millivolts/°C. With a 12-bit A/D converter and a 10-volt full scale, this corresponds to 10 bits/°C. The temperature measurement feature allows the cold junction error to be subtracted out in software. Thermocouple linearization routines written in BASICA are available for J, K, T, E, S, R, and B thermocouples.

A problem can occur when monitoring thermocouples. For accurate thermocouple measurement, it is necessary to know the temperature of the connection of the thermocouple wire to the input terminal. This cold junction temperature error is then compensated for in the software. However, each cold junction temperature measurement device requires an A/D input channel. If each MUX is also monitoring the cold junction temperature, then some of the available input channels are taken, and the maximum number of thermocouples that can be monitored is reduced. If all boards are at approximately the same temperature, the cold junction monitoring device can be connected on one bank, with only a slight loss of accuracy.

Thermocouple gain settings

The following list shows recommended gain settings for thermocouples over their operating range.

Thermocouple Type	Maximum Output (mV)	Maximum °C	Suitable Gain
J	43	760	100
K	55	1,370	50
T	21	400	200
E	76	10,000	50
S	19	1,760	200
R	21	1,760	200
B	14	1,760	200

Higher gains can be used for less than the full-scale span. Gains are based on ± 5-volt output.

Typical input gains include X1, 2, 10, 50, 100, 200, and 1000. The input impedance is usually greater than 5 megohms, and the bandwidth is greater than 5 kHz.

ANALOG BOARD SPECIFICATIONS

The following factors are important in selecting the right A/D board for a particular application:

1. Number of input channels.
2. Input scan rate in samples/second.
3. Resolution or accuracy, usually referred to in bits of resolution.
4. Input range specified in full-scale volts.
5. Single ended or differential inputs.
6. Signal conditioning available.
7. Other features, such as D/A channels, digital I/O, and counter/timers.

A/D SAMPLE RATE

Most suppliers specify the maximum sample rate as total board throughput. However, multichannel analog input boards typically use a single A/D converter and an input multiplexer. This circuit configuration is shown in Figure 8-7.

When a single channel is being sampled, the maximum throughput applies. If more than one channel is being sampled, the maximum sample rate per channel can be determined as follows:

$$\text{Maximum Rate/Channel} = \frac{\text{Maximum A/D Board Sample Rate}}{\text{No. of Channels Sampled}}$$

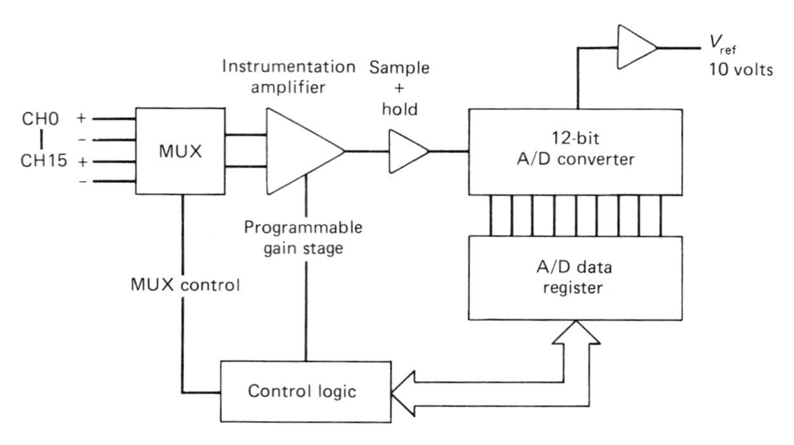

Figure 8-7 Typical A/D input stage.

For example, suppose that the sample rate is 50 kHz. One channel can be sampled at the full 50 kHz. Two channels can be sampled only at 50/2 or 25 kHz each. Four channels gives 50/4 or 12.5 kHz.

INPUT RANGE AND RESOLUTION

Input resolution of a data acquisition system is usually specified in bits. The conversion from bits of resolution to actual resolution is

$$\text{Resolution} = \text{One Part in } 2 \text{ (No. of bits)}$$

For example, a 12-bit A/D converter's resolution is found as follows:

$$\text{Resolution} = \text{One Part in } 2^{12}\text{, or } 4096$$

The input range is typically specified as a voltage range. The specifications may be either unipolar (0 to +5 volts, 0 to +10 volts) or bipolar (−5 to +5 volts, −1 to +1 volt). To determine the minimum A/D resolution in volts, divide the full-scale input voltage by the resolution found above. The full scale range of a 0 to N volt input is N volts, while the full-scale input range of a $-N$ to $+N$ volt input is $2 * N$. Thus a 0 to 5-volt input has a 5-volt full-scale range, while a −5 to +5 volt input has a 10-volt full-scale range. The input resolution in volts can be determined as follows:

$$\text{Voltage Resolution} = \frac{\text{Full-Scale Input Range}}{\text{Parts of Resolution}}$$

For example, a 12-bit A/D converter with a −5 to +5 volt input range will give

$$\frac{10 \text{ (Volts Full Scale)}}{2^{12}} = 0.00244 \text{ volts}$$

This says that 2.44 millivolts is the smallest input voltage change that the system can detect.

Another consideration is input gain. Many A/D systems offer inputs with a variety of input gains. To determine the overall minimum detectable input voltage, divide the voltage resolution found above by the gain. The combined equation to determine the overall input resolution is

$$\text{Absolute Resolution (in Volts)} = \frac{\text{(Full Scale Input)/(Gain)}}{2 \text{ (No. of Bits of Resolution)}}$$

For example, a 12-bit A/D converter with a −10 to +10 volt full-scale input range and an input gain of 100 has a voltage resolution of

$$\frac{\text{(20 Volts Full Scale)/(100)}}{2^{(12 \text{ bits})}} = 48 \text{ Microvolts}$$

Voltages larger than the specified input range of an analog input can be measured by using a voltage divider. This voltage division is accomplished by two resistors, as shown in Figure 8-8a. For example, suppose that a 0- to 50-volt input source is to be measured by a 0- to 10-volt input. Let $R2$ = 10 kohm; in

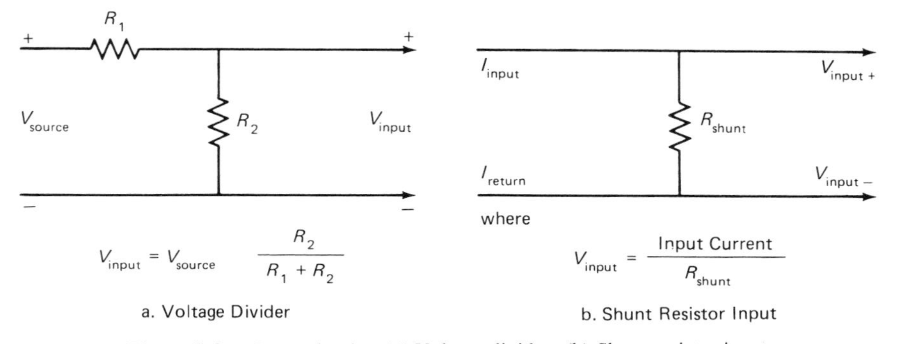

Figure 8-8 Input circuits. (a) Voltage divider; (b) Shunt resistor input.

most cases you can use any common value between 10 and 100 kohm. Then

$$V_{\text{Input}} = 10 \text{ volts} \quad \text{and} \quad V_{\text{Source}} = 50 \text{ volts}$$

Solving for $R1$ gives 40 kohm.

Most A/D systems measure voltage, so it may be necessary to convert a current source into a voltage source by using a shunt resistor. Figure 8-8b shows a typical shunt resistor input circuit.

The most common current input signal is 4–20 mA. The 4-to 20-mA loop is often used in process monitoring because of its high noise immunity. The shunt resistor required to connect a 4-to 20-mA loop to a 0–5 or +5 volt input can be found by applying the approach used above. Thus, we find:

$$R_{\text{shunt}} = \frac{5 \text{ volts}}{20 \text{ mA}} = 250 \text{ ohms}$$

This shows that with a 250-ohm shunt the 4- to 20-mA input will be converted into a 1- to 5-volt input. Precision resistors should always be used, since the accuracy of the measurement will be directly related to the resistance accuracy.

Many screw terminal accessory boxes and system expansion multiplexers use plated-through holes that allow the installation of shunt resistors. Some signal conditioning modules are designed for monitoring 0- to 20-mA and 4- to 20-mA inputs.

9

Interfacing Displacement and Proximity Sensors

This chapter is concerned with the interfacing of resistance, capacitive, inductive, and digital displacement sensors. The use of the 8255 programmable peripheral interface is investigated, and its operation in digital I/O board applications is explained.

The use of high-speed parallel digital interface boards for digital encoder applications is discussed. The characteristics of the different operational modes are explained. We also investigate digital I/O boards and their operation, as well as high-speed parallel digital interface boards, and explain the use of I/O mapping for these boards.

APPLICATION INTERFACING

Sensors for the measurement of displacement and proximity are an important part of many automation and control applications. Resistive, capacitive, inductive, or optical techniques can be used to measure linear or angular displacements between the point or object being sensed and a reference or fixed point or object. Proximity sensors measure linear or angular motion without any mechanical linkage.

The output from a displacement or proximity sensor may be the analog or digital voltage equivalent of the absolute distance being sensed, or it may be a function of the distance from a given starting point. Many of the common displacement and proximity techniques are used as primary sensors in other transducers, such as those used to measure pressure.

TABLE 9-1 Disadvantages of Resistance Sensors

Finite resolution for wire-wound devices
Friction and limited life due to contact wear
Increasing electrical noise due to wear
Sensitivity to shock

RESISTANCE SENSORS

Variable resistors can be used as voltage or current dividers. The displacement information is a function of the wiper or movable arm of the resistor as it slides over the resistance element.

Sensors are available for both linear and angular measurements, including fractional and multiturn configurations for rotary applications. The potentiometric displacement transducer is a low-cost device, but it is subject to wear due to the mechanical wiper. The major disadvantages of resistance sensors are summarized in Table 9-1.

Resistance elements can be wirewound, carbon ribbon, or deposited conductive film. Excitation can be either ac or dc. Usually no output amplifiers are required, and the output can take on a variety of functions, such as linear, sine, cosine, exponential, or logarithmic. The characteristics of resistance sensors are summarized in Table 9-2.

The interfacing of resistive displacement sensors is done using the same techniques discussed for potentiometric pressure transducers in Chapter 8. The essential components of the interface are shown in Figure 9-1.

CAPACITIVE SENSORS

Capacitive sensors are typically used for linear instead of angular proximity measurement. In these devices, either the dielectric or one of the capacitor plates is movable.

The capacitive proximity sensor sometimes uses the measured object as one

TABLE 9-2 Typical Resistance Sensor Characteristics

Resolution 0.2%
Linearity ±1%
Repeatability ±0.4%
Hysteresis 0.5–1%
Temperature error ±0.8.

Note: Higher-precision instruments are available with total errors in the range of ±.5%.

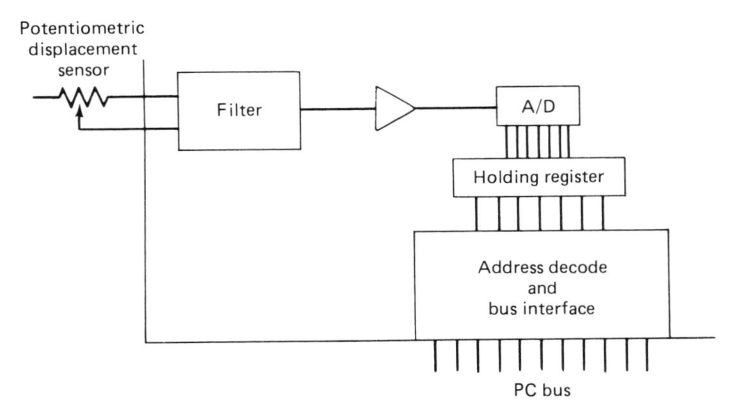

Figure 9-1 Interfacing resistance displacement sensors using an I/O expansion board.

plate, while the sensor acts as the other plate. The capacitance changes as a function of the area of the plates, the dielectric, or the distance between the plates.

Capacitive transducers are sometimes packaged with internal signal-conversion circuitry which provides a dc output. Accuracy for small displacements can be in the range of .25%, and accuracies of up to .05% are possible for high-cost devices.

Capacitive sensors tend to be accurate for small displacements. They are relatively small devices with an excellent frequency response. They suffer from their sensitivity to temperature, as well as the need for additional electronics to produce a usable output.

Capacitive sensors have good linearity and output resolution. Their drawbacks are the temperature and output sensitivity, which require the amplification circuits to be located close to the transducer.

In a typical capacitor transducer circuit, an ac voltage is applied across the plates and an emitter follower or other amplifier is used to couple the detected changes. The capacitor can also be made part of an oscillator circuit in which it causes a change in the output frequency. The interface to the PC can use an expansion board, as discussed in Chapter 7.

INDUCTIVE SENSORS

Inductive sensors may use a number of coil configurations. A single coil can be utilized in which a change in the self-inductance of the coil is monitored. Multiple-coil sensors employ the change in magnetic coupling or reluctance between the coils. Single-coil displacement units allow a movable core to change the self-inductance, while single-coil proximity sensors use the magnetic properties of the object itself to modify the self-inductance. A small change in inductance is usually sensed with a bridge circuit or oscillator.

Multiple-coil inductive sensors use the differential transformer technique. The linear variable differential transformer (LVDT) uses three windings and a movable slug or core to sense linear displacement. A typical LVDT configuration is shown in Figure 9-2 (upper left).

In an LVDT, the secondary windings are wound so as to produce opposing voltages and are connected in series. When the core is in a neutral or zero position, the voltages induced in the secondary windings are equal and opposite, and the net output voltage is minimal.

When the displacement of the core increases, the magnetic coupling between the primary coil and one of the secondary coils increases, and the coupling between the primary coil and the other secondary coil decreases. The net voltage increases as the core is moved away from the center position, and the phase angle increases or decreases, depending on the direction.

Differential transformers are also available in an angular arrangement in which the core rotates about an axis. Variations in the winding configurations are used in synchros, resolvers, and microsyns.

Figure 9-2 also shows the interface to the PC bus using an I/O expansion board and a demodulator circuit that can produce a dc output from the transformer windings. Typical input frequencies range from 60 Hz to 30 kHz. Some units have built-in conversion electronics to allow dc input and output.

Inductance bridge sensors are designed to use two coils with a moving core. The core changes the inductance of the coils which form one-half of an ac bridge. These sensors are available in both linear and angular configurations.

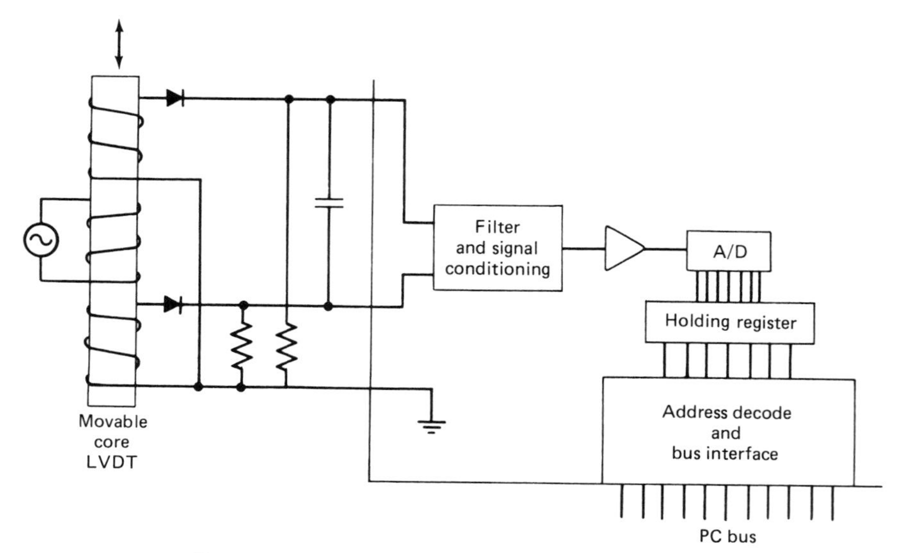

Figure 9-2 Displacement measurement using an LVDT.

The temperature error for variable reluctance transducers is typically 2% for a 100°F change. Transducers without dc-dc conversion circuitry can generally operate up to 350°F, while transducers with conversion circuitry are usually limited to −65 to +200°F. Typical linearity is ±0.5%, while repeatability and hysteresis can be less than 0.2%. Variable reluctance transducers have good shock and vibration characteristics, along with a good dynamic response.

DIGITAL DISPLACEMENT SENSORS

Digital position transducers are popular in control and automation applications, since they simplify interfacing to displays and data acquisition equipment. These devices provide either digital frequency or digital-coded output which is a function of either displacement or proximity.

Transducers with a frequency output use a frequency control that is a function of sensing movement. Transducers with digital-coded output detect position and convert this deflection into a digitally coded word.

The digital output or series of pulses may be produced in displacement and proximity sensors using changes in electrical conduction, induction, or photoelectric conduction. Conducting encoders use brushes or wipers to detect the position of a coded disk or plate. When a single sensing track is used, a series of pulses is produced as the disk or plate is moved. Direction sensing is accomplished by adding another track which is offset to produce the sequencing logic. Counters are needed to count the pulses and perform the conversion to angular or linear measurement. Multiple track encoders provide digital or binary-coded output which is a function of the absolute angular or linear position.

Rotary encoders evolved from complex rotary switches which produced multiple outputs or combinations of outputs as the switch was rotated. The switch outputs evolved into coded patterns which indicate the position of the shaft.

The early shaft encoders used carbon or metal wipers and brushes which made contact on conductive disks. The disks were eventually plated with precious metal alloys for conduction. The metal wipers were also plated with precious metal alloys to reduce arcing, corrosion, and noise. The use of precious metal alloys is costly, and these contacting encoders still had a limited life. Thus, they were used primarily in applications where replacement was easy and cost was not critical.

Magnetic displacement sensors use linear or rotary gears of ferromagnetic material which provide pulses due to a change in linear or angular position. Direction sensing is accomplished by shaping the gear teeth in an asymmetrical pattern in order to modify the output waveform.

Photoelectric encoders use a light source and a detector with discs or plates of transparent and opaque windows. The switching function is similar to that of conducting encoders, except that switching is accomplishing by breaking the path of the light beam between the light source and the detector. Optical encoders are an improvement over mechanical contact devices, since switching is done using a light beam. The transparent and opaque windows indicate either the relative or absolute position of a rotary or linear shaft.

Early optical encoders used a hot filament bulb for the source with or without a lens, and the light detector was an open semiconductor chip. The hot filament bulb was subject to premature failure due to vibration and voltage surges. Most of these units were designed for easy bulb replacement and were used for non-critical applications. Accuracy was a function of brightness, which increased the size and power of the bulb and reduced the reliability.

More recent devices use LEDs, which provide a light source which is not damaged by vibration. Their operating life exceeds several hundred thousand hours. Power consumption as well as size have been reduced.

Position encoders can be incremental, which generally use one or two tracks with equal spaces of transparent and opaque windows. This produces a series of pulses as a function of position, and direction is detected by comparing the two outputs from multiple-track devices. Absolute position linear encoders requires no signal processing to determine linear position. It utilizes a multiple track and arrays of emitters and detectors.

Figure 9-3 illustrates the type of interface required to a PC for an incremental position encoder. This type of counter/timer expansion board is discussed in Chapter 7. An absolute position encoder requires a parallel expansion board for interfacing to a PC, as shown in Figure 9-4.

The interface to the PC can be accomplished using the expansion boards that will now be discussed. These are the parallel digital I/O interface boards that use parallel I/O chips like the 8255.

A parallel digital I/O card can provide TTL/DTL-compatible digital I/O lines, interrupt input and enable lines, and external connections to the PC's bus power supplies. It is a useful interface for parallel I/O devices such as encoders, displays, and user-constructed systems and equipment.

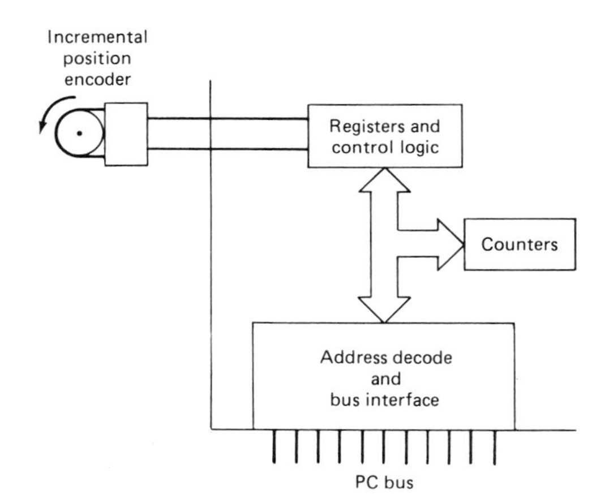

Figure 9-3 Incremental position encoder interface using a timer/counter I/O expansion board.

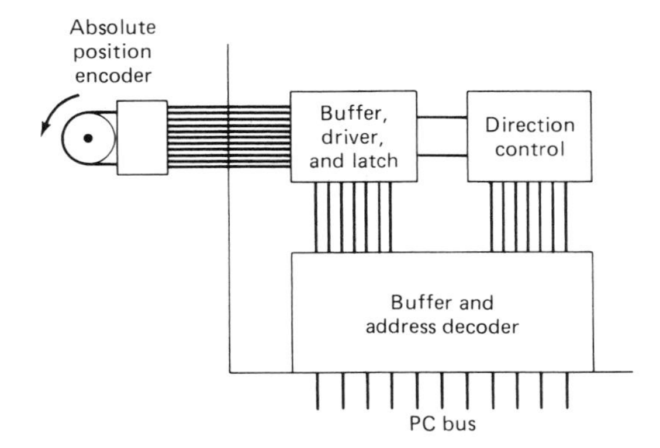

Figure 9-4 Absolute position encoder interface using a parallel I/O expansion board.

THE 8255 PROGRAMMABLE PERIPHERAL INTERFACE

This interface is a parallel I/O chip with four registers. The interface to the microprocessor is made up of a chip select pin (CS); two address pin (A0 and A1); the three control pins READ (RD), WRITE (WR), and RESET; and eight bidirectional data pins (D0 through D7).

There are four groups of I/O pins: Port A, Port B, Port C Upper, and Port C Lower. Ports A and B are 8 bits wide, while Ports C Upper and Lower are 4 bits each. This provides a total of 24 I/O pins. Operation of the 8255 registers depends on the state of the control inputs, as shown in Table 9-3. The three modes of operation are as follows:

TABLE 9-3 Register Access Codes

Pin					
A0	A1	$\overline{RD}$	$\overline{WR}$	$\overline{CS}$	Operation
0	0	0	1	0	Read Port A pins
0	1	0	1	0	Read Port B pins
1	0	0	1	0	Read Port C pins
0	0	1	0	0	Write to Port A
0	1	1	0	0	Write to Port B
1	0	1	0	0	Write to Port C
1	1	1	0	0	Write to control register
X	X	X	X	1	Not recognized
1	1	0	1	0	Illegal
X	X	1	1	0	Not recognized

Note: X indicates that the pin may assume either level.

1. Mode 0 is the basic input and output mode for all 24 I/O pins; it is also called the *bit I/O mode*.
2. Mode 1 provides strobed I/O. Port C is used for control and status.
3. Mode 2 is the bidirectional data bus mode. Five bits on Port C are used for handshaking.

The following rules apply to the way the ports may be set:

1. Ports A and B can be set to the different modes as needed.
2. The Port C Upper configuration depends on how Port A is set.
3. The Port C Lower configuration depends on how Port B is set.

Programming is done by sending a control word from the microprocessor through the 8255 data bus. In the mode definition control word, bit 7 must be written as a 1 to set the mode active flag. Bits 5 and 6 are used to set the Port A mode. Bit 4 is used as a Port A data-direction bit; it determines whether the Port A pins are inputs from or outputs to the 8255A. Bit 3 determines the direction of the Port C Upper pins. Bit 2 of the 8255 control register is the mode-select bit for Port B. Port B cannot be configured to mode 2, the bidirectional-bus mode. Bit 1 determines the direction of the Port B pins, and bit 0 determines the direction of the Port C Lower pins.

In addition to this control word, an 8255 bit set/reset control word is defined. In this configuration, Port C pins are set or reset for status and control for Ports A and B. Bit 7 must be 0 to differentiate this configuration from the mode definition format. Bits 4, 5, and 6 are not used. Bits 1 through 3 specify which Port C bit is to be used, and bit 0 sets the state of the bit.

MODE 0 CHARACTERISTICS

In this mode, the 8255 operates in a bit I/O configuration. The pins are defined as outputs or inputs by the microprocessor and are changed by the microprocessor with another instruction. The current pin voltages are read by the microprocessor. The outputs in mode 0 are latched and single-buffered. The inputs are not latched. All bits of Port A must be set to be either inputs or outputs. This also holds for the pins of Ports B, C Upper, and C Lower. The data direction must be the same for all signal lines in a given port. Ports A or B cannot be set as six inputs and two outputs. Port C is split and can be set as eight outputs, eight inputs, or four inputs and four outputs. Mode 0 operation is like a zero-wire handshake, since no timing information is passed between the CPU and peripheral.

MODE 1 CHARACTERISTICS

In this strobed-I/O mode, the data ports are A and B, and the C ports serve as control and status for the strobed handshake. Port A uses Port C bits 3, 4, and

5 for handshaking, and Port B uses Port C bits 0, 1, and 2. The remaining Port C, bits 6 and 7, are available for bit I/O.

For the input handshake, the three lines used are called *strobe* ($\overline{STB}$), *Input Buffer Full (IBF)*, and *Interrupt Request (INTR)*. For Port A, $\overline{STB}$, IBF, and INTR correspond to Port C bits 4, 5, and 3, while for Port B they correspond to Port C bits 2, 1, and 0. An external device places a byte of data into the input latch of the data port by setting $\overline{STB}$ low. This causes the data on the data lines to be latched and IBF to be set.

$\overline{STB}$ is low true, while IBF is high true. $\overline{STB}$ is negated by raising it to a high-logic level, while INTR is raised to a high-logic level to signal the acquisition of a data byte. INTR is negated at the start of a read cycle by the microprocessor to the port input latch. IBF is negated by the 8255 at the end of this read cycle.

This handshake mechanism can be used to implement a full interlocked handshake with a peripheral. The peripheral must wait for IBF to be in a false state. Then it can place data on the data lines and set $\overline{STB}$. After the 8255 accepts the data and sets IBF, the peripheral will negate $\overline{STB}$ and wait for IBF to be negated for the next transfer.

In the output handshake, the microprocessor places a data byte at the output-data pins. This causes a signal called *Output Buffer Full* ($\overline{OBF}$) to be set. The peripheral uses a pin called *Acknowledge* ($\overline{ACK}$) to indicate that the data have been accepted. An Interrupt-Request (INTR) is used to show that the peripheral has accepted the data by setting $\overline{ACK}$.

The sequence is started by a $\overline{WR}$ (write) control signal, which causes INTR to be negated. The end of the write cycle causes $\overline{OBF}$ to be set, signaling the peripheral that a byte of data is ready. Then the peripheral accepts the data by setting $\overline{ACK}$, and the 8255 negates $\overline{OBF}$. When the peripheral returns $\overline{ACK}$ to false, the 8255 sets INTR, requesting another data byte.

The INTR line for both input and output operations does not have to be used as an interrupt request. The two INTR lines can be buffered so that the microprocessor can read the state of the lines. The 8255 can then be operated as a polled device. The microprocessor will periodically check the status of the INTR lines to determine if one of the data ports is requesting service. Interrupt-enable A (INTE A) is set and reset with Port C bit 6, and interrupt-enable B (INTE B) is set and reset with Port C bit 2.

Mode 1 operation suggests a two-wire handshake. The peripheral places data on the data lines, where they are strobed into the input latch with the $\overline{STB}$ signal. The 8255 indicates that it is ready for another transfer with the IBF line. For a one-wire handshake, the peripheral can ignore the IBF signal, but the peripheral must not try to transfer information faster than the computer can accept it. This is an implicit return handshake. Another approach is to let the 8255 handshake with itself. Connecting $\overline{OBF}$ with ACK will allow the 8255 to receive ACK as soon as it sets $\overline{OBF}$.

MODE 2 CHARACTERISTICS

Mode 2 is the most complex mode. Only Port A may be operated in mode 2. Port A becomes a bidirectional data port, and 5 bits of Port C become the handshake lines. Port B can be operated in either mode 0 or mode 1 while Port A is operating in mode 2. The rest of the Port C bits are available either for bit I/O or as handshake lines for Port B.

An INTR is used as a handshake line in both input and output operations in mode 2. It is used when the data port requires service. The other handshake lines are $\overline{\text{OBF}}$ and $\overline{\text{ACK}}$.

The input handshake lines are $\overline{\text{STB}}$ and $\overline{\text{IBF}}$. INTR uses Port C, bit 3. $\overline{\text{OBF}}$ and $\overline{\text{ACK}}$ use Port C, bits 4 and 5.

The handshake for mode 2 is similar to the mode 1 transfer. A major difference is that data flow in both directions on the data pins of Port A. In order to control data direction, the 8255 does not move data held in the output latch to the Port A data lines until the peripheral sets $\overline{\text{ACK}}$. When the 8255 receives the $\overline{\text{ACK}}$, it moves the contents of the output latch to the Port A data lines within 300 nanoseconds.

The INTR line can be enabled for both input and output transfers. INTE 1 enables the INTR line for output transfers, and INTE 2 enables it for input transfers. When an interrupt occurs, the microprocessor must determine the cause of the request. The input latch may be full or the output latch may be empty, or both conditions may occur. The type of request is determined by reading Port C to determine the status of the IBF and $\overline{\text{OBF}}$ lines. If IBF is high, there are data in the input register. If $\overline{\text{OBF}}$ is high, the output data latch is empty.

Since mode 2 handshakes are similar to mode 1 handshakes, the same uses for one- and two-wire protocols exist.

DIGITAL I/O BOARD OPERATION

Digital I/O lines can be provided through an 8255 programmable peripheral interface chip which provides three ports: an 8-bit PA port, an 8-bit PB port, and an 8-bit PC port. The PC port can also be used as two half ports of 4 bits: PC upper (PC4-7) and PC lower (PC0-3). Each port and half port can be configured as an input or an output by software control according to the contents of a write-only control register.

The PA, PB, and PC ports can be read as well as written to. Other configurations are possible for unidirectional and bidirectional strobed I/O where the PC ports are used for control of data transfer and interrupt generation.

Interrupts are controlled by the 8259 interrupt controller in the PC. This is set by BIOS system initialization. Users program the 8259 to respond to their requirements and set up the desired interrupts. Interrupt handling is done using a tristate driver with a separate enable, interrupt enable-active low. This may be connected to any of the interrupt levels 2–7 available on the PC bus by jumpers on the board.

The 8255 uses four I/O address locations which are decoded within the I/O address space of the PC. The base address is usually set by a DIP switch on the board, and the address can be placed anywhere in I/O address space. Base addresses below FF hex (255 decimal) should be avoided, since this address range is used by the internal I/O of the computer.

Programming

Programming is usually done with assembly language or BASIC. The 8255 is configured in the initialization section of the program by writing to the control register. During a power-up or reset, all ports are configured as inputs. The configuration is set by writing the appropriate control code. There are three possible operating modes, as shown below. The PA/PC4-7 and PB/PC0-3 groups can be in different modes at the same time.

Mode 0—Basic I/O; all ports are I/O ports.
Mode 1—Strobed I/O; part of the PC port controls data transfer.
Mode 2—Bidirectional I/O on the PA port only; part of the PC port controls data transfer.

For example if:

PA input, PB output, PC0-3 input, PC4-7 output:

```
CONTROL WORD = 1001 0001 BINARY OR 91 HEX
```

Strobed output on PB, PA output, PC0-3 control, PC4-7 input:

```
CONTROL WORD = 1000 1100 BINARY OR 8C HEX
```

To program, write to the control register to set the configuration in BASIC:

```
xxxx OUT (Base address +3), &H91
```

To access the ports as required, for example, to read PA:

```
xxxx X% = INP (Base address + 0)
```

To write to PB:

```
yyyy OUT (Base address 1), DATA
```

To read PC:

```
zzzz Z% = INP (Base address + 2)
```

Once the configuration has been set in the initialization, the 8255 will stay in that configuration until a write occurs to the control register. In the 8255 all port registers are cleared by a write to the control register. Repeated changes of configuration require some provision for restoring data to cleared ports.

The following example shows how the input bits of the PC port can be used to monitor the status of two points. At a set condition, a binary value is sent to the PB port.

```
10      OUT &H303, &H89
20      K%=INP(&H302)
30      IF (K% AND 5)=5 THEN OUT &H301,9
40      IF (K% AND 5)=0 THEN OUT &H301,0
50      GOTO 20
```

When input port PC bits 0 and 2 are true, then output A is binary 9 (bits 0 and 3) from the PB port; otherwise output A is zero.

The outputs and inputs are usually TTL/DTL compatible. Outputs will usually drive one standard TTL load (74 series) or a four-LSTTL (74LS) load. CMOS compatibility can be obtained by connecting a 10k ohm pull-up resistor from the input or output to +5 volts. Power consumption is typically less than 170 watts at +5 volts. There are also high-speed parallel I/O boards that can be used for interfacing absolute encoders.

HIGH-SPEED PARALLEL DIGITAL INTERFACE BOARDS

High-speed digital I/O interface boards will use 8- or 16-bit DMA capability. Data transfer rates of 250,000 bytes/second or 125,000 words/second are possible with a 4.77-MHz 8088-based PC, and higher speeds are obtainable on 8088-2-, 8086-, 80286-, and 80386-based machines.

Normal processor I/O transfers may also be made through the data ports. An on-board counter/timer permits the user to set data transfer rates, or the transfer may be triggered externally.

A BASIC callable software driver is usually provided to allow the user to perform DMA controller setup, timer setup, and data block transfers. If the source code is available, the driver can be modified to run with higher-level languages.

These boards usually provide two 8-bit I/O ports, A and B. Each port can be set as an input or output under software control. The ports can be addressed as normal I/O locations using programmed transfer or via an 8237 DMA controller. DMA transfers may be 8-bit (byte) through Port A only, or 16-bit (word) using both ports A and B. In word mode, double buffering provides simultaneous update of ports A and B. Both ports are addressable at any time as normal I/O locations. External signals provide the current direction of the ports.

DMA transfers may be initiated by either an external or an internal timer. Suppose that the internal timer consists of a 10-MHz xtal oscillator divided through two sections of an 8254 counter. This will provide a clock rate ranging from 2.5 MHz to 0.0023 Hz, about 8 pulses/hour.

The choice of external signal or internal clock is made using software control. On receipt of a positive edge on a XFER REQUEST input, the XFER ACKNOWLEDGE output goes low. Completion of the transfer to/from memory is acknowledged by the XFER ACKNOWLEDGE output returning to the high state.

The operating DMA level is also selected using software control. The user is required to set up the 8237 DMA controller on the system board before a transfer. A BASIC callable subroutine can be used to do this. Software control is also used to select the active interrupt level (2–7) and between a positive or negative edge external interrupt on the INTERRUPT pin, or if the interrupt is initiated from an internal timer or generated by the 8237 DMA controller.

Programming

Programming can be done in BASIC using a software driver. The driver will handle initialization, as well as controlling the actual data transfer. The BASIC driver can be modified to be linked to other upper-level languages or assembly code. The I/O structure of this type of board will now be described.

Register Functions: Port A

Port A data correspond directly to the data bus. In word mode, Port A is the least significant byte.

D7	D6	D5	D4	D3	D2	D1	D0
A7	A6	A5	A4	A3	A2	A1	A0

Register Functions: Port B

Port B data also correspond directly to the data bus. In word mode, Port B is the most significant byte.

D7	D6	D5	D4	D3	D2	D1	D0
B7	B6	B5	B4	B3	B2	B1	B0

DMA Control

The DMA control register bits have the following functions:

D7	D6	D5	D4	D3	D2	D1	D0
DMA Enable	DMA Level	AUX2	AUX1	Xfer Source	Byte /Word	B Dir	A Dir
				0 = External (XFER REQ 1) 1 = Internal 8254 timer			

Interrupt Control
The interrupt control register bits have the following functions:

D7	D6	D5	D4	D3	D2	D1	D0
INT Enable		Interrupt Level		AUX3	INT	Source	Slope
0 = disable 1 = enable		000 = inactive 001 = inactive 010 = level 2 011 = level 3 100 = level 4 101 = level 5 110 = level 6 111 = level 7		00 = external input 01 = 8237 terminal 10 = 8254 timer 11 = 8237 terminal		0 = +edge 1 = −edge	

I/O MAPPING

Suppose that the board uses eight consecutive locations in I/O address space. This does not interfere with memory addressing. The base address can be set by a DIP switch and may be on any 8-bit boundary in the decoded address range of 0 to 3FB hex. One useable range is 100 3F8H. The addresses are mapped as follows:

Address	Function	Type
Base +0	Port A	R/W
+1	Port B	R/W
+2	DMA control	R/W
+3	Interrupt control	R/W
+4	Counter 0	R/W
+5	Counter 1	R/W
+6	Counter 2	R/W
+7	Control	W
	Status	R

R = read; W = write.

The first four addresses correspond to the digital I/O and control, and the last four addresses correspond to the 8254 timer.

Power supply requirements are normally +5 V at less than 1 amp. Some boards require two-thirds of a slot and will not fit in half slots.

10

Interfacing Flow Transducers

This chapter provides an introduction to flow measurement, including explanations of the differential pressure flow techniques and the use of orifice plates, Venturi tubes, and flow nozzles. Turbine flowmeters, variable area meters, fluid characteristic sensors, and electromagnetic flowmeters are investigated, and interfacing techniques are explained.

INTRODUCTION TO DIFFERENTIAL PRESSURE FLOW TRANSDUCERS

Control and automation of almost every type of chemical process require the measurement of flow. The medium may be a gas, water, air, or processed liquids, or gases which form the product itself.

Some flow sensors respond directly to the flow rate of a fluid fall. These types of sensors fall into three general groups:

1. A restriction in a pipe or duct is used to produce a differential pressure which is proportional to the flow rate. This pressure is measured using a pressure transducer system calibrated in flow units.
2. A mechanical member is used to sense the flow of a moving fluid by rotation or deflection in a tapered tube.
3. The physical characteristics of the fluid are used to sense the flow.

The sensing elements in the first group are called *head meters*, since the differential pressure between the two sensing ports can be equated to the head or the height of a liquid column. These flow-sensing elements provide a constant area for flow passage.

The head meter type of flow measurement is simple, reliable, and inexpensive and offers good accuracy. Since it is a direct measurement of flow rate, it is well suited to automatic flow control applications. The interface to the PC will depend on the type of pressure sensor used. Almost all of the types of pressure sensors and interfaces discussed in Chapter 7 can be used for head measurement.

ORIFICE PLATES

Orifice plates provide a simple restriction for the differential pressure measurements of liquids or gases. These restricting plates are mounted in the flow line, perpendicular to the flow.

The flat orifice plate has an opening smaller than the inside diameter of the piping, as shown in Figure 10-1. Taps are made at two points in the line. One is used to measure the pressure upstream from the orifice, and the other is used to measure the pressure downstream at the point of lowest pressure. The pressure difference between the two taps is a measure of the rate of flow. Orifice plates are not normally used for slurries and dirty fluids because of accumulations and rapid wear at the orifice edges.

An orifice can be properly sized with tables which include compressibility factors, Reynolds curves, and thermal expansion factors. It is also common to depend on the orifice manufacturer to furnish a proper orifice from specifications of the application.

The flow rangeability of a particular orifice plate tends to be low, on the order of 3 or 4 : 1. This means that the flow may not change by more than a factor of 4. Range changes are effected by changing the plate size.

There are also limitations on the minimum line size and minimum flow rates for accuracy when orifice plates are used. Plates with a quadrant edge are available for low flow rates and flows involving viscous fluids. An accuracy of 1% is possible. Table 10-1 summarizes the basic characteristics of orifice plates.

VENTURI TUBES AND FLOW NOZZLES

This is another class of flow-sensing elements which operates in a similar manner to the orifice plate. The Venturi tube construction, as shown in Figure

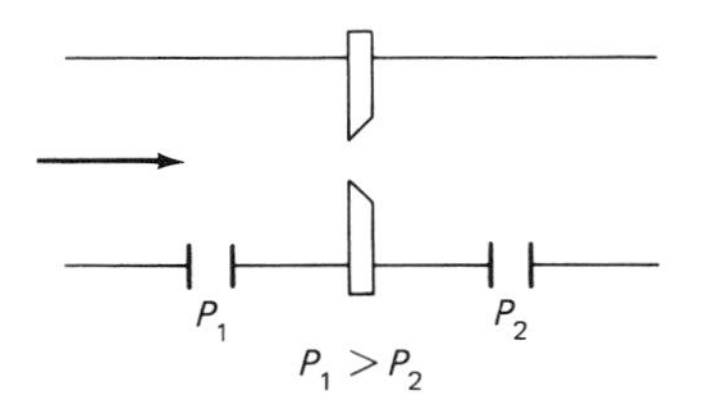

Figure 10-1 Orifice plate flow sensor.

TABLE 10-1 Characteristics of Orifice Plates

Low cost of the restriction itself
Capacity changes made by switching plate size
Large amount of coefficient data available
Applicable to a wide range of temperatures and pressures
Straight approach of piping required
Low rangeability for a particular plate
Upstream edge wear causes inaccuracies to develop
Unsuitable for slurry applications

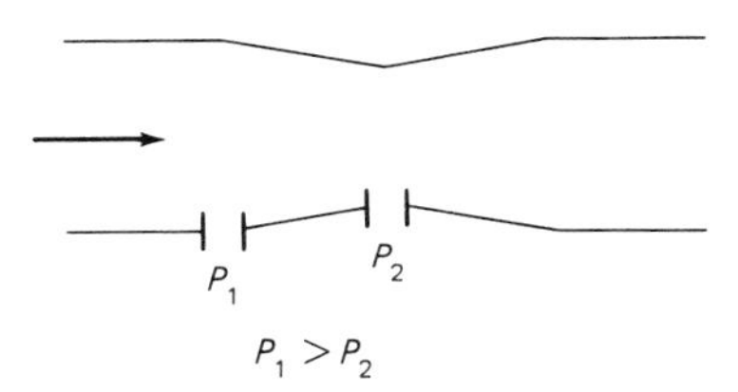

Figure 10-2 Venturi tube flow sensor.

10-2, consists of a short, constricted portion called a *throat* between two tapered sections.

It is usually installed between flanges in the line. In operation it tends to accelerate the fluid and lower its pressure. Venturi tubes are normally used for liquids. They can be used for gas flow measurement under the proper conditions.

An orifice presents a restriction at a point in the line; the Venturi tube spreads this restriction over a longer distance. The flow through the center section is at a higher velocity than at the end sections.

Taps at the entrance and throat are used to measure the pressure difference. An accuracy of 1% can be obtained. Table 10-2 summarizes the basic charac-

TABLE 10-2 Characteristics of Venturi Tubes

High capacities possible
Pressure recovery better than that of orifice plates
Fluids-containing suspended solids can be used
Resists wear due to abrasion
Larger sizes require considerable room
More costly than orifice plates
Produces a lower differential pressure than an orifice for the same flow and throat size
Low rangeability of 3–4 : 1.
Difficult to modify for major changes in flow range

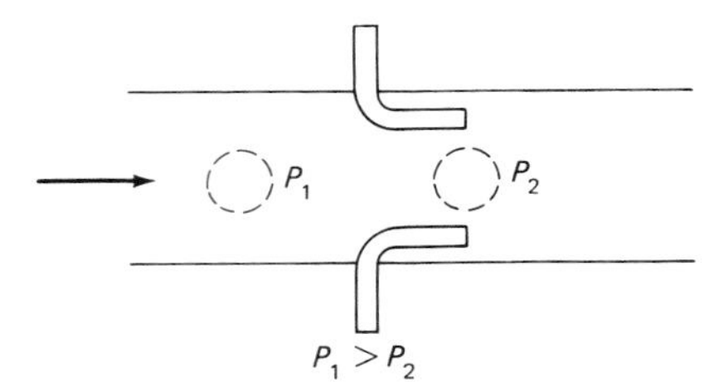

Figure 10-3 Flow nozzle sensor.

teristics of Venturi tubes. Similar to the Venturi tube are the Dall and Foster types.

The flow nozzle is curved, as shown in Figure 10-3. It is inserted between two sections of piping. The curvature of the inside contour exhibits a tangency to the throat that occurs without a sudden change of contour. The outlet end of the nozzle is usually beveled or recessed. Table 10-3 summarizes the basic characteristics of flow nozzles.

The pitot tube is another variation of a curved nozzle. It is not common in industrial systems. It is sometimes used for spot checking flows. A typical pitot tube is shown in Figure 10-4. One pressure opening faces the flowing fluid. This opening intercepts a small portion of the flow and responds mainly to the impact pressure. The other opening is perpendicular to the axis of flow and measures the static pressure.

TABLE 10-3 Characteristics of Flow Nozzles

Suitable for fluids with small amounts of solids
More rugged than orifice plates
Useful for high-velocity measurement
Pressure recovery better than that of orifice plates
Capacity 65% greater than that of orifice plates of the same diameter
Less susceptible to wear than orifice plates
More easily installed than Venturi tubes
Produces higher differential pressures than Venturi tubes

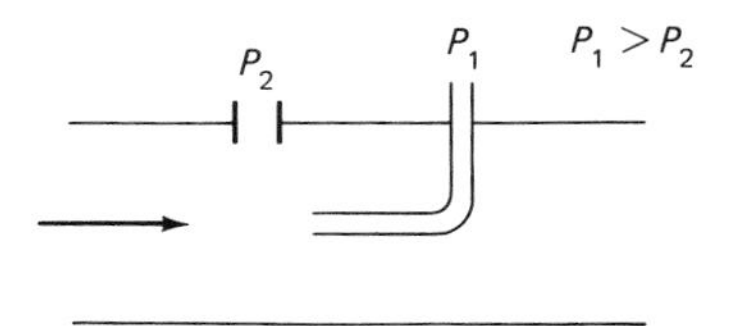

Figure 10-4 Pitot tube flow sensor.

TURBINE FLOWMETERS

The turbine flowmeter uses the movement of the fluid to turn a turbine wheel. The speed of the rotor is proportional to the flow rate. The meter's output is a precise number of pulses, which depends on the volume of the fluid that passes between the rotor blades in a unit of time. As each blade passes the coil's pole piece, a voltage is induced in the coil. The transducer output is an ac voltage with a frequency proportional to the flow rate. The number of cycles per revolution is a function of the number of blades on the rotor, and the revolutions per unit time are a function of the flow rate. The rotor blades can be shaped to produce a sinusoidal output voltage with little harmonic distortion.

The transducer output frequency band for a range of flow rates may be selected to coincide with a particular frequency band for modulation purposes. This is set by the number of rotor blades and their pitch.

Turbine meters are widely used in many processing industries because of their range and performance characteristics. They can also be used for gas flow measurements.

Turbine meters are usually selected with a capacity ranging from 30% to 50% above the expected maximum flow rate. Operation below maximum capacity provides greater reliability.

Turbine flowmeters have high performance. Each meter is calibrated individually for one or a number of fluid conditions before shipment. Turbine flowmeters are among the most accurate flow monitoring devices in common use.

The output relationship is linear within the particular transducer's range for flow rate and viscosity. A linearity of $\pm 0.5\%$ of flow rate and a repeatability of 0.02% of the rate can be achieved.

Turbine meters can be used in control and automation applications with flow rates ranging from fractions of gallons per minute to tens of thousands of gallons per minute. Pressures are allowed up to 7500 psi, and temperatures can range from $-400°F$ to $1000°F$. These meters can be used with most reasonably clean fluids and have a rangeability of 20 to 1.

Bearing friction, fluid, and magnetic drag and swirl in the fluid stream can produce errors in these transducers. These causes of error are minimized in the mechanical design. Swirl effects are reduced by the use of flow straighteners. Magnetic drag is not usually a problem except in miniature transducers. Thrust bearing friction can be reduced using the hydrodynamic forces to balance the axial loads on the rotor.

The transducer case is made of nonmagnetic metal and is usually threaded for the detector coil. The amplitude of the output voltage from the coil can be changed by adjusting the distance between the pole piece and the turbine blade tips.

Most turbine flowmeters use a permanent magnet coil. The turbine blades are made of a ferroelectric material. It is possible to achieve an accuracy of

TABLE 10-4 Characteristics of Turbine Flowmeters

High-pressure, high-flow measurement capabilities over a wide temperature range
High accuracy and repeatability
Good flow range
Good response times to flow changes
Short piping approaches required; flange ends allow quick replacement
Can be converted to mass flow with compensating hardware or software
Abrasive materials can wear out the internal bearings
Pulsating flows or water hammer can damage moving parts
High-viscosity liquids can cause measurement errors
More expensive than similar head meter installations

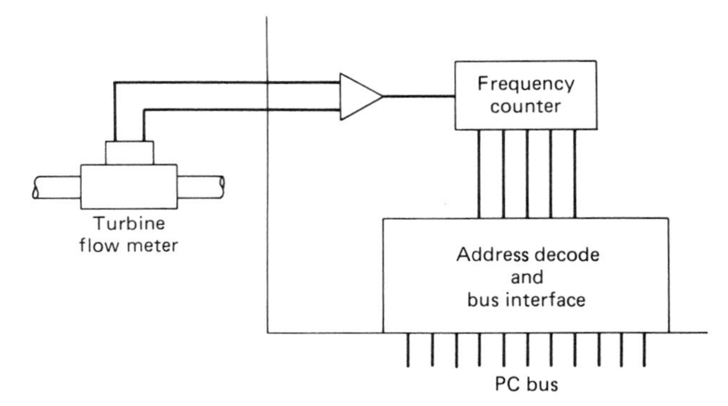

Figure 10-5 Turbine flowmeter interface using an I/O expansion board.

0.5% in the linear portion of the flow coefficient curve. Table 10-4 summarizes the basic characteristics of turbine flowmeters.

The interface to a PC requires a frequency counter, as shown in Figure 10-5. This can be accomplished using an I/O expansion board, as discussed in Chapter 8. Signal conditioning may also be needed to condition the turbine output.

VARIABLE AREA METERS

Variable-area meters can employ a float in a tapered section of tubing. These meters are called *rotameters* and use a spring-restrained plug or vane. The displacement due to the flow of elements causes the area of the flow passage to vary while the differential pressure or head remains constant. The displacement is then measured to provide an output proportional to the flow rate. Rotating flow-sensing elements include spring-restrained turbines installed in pipe sections and spring-restrained propellers installed in flow streams. These turn at

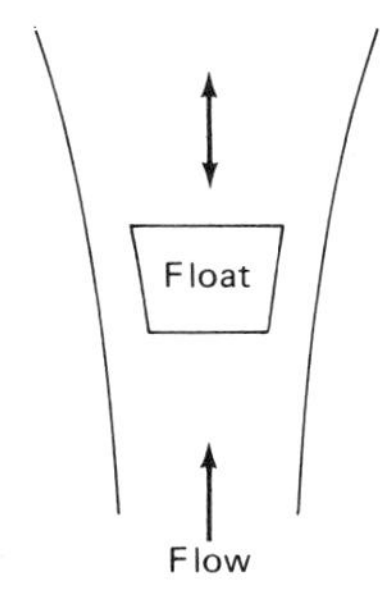

Figure 10-6 Rotameter (variable area) flow sensor.

TABLE 10-5 Characteristics of Rotameters

Low pressure drop
Allow some solids in the fluids
Provide an average reading for pulsating flows
Require vertical installation
High cost for sizes larger than 2 inches
Flow-induced deflections can be sensed by displacement sensors or strain gauge bridges

an angular speed proportional to the flow rate and are displaced along the axis of rotation.

Rotameters are designed so that the stream flows through tapered vertical tube using a float, as shown in Figure 10-6. The float rises as a function of the rate of flow and stabilizes for a constant flow rate.

The density and viscosity of the liquid or gas being measured by the rotameter can affect the position of the float. Rangeability is 12 : 1, and an accuracy of 1% is possible. A range of sizes allow capacities from .5 cc to 5000 gallons/minute. Units are available for temperatures of up to 1000°F and pressures of up to 2500 psi. Table 10-5 summarizes the basic characteristics of rotameter flow sensors. Since the flow-induced deflections can be measured by different types of displacement sensors, the same methods discussed in Chapter 9 can be used to provide the interface to the PC.

FLUID CHARACTERISTIC SENSORS

One class of flow sensors uses the characteristics of moving fluids to measure flow. One technique is to use a heated wire as a hot-wire anemometer sensor. The heated wire will transfer more heat as the velocity of the surrounding fluid increases. This results in a cooling effect that causes the resistance of the wire to decrease. Another type of sensor uses a small, slightly conductive fluid that flows through a transverse magnetic field. It provides a voltage proportional to the flow velocity.

In still another type of flow sensor, the boundary layer of a moving fluid is heated by a small heating element. The convective heat transfer is measured by a temperature sensor located at a point downstream from the heater. The temperature changes with a change in flow velocity.

Hot-wire anemometers use a wire element which is located normal to the flowing stream. The wire is heated electrically, and cooling due to the flow changes the resistance of the wire. The resistance change is calibrated to indicate the flow velocity. Some hot-wire anemometer circuits maintain the wire temperature constant and measure the current required to maintain this temperature. The measured current is then calibrated to indicate flow. Other circuits use a nonlinear amplifier to maintain a constant current through one side of a bridge circuit. The interface to a PC can be done using an I/O expansion board, as discussed in Chapters 7 and 8. The essential elements are shown in Figure 10-7.

Hot-wire anemometers are typically used for measurement of the low-flow rates of gases. They are also used for determining gas velocity. Anemometers have good sensitivity to flow changes.

Hot-wire anemometers can be used to measure mass flow, provided that the product of the thermal conductivity, specific heat, and density remains almost constant. This is true for many gases at low pressure.

The measurement of flow rate can be done with heat transfer methods using two types of flow instruments. The thermal flowmeter and the boundary layer flowmeter use an electrical heater to increase the heat of a portion of the moving fluid, as shown in Figure 10-8.

Temperature sensors measure the temperature of the fluid at upstream and downstream from the heater. The heat transferred between the heater and the

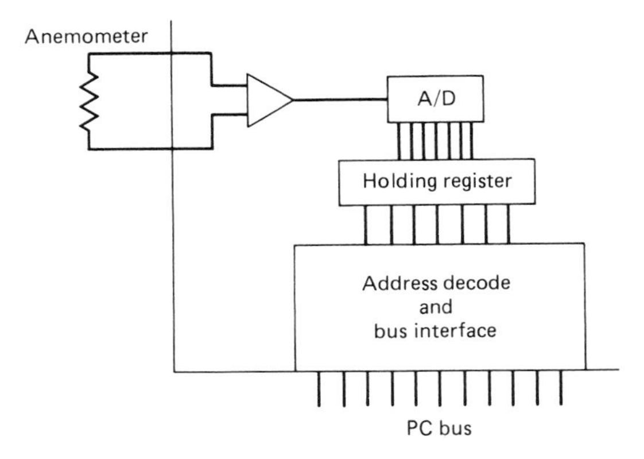

Figure 10-7 Anemometer interface using an I/O expansion board.

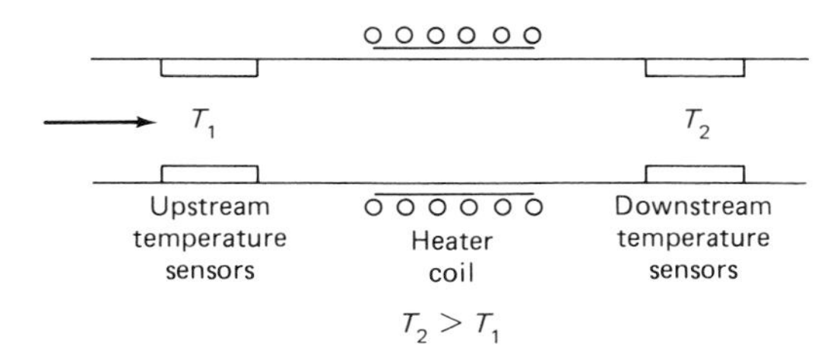

Figure 10-8 Heat-transfer technique for flow measurement.

downstream sensor is found either by measuring the temperature difference between the two sensor areas, using a constant heat input, or by measuring the change in heater current to keep this temperature difference constant.

Early thermal flowmeters used a heater immersed in the fluid. Since the entire core of the fluid was heated, the heater power required could be in the kilowatt range. This high power requirement limited the application of these early thermal flowmeters.

The power requirement is reduced with the boundary layer flowmeter. In these devices, only the thin boundary layer of the fluid adjacent to the pipe wall is heated. The temperature sensors are mounted flush with the inside pipe surface, and the heater either surrounds the outside of the pipe or is embedded in the wall of the pipe. The heat input varies with the mass flow rate and the temperature differential. The heater power requirement is usually less than 50 W for boundary layer units.

The forced vortex flowmeter, also known as the *processing vortex meter* or *swirl meter*, uses vortex procession of the fluid. The principle of vortex procession is based on the fact that if a rotating body of fluid enters an enlargement, as shown in Figure 10-9, under the proper conditions, the flowing fluid will oscillate with a well-defined frequency that is proportional to flow.

The frequency of the oscillations can be detected with a pair of temperature sensors. The output will be a train of pulses with a frequency that is proportional

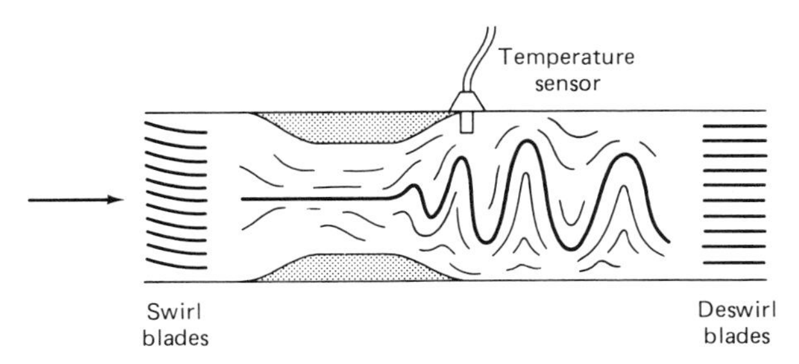

Figure 10-9 Forced vortex flowmeter.

to the flow rate. This train of pulses is a function of the fluid generating a hydrodynamic procession of the flow.

Swirling takes place after the fluid is sent through a special set of blades. Downstream of these blades is a Venturi-shaped contraction and expansion of the flow passage. Following this region is an area where the cross section enlarges and the swirling flow processes. The flow pattern leaves the axial path of the flowmeter's center line and assumes a helical path. The frequency of this procession is proportional to the flow rate.

The oscillations of the fluid cause variations in fluid temperature, which can be detected by a platinum film sensor. Deswirl blades are used to straighten the flow as it leaves the flowmeter. This type of flowmeter has a wide linear range of 100 : 1 and provides an output within $\pm 1\%$ of the rate for an output frequency of 10 Hz to 2 kHz.

The vortex shedding flowmeter uses a similar design but depends on hydrodynamic principles. A nonstreamlined obstruction is placed in the pipe so that the fluid does not flow by it smoothly (Figure 10-10). Eddies or vortices are formed. These grow larger and eventually detach themselves from the obstruction. This detachment, called *shedding*, occurs alternately at each side of the obstruction. The vortices form tails downstream of the obstruction. The shedding frequency of the vortices is directly proportional to the flow rate and can be measured with temperature sensors. The action of the obstruction is similar to that of an orifice plate, and the output signal is like that of a turbine-type flowmeter.

The vortex shedding flowmeter senses flow velocity; thus, it acts as a volumetric measuring device. It can be used as a mass flowmeter if the fluid density is known. The basic characteristics of vortex shedding flowmeters are shown in Table 10-6.

All of these heat-transfer flow measurement techniques can use the I/O expansion board for interfacing to a PC discussed in Chapter 8. The essential requirements are shown in Figure 10-11. After signal conditioning, an A/D conversion is made and the result is stored in a holding register before it is passed to the PC bus.

Another flow-sensing method is based on the use of radioisotopes and de-

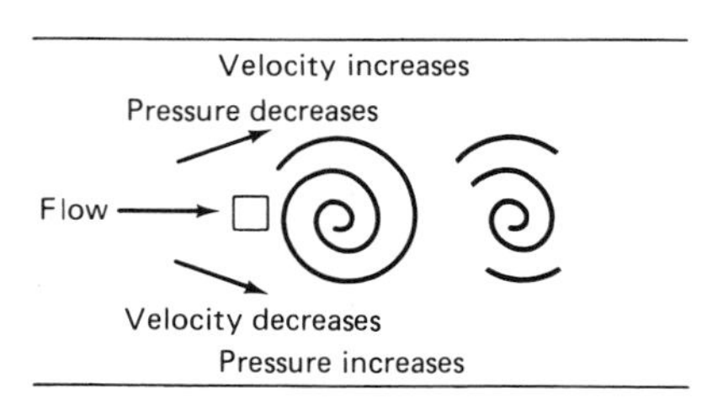

Figure 10-10 Vortex shedding.

TABLE 10-6 Characteristics of Vortex Shedding Flowmeters

Wide range
Universal calibration
No moving parts
Cannot be used in laminar flow
Reynolds number of fluid must be at least 3000
Can be used in both liquid and gaseous fluids

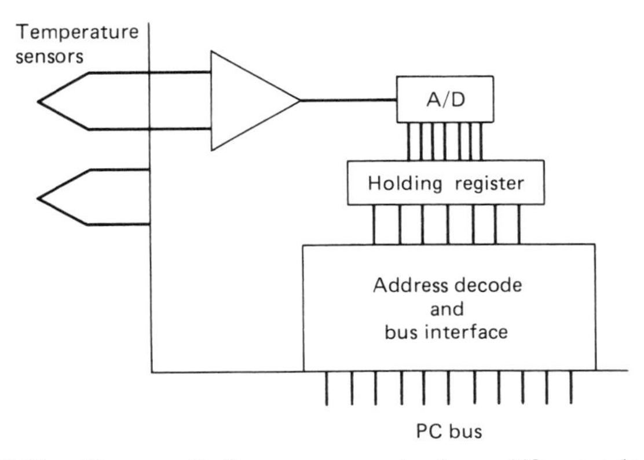

Figure 10-11 Heat-transfer flow measurement using an I/O expansion board.

tectors of the nuclear radiation that is imparted to the fluid. A source is placed outside the pipe upstream of the detector. Neutrons from the source collide with the moving fluid and cause particle and electromagnetic radiation to be emitted. The particles or radiation effects are detected as they flow through the pipe. The amount of radiation detected depends on the rate of flow of the fluid.

A timer/counter expansion board can be used to interface this system to a PC. This type of expansion board is discussed in Chapter 7. The essential requirements for the flow sensing and interface are shown in Figure 10-12.

ELECTROMAGNETIC FLOWMETERS

Electromagnetic flowmeters depend on Faraday's law, which states that if relative motion takes place at right angles between a conductor and a magnetic field, it will induce a voltage in the conductor. This voltage is a function of the relative velocity of the conductor and the intensity of the magnetic field.

An electromagnetic flowmeter, as shown in Figure 10-13, is made from nonmagnetic materials and uses a conductive liquid. Magnetic coils provide the magnetic field, and as the liquid moves through this field, a voltage proportional to the fluid flow rate is generated.

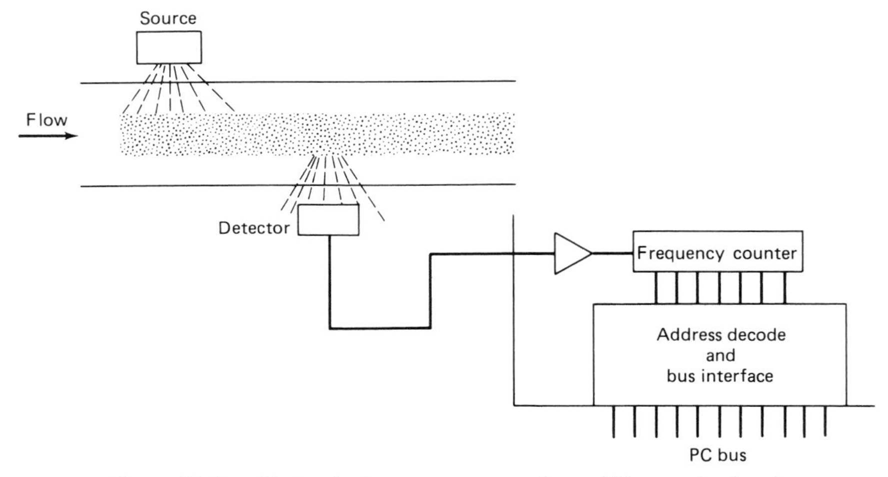

Figure 10-12 Nucleonic flow measurement using an I/O expansion board.

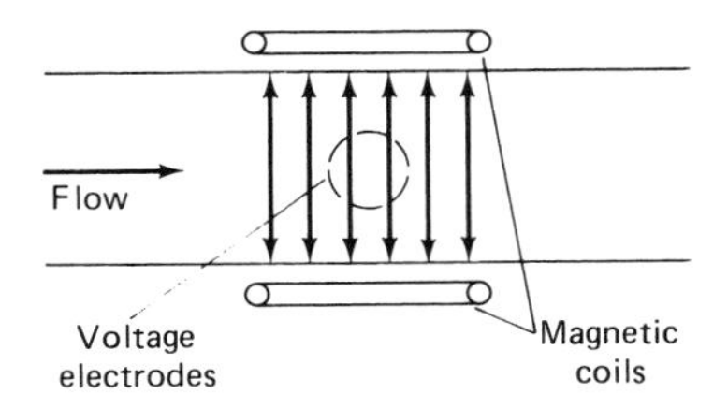

Figure 10-13 Electromagnetic flowmeter.

Electromagnetic flowmeters have an output that ranges from microvolts to millivolts. Flow sensing in this type of flowmeter is not affected by changes in liquid density or viscosity. Turbulence of the liquid and variations in piping have only limited effects. The main requirement is that the conductivity of the fluid be greater than about 10^{-8} mho/cm^3.

Electromagnetic flowmeters are calibrated to detect the liquid velocity. Therefore, the entire cross-sectional area of the pipe must be full and there should be no gas bubbles in the liquid, since these will also cause errors.

Mixed-phase fluid may cause conductivity variations and major errors. DC coil excitation can cause electrolysis problems; therefore, it is not used. The linearity of the output voltage with flow rate is an advantage in many process automation applications. This type of flowmeter can use the basic I/O expansion board for interfacing to a PC, as discussed in Chapter 8.

11

Distributed Control and Local Networks

Networks provide many advantages in PC applications. This chapter explains PC network configurations and the general network organization needed for distributed process control. PC control system configurations are compared in terms of their ability to increase reliability and reduce failures.

This chapter shows how to perform a network needs evaluation. Private branch exchange, baseband, and broadband network technologies are compared, and tips on network implementation are given.

There are many factors to consider when a network implementation is planned; these are examined in this chapter. The concept of network protocols is introduced in this chapter and expanded in the next.

PC CONTROL SYSTEM CONFIGURATIONS

The single-computer configuration has been used in many installations. This has had a major influence on many of the technical and human aspects of computer control. The large central computer originally had several advantages:

1. More and better software was generally available.
2. The large computer was equipped to handle complex problems such as the solution to difficult algorithms.
3. In applications that required many calculations and a large memory capacity, the larger computer was generally more suitable.

Applications that did not require complex calculations were handled by smaller systems. Many control tasks in discrete-parts manufacturing involve simple binary inputs and outputs. These tasks may be performed with a microcomputer or PLC.

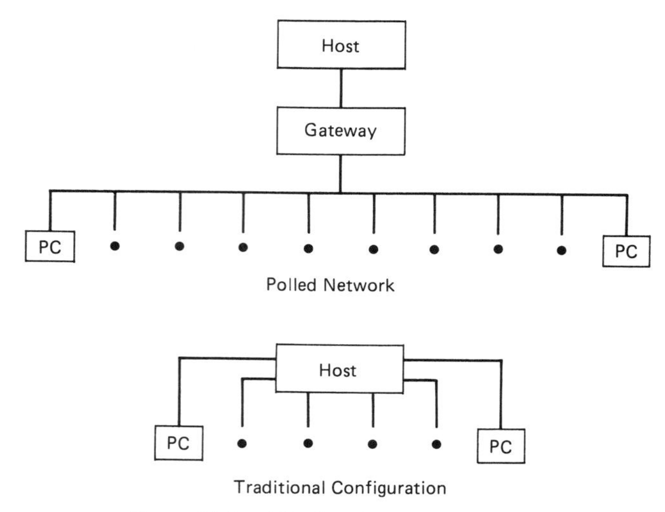

Figure 11-1 Distributed processing systems.

It has become increasingly apparent that most control systems can be divided into smaller and simpler tasks. PCs can then assume the divided responsibility of the complete control system, with one PC controlling the overall flow of the tasks, as shown in Figure 11-1. This technique is generally known as *distributed processing*. The PCs may not always be dedicated to a single function, but they can be assigned tasks by a master processor which operates the complete network.

One advantage of distributed processing is the division of labor, since the remote units offload the central or master processor to improve the performance of the total system. The resulting improvement in time response can result in less overhead in the system and improve the execution of functions without waiting for the availability of a central computer.

Distributed systems also provide a modularity which is not available in many centralized systems. As sections of a plant are automated, more remote units can be added.

RELIABILITY IMPROVEMENT AND FAILURE REDUCTION

Distributed processing can also improve system reliability and failure tolerance. The remote units allow some operations to continue independently of the central processor for short periods of time. Thus, it is possible to tolerate a short outage in the central computer or in a communications link. Another computer can also take over control, for a limited time outage of remote units.

A factory network is often required to operate in harsh, noisy industrial floor environments. In a distributed system or network, reliability can be achieved by the following mechanisms:

1. Redundancy, which can include fully redundant interfaces and cabling.
2. Fail-safe features that are designed to prevent any single failure in the system from bringing the entire system down.
3. Self-repairing mechanisms that automatically disconnect failed devices or add new ones.

In a distributed configuration such as this, a master processor assumes the control task. Then, when a failure occurs, a backup processor takes over control. The failed processor can then be taken offline, repaired, and returned to service without interrupting normal operations.

To guarantee that higher reliability will be achieved, the system should be analyzed for the effects of each type of failure. The software must be able to take advantage of the higher potential reliability of the duplicated hardware.

The distributed processors provide greater computing capacity, and this excess power can often be used for other optional tasks. In order for the other processors to keep up-to-date with the status of the program, they need access to current processing status information. When a shared-file system is used, this information may be gained by using an access mechanism such as token passing. A token-passing system requires that each station on the network have the token before another device talks to the node (Figure 11-2). The method used to determine which node has the token may be deterministic or it may be sequential; in the latter system, the token is passed from node to node in a strict order.

The requirement to follow the token-passing order can reduce the response too severely for some dynamic applications. In a typical system with a number of processors, the tokens or requests from one processor to the others are posted in a shared file. Then a scheduling program periodically checks to determine if

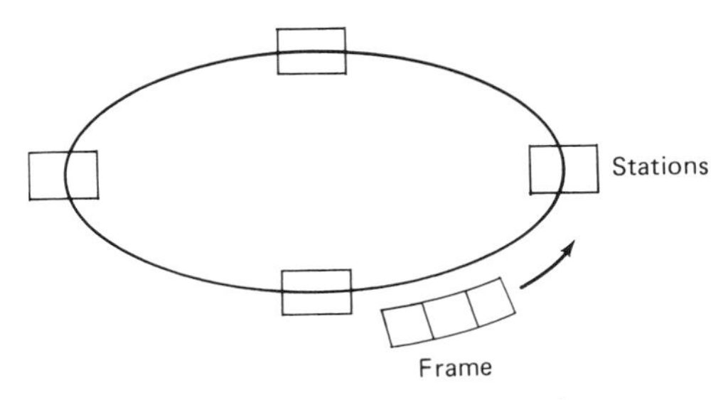

Figure 11-2 Token ring operation.

a request is waiting. The request can then be scheduled according to the system's priority.

The primary devices are in the token-pass sequence and initiate communications. The secondary devices respond only when queried, and the demand devices are secondary devices that are allowed to behave like primary devices only if a specific event occurs.

When the disk files are serial devices, so that two processors cannot be used at the same time, the contention problem must be solved by circuits or by another processor (usually called an *arbiter*) which may be part of the file interface.

Operation of the factory network includes functions such as the following:

1. Adding users.
2. Deleting users.
3. Assigning specific network resources to users.
4. Configuring the network.
5. Daily monitoring.
6. Locating and isolating network problems.
7. Control of access to the network.

A clear understanding of these functional areas will help factory network conceptualization, design implementation, and operation.

Training of the staff to operate and maintain the network is usually a major consideration. A network operations plan should be used to define the policies and procedures required for network operations, including network resource allocation, system configuration, security, network diagnosis and fault location, and network performance monitoring. A traffic analysis phase can be used to quantify the data transfer requirements of the application. This analysis should quantify the sources and destinations of transmitted data, the frequency of data transfer over the network, and the volume of data transfer.

PCs provide many networking alternatives. In a typical application, an industrial PC can be used to control the operations of a series of welding robots in a self-contained manufacturing area commonly known as a *cell*. In this system, a PC located in the factory's computer room can download parts of programs over the network to the cell controller, which can, in turn, pass these programs to the welding robots.

High-speed communications allows the linking of a large variety of PLCs, PCs, and communication terminal devices in control and automation environments. The roles of industrial control devices are changing as these networks permit fast communication between the control devices and allow the sharing of peripherals. This reduces the amount of hardware that must be dedicated to single applications and results in lower cabling costs.

GENERAL NETWORK ORGANIZATION

Factory networks can be organized in a number of ways. Normally, the computers are loosely coupled and capable of stand-alone operation. The general network control scheme may be defined as master-slave, hierarchial, or peer-connected. This level of the configuration determines how the responsibilities are distributed among the processing units.

In a master-slave network configuration, a host processor is connected to a satellite processor (Figure 11-3). The communication line between the processor is known as a *link*. The link may be a communications channel, such as a coax cable or telephone line. Each device in the network is usually referred to as a *node*. A master-slave network with multiple slaves may also be called a *multipoint* or *multidrop* configuration.

In the master-slave network, a single master processor has control and determines which slave computer will operate on a task. Communication among the slave computers is generally under the control of the master. After the slaves are assigned their tasks by the master, they operate asynchronously with respect to the master until their tasks are completed or until they require service from the master.

The hierarchial network is hybrid and uses a multilevel master-slave configuration. The different levels in the hierarchy can be assigned responsibilities for specific functions. The upper level in the hierarchy makes the major decisions, while the lower levels have the responsibility for controlling specific operations.

A peer or peer-to-peer network allows any device to communicate with any other device, with no master or bus-arbitration devices required. This single-level control scheme is in contrast to the master-slave and hierarchial networks, which are based on top-down control. The peer network depends on mutually cooperating computers with no defined masters or slaves. This type of network is popular for PCs. It requires that the operating system of each computer be aware of the status of the other computers in the network.

A scheduling scheme such as token passing may be used to distribute the tasks. As a job is passed to a computer in the network, the originating computer

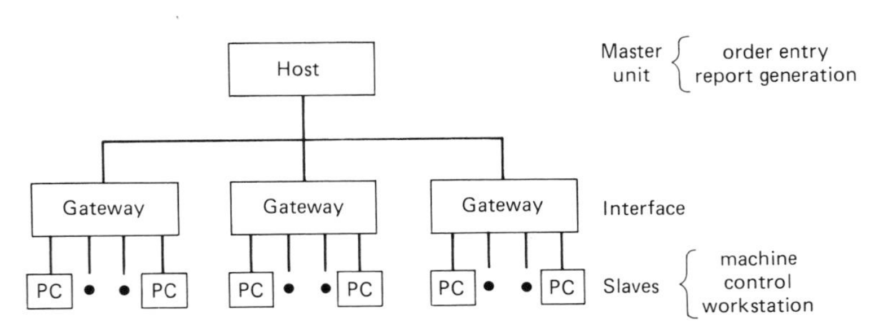

Figure 11-3 Distributed network in a master-slave configuration.

moves on to a new task. A computer which is busy may pass the task on to an available computer, which then executes it.

The time response in peer networks is difficult to predict, since one computer does not know the workload of another and since there is no permanent master to control the tasks. This is a disadvantage.

Peer networks have the following advantages: The facilities are not required to be available on the originating computer. In addition, the processors can share the computing load in a dynamic fashion, allowing more efficient use of the total facility.

Besides the computer control scheme, one must also consider the physical configuration of the network. The star configuration, for example, uses a host or master processor located at a central point in the system (Figure 11-4a). Each processor communicates directly with the host, and any communication between the satellite processors must go through the host. This physical network is also known as a *radial* or *centralized* configuration. Since the host controls all communications, the star network is limited by the speed of the host. When several satellites attempt to communicate with the host at the same time, the host must be able to control the data flow among satellites properly to be effective.

In the multidrop configuration, the host controls the flow of data between any two nodes, and any satellite can communicate with the host or with any other satellite at any one time. The multidrop configuration is shown in Figure 11-4b. It is also called a *data bus*, *data highway*, or *multipoint* configuration.

The loop or ring configuration shown in Figure 11-4c is often used in remote multiplexing systems. If a single link breaks, the nodes can still communicate. The loop can begin and end at a single loop controller, or each computer may be given control of the communications. Messages between computers in the loop are usually in the form of a string of words, with some bits or words containing information on the originator and the addresses. When a computer in the ring recognizes a message addressed to it, it accepts the message. When a processor receives and verifies a message, the starting and ending addresses of that message are passed to the destination processor. This processor then encodes and retransmits the message to the next point in the network.

A supervisory processor or controller maintains network information and data link assignments based on the equipment conditions, the message load, and the most direct route to the final destination. In the ring configuration, control can be difficult due to the way the messages must pass through the computers; thus, higher data transfer rates are needed.

In the point-to-point network, every processor has a direct access link to every other processor (Figure 11-4d). For n processors, $n(n - 1)/2$ interconnections are required. Using this relation for a 3-processor system, 3 communication links are needed. For a 5-processor system, the number of links required is 10; for a 10-processor system, the number of links increases to 45.

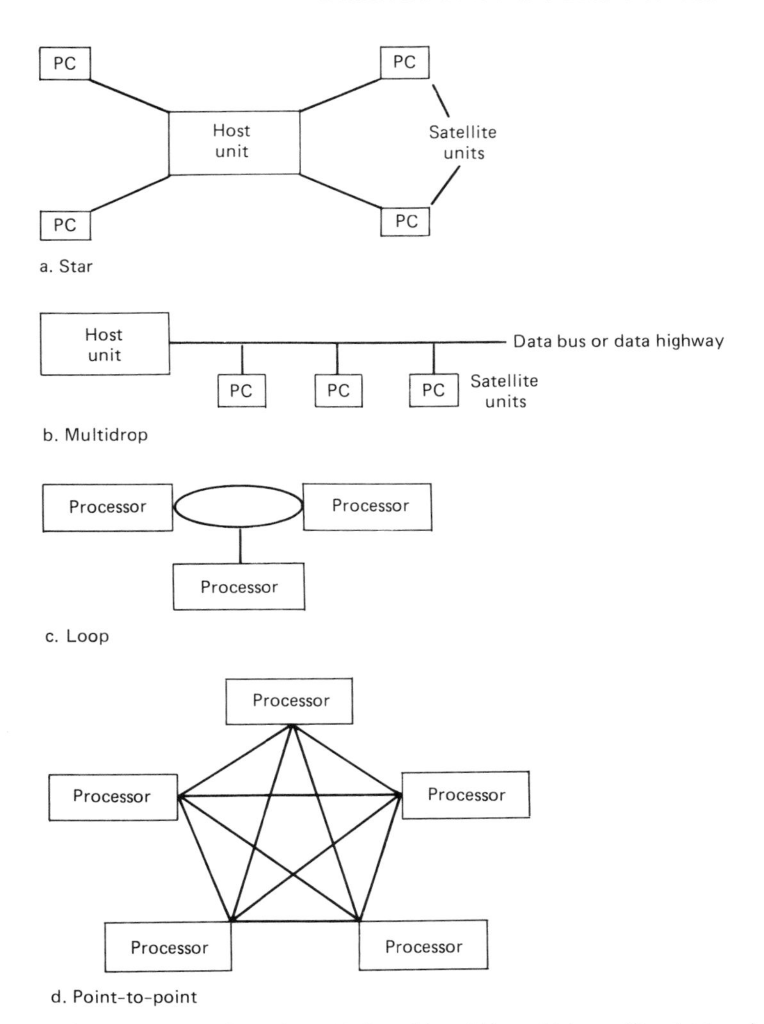

Figure 11-4 Network configurations. (a) Star; (b) multidrop; (c) loop; (d) point-to-point.

The point-to-point configuration permits a faster response or a lower grade of communications lines. The use of alternate paths allows messages to be forwarded even if some of the links are broken. Then, if one of the processors becomes inoperable, the other processors still have access to the data and can maintain operations.

In a multiprocessor configuration, it must be determined when another processor should be activated to take over the master role. One technique requires the primary processor to set a timer in the backup unit on a periodic basis. Failure to set this timer causes the backup unit to assume control and disable the primary unit. This type of timer is called a *hardware* or *watchdog* timer. The switchover from the primary to a backup computer should result in only a small control deviation in most applications.

A basic method of connecting processors is through the I/O channels. This can be done through either serial or parallel data transfer. The transfer of information will be at the data rate of the slower processor. These I/O-connected systems can use a number of operating modes. Each processor can control a portion of the tasks, and the necessary coordination can take place through the interconnections.

DISTRIBUTED PROCESS CONSIDERATIONS

The use of computer networks in general data processing has increased greatly in recent years. The increased interest in factory networks such as the Manufacturing Automation Protocol (MAP), along with the increased popularity of remote multiplexing, has resulted in the wider acceptance of distributed control schemes for process control industry.

MAP is supported by many of the nation's largest process manufacturing corporations. Several are members of the MAP Process Industry Initiative and are also represented on its public relations subcommittee. The subcommittee's mission is to develop awareness of the emerging MAP standard in the process manufacturing industry. The MAP Process Industry Initiative is a working group established under the MAP/Technical and Office Protocol Steering Committee. The user/vendor group includes companies such as Proctor & Gamble, E.I. DuPont, Exxon, Weyerhaeuser, InLand Steel, Tennessee Eastman, Foxboro, Honeywell, and Fisher Controls International.

In many process industries, distributed control has been necessary for years. Now the smaller computer makes distributed control ideal for discrete-parts manufacturing, where several separate stand-alone control devices can be installed at various points in the production process. MAP is discussed in more detail in Chapter 12.

Distributed control differs from distributed processing in that it also makes use of remote multiplexing. The processors communicate with field-located multiplexers, and each processor is usually dedicated to the same task.

The two basic techniques for distributed control are the loop approach and the unit approach. The loop approach uses a number of satellite processors which perform a fixed number of functions.

A single PLC can be used to perform one function for a number of different control loops. The processor can be dedicated to a single control loop, and the controllers can be interconnected electrically by data buses. Depending on the complexity of the system and the nature of the application, they can operate independently, communicate with each other, and even assume control of each other.

The ladder logic can be entered step by step at individual PLCs. The small display panels usually found on the PLC may be difficult to read, so the PC's CRT displays are desirable to show several rungs of ladder logic at once.

The unit approach is based on a separate control system for each unit in the system, with a processor assigned to each unit operation. Some advantages of distributed control can be achieved by either approach, but others can be gained only by using the unit approach. One major advantage is a reduction in the cost of installed wiring. If the I/O equipment is located throughout the physical system, the wire runs will be shorter. Besides the savings in installation and wire costs, there will be fewer problems from interference with low-level signals.

The unit approach tends to be modular, and computer control can be added to the units in the system one at a time, without disrupting control of those units already on computer control. The most likely unit for computer control can be established after conducting an analysis of the system.

A system study can be conducted to determine which unit may be a control system or process bottleneck, a large energy user, or difficult to operate without computer control.

A frequently used local protocol is RS-232C, but RS-422 is also popular because of its immunity to RFI and the low cost of the twisted pair of wires. The high-level, or supervisory net, will more likely use a version of IEEE-802.

The reliability and maintainability of process control systems with distributed control are improved. The cost of microcomputer hardware is low, so processors in a unit configuration can be backed up with a spare (duplexed processing); thus, a single processor failure will have no effect on the operation. Since most microcomputers use a single printed circuit board for the CPU, the system is more easily repaired. It is often a matter of replacing a plug-in board if the problem is not in the CPU board.

NETWORKS

For control applications, computer networks can be viewed as an extension of the distributed processing concept. The difference between a distributed processing system and a computer network tends to be a matter of degree.

Computer networks consist of two or more computer systems which are separated. The distance can be a few feet within the same room or thousands miles between facilities which are connected by common-carrier lines.

These networks may connect PLCs, PCs, printers, badge readers, I/O devices, time/attendance stations, and other production monitoring equipment.

Many different steps are involved in the design, implementation, and operation of a factory network. First, it is necessary to consider the environment in which the system will operate. This environment includes the following factors:

1. Manufacturing objectives and constraints.
2. Functional consideration of users.
3. Characteristics of the environment.
4. Performance specifications.
5. Reliability goals.
6. Network security.

A complete consideration of the application includes the type of information to be processed, the geographic location and experience of the users, and the existing hardware and software resources.

The network architecture identifies the network components and their interrelationships. A wide array of technologies is available for the factory network. Selection of the appropriate network architecture is crucial to the successful implementation of the manufacturing network. The major components of the network are network stations, applications software, network transmission media, networking hardware and software, network servers, interconnection devices, and the network's management system. After the network architecture is defined, a strategy for the implementation and operation of the network must be developed. This plan will usually consist of four steps:

1. Network acquisition.
2. Network development.
3. System installation.
4. Staff training.

The network acquisition phase involves acquiring the hardware and software. This includes determining component costs and developing schedules for component procurement. Installation includes the procedures, costs, and schedules for network wiring, the layout of the transmission media, and the related hardware and software.

INTRODUCTION TO PC NETWORK TECHNOLOGY

Network technologies can be grouped into three basic types: private branch exchange (PBX), cable television broadband, and the baseband types such as Ethernet. PBX technologies operate in the telephone bandwidth of 4.4 kHz. If digital transmission is used, PBX networks can support RS-232C or V24 types of terminal communications. PBX is most suitable for terminal-to-computer networking.

Broadband networks use cable television technology and provide up to 400 MHz of bandwidth. The bandwidth is usually frequency-divided into small channels, which makes it suitable for bundling together a number of lower-speed lines. Some of the faster broadband networks provide channels of up to 128 kbits/second.

Digital baseband networks generally operate at about 10 megabaud and can provide their entire bandwidth, if required, between any two points. These networks are most useful for handling the bursts of data generated by communications between computers.

The different network technologies can coexist, with each serving a different set of functions. Interfaces are available between the various technologies, such as gateways to Ethernet. The International Standards Organization's (ISO) seven-level network model provides a way of defining the compatibility of the layers of network connection and protocol.

Protocols such as IBM's SNA, DEC's DECnet, and Xerox's XNS are all higher-level architectures for levels 3–7. An SNA network may include Ethernet as a low-level transmission medium.

A variety of local digital networks at levels 1 and 2, such as Ethernet, RS-232C, 3270, or IBM's token-passing network, can coexist with SNA. However, many problems can occur. Consider the example of one vendor's Ethernet sharing a cable with another vendor's Ethernet. While the microcomputers connected by one vendor may communicate with each other, they may not be able to utilize the Ethernet file or printer servers of the other vendor, even though both vendors are placing data on the same cable.

NETWORK NEEDS EVALUATION

A needs assessment can be used to quantify the present requirements of the factory network, as well as those projected for the next several years. This assessment analysis should concentrate on two important areas: (1) quantification of the support requirements for the operators and (2) prioritization of the desired needs that must be serviced by the network.

Another task is to determine which technology to implement. This requires an understanding of network technology. The viable local communications solutions may include using facilities provided by the telephone company or installing a new local area network (LAN). The three basic types of LAN technology are as follows:

1. The PBX for data-only or integrated voice/data.
2. Baseband signaling using twisted-pair wiring or coax cable.
3. Broadband signaling using coax or fiberoptic cable.

The following important technical factors also need to be considered:

1. Configuration alternatives.

2. Types of PCs supported.
3. Typical device speeds and maximum network speed.
4. Signaling medium and capacity per cable.
5. Maximum network distance.
6. Support for gateway and bridge processors to other networks.
7. Support for video and voice, as well as data communications.
8. Installation.

PBX NETWORK TECHNOLOGY

PBX-based LANs use star or tree configurations and normally take advantage of in-place wiring for the signaling medium. These networks are suitable for applications that have been renting telephone equipment. There is an opportunity to utilize integrated voice and data workstations, and a variety of data devices can be supported by purchasing new PBX facilities.

The PBX acts like an automated switching matrix similar to a telephone switchboard, except that the PBX does not have physical wire connections between users. Time division sampling techniques are used to sample each conversation and convert the analog signal into a digital representation. By synchronizing the paths of the digital signals from the various extensions, the PBX establishes a virtual circuit connection.

PBX networks typically support microcomputers and other devices operating at data rates of 300 to 9600 bits/second (bps). Some networks support device speeds of up to 3 megabits/second (Mbps). The maximum network distance for microcomputers is 20 meters without modems or repeaters. A limited-distance data set is normally used to connect the microcomputer to the PBX switch.

Microcomputers may be connected to central data bases using terminal emulation software. These microcomputers can also be used as integrated voice/data workstations. They support voice, data, and facsimile communications but generally not video. Each network cable supports one voice channel with control signals and one data channel.

The cost per port for a typical voice/data PBX can range from several hundred to well over $1000 each for voice and data transmission. The cost per port for a typical data-only PBX is about half as much.

BASEBAND NETWORK TECHNOLOGY

Baseband network technology uses signaling techniques in which each node or station sends digital signals on a communications channel shared by all of the network stations. Each station will listen to all of the network traffic but respond only when it recognizes those messages that are addressed to it.

Baseband networks are available which support the star, ring, and bus configurations. These networks use twisted-pair telephone cable, 50-ohm coax cable, or Ethernet coax cable.

Baseband networks are often used for networking a small group of microcomputer workstations. The microcomputers are typically provided with baseband transceivers in the form of expansion boards. A microprocessor on the board is usually dedicated to managing the network protocol.

These systems usually operate at speeds ranging from 2.4 to 56 kbps, with a maximum network speed of 10 Mbps. They can support one data channel with 10 to 100 devices sharing the channel by circulating the data, which may be in the form of packets.

The total transmission distance is usually less than 20 kilometers, with each segment separated by less than 1500 meters. The microcomputers are usually closer than this because of the high cost associated with the use of baseband repeaters.

Some networks support digitized voice as well as data transmission. They do not usually support analog facsimile devices or video transmission, although some of them support digital facsimile devices.

BROADBAND NETWORK TECHNOLOGY

Broadband networks use tree and bus configurations with single or dual 75-ohm coax cable. They normally operate with devices at speeds ranging from 2.4 kbps to 3.0 Mbps. The maximum channel speed is typically 5 Mbps.

Frequency division is used to allow the simultaneous transmission of many data channels on a single physical channel. The different device groups connected to the network are assigned frequency ranges called *bands*. The transmission mode is analog, and digital signals require conversion using modems.

These networks can provide the high bandwidth necessary to support hundreds of voice, video, and data channels in the network. Broadband hardware is based on community antenna television (CATV) technology which is relatively mature. The maximum geographic distance is 50 kilometers. Broadband is especially suitable for applications with a number of buildings in an open environment. The cost per port ranges from $600 to $1000.

COMPARING NETWORK TECHNOLOGIES

PBX local networks offer a mature technology, and existing resources can often be used. Many facilities have a PBX, and most of these can be used to provide some data transmission between locations. These installations allow flexibility in providing access to data channels through telephone jacks. Disadvantages include limited device capacities, relatively slow data transfer rates, and high costs for high-traffic data applications.

Baseband LANs allow a greater variety of traffic types and applications than is possible with PBX networks. A typical application is a cluster of microcomputers sharing a data base.

Baseband networks allow the sharing of information among users on a highly interactive basis. The cost of baseband technology is similar to that of broadband technology, depending upon the data rates and the network configuration. Baseband networks do not support video and usually do not support digitized voice. They are limited in maximum cable distances and tend to have less growth capacity than broadband networks, which are better for higher-volume and large-bandwidth applications.

Broadband technology includes integrated voice and data communications, as well as video requirements. Broadband networks, with their large bandwidths and high transfer rates, permit multiple data channels for data and video. Disadvantages include high cost, the need to add stations, especially in densely populated areas, the analog signaling requirement, and the initial extra expense for installation. Network technology is discussed in more detail in Chapter 12.

NETWORK IMPLEMENTATION CONSIDERATIONS

In many plants, a great deal of equipment operates as islands which are separate from the mainstream of product design and production. This tends to limit our view of what is possible in the integration of equipment in the total factory environment.

The technology exists today for automating all types of factory operations. A comparison of today's manufacturing equipment with the equipment and systems of only a few years ago illustrates the rapid changes of computerization in the factory.

As automation is implemented throughout the factory, the full potential for increased productivity cannot be fully realized unless all of the data inputs are automatically acquired and all of the data exchanges occur without human intervention. Inherent in such a system is the computer analysis of manufacturing data to reduce rather than expand paper output. The human operator in these systems should be concerned only with monitoring performance, with an occasional intervention, when an unforeseen event occurs. Manual data entry is prone to error and can easily become a plant bottleneck, limiting production and productivity for the entire facility. This can be eliminated if the input data exchange to all factory automation equipment is automated.

The test and measurement functions of the system always become more critical as the product reaches the finished product stage.

In a complete factory implementation, automation equipment allows computer-aided testing (CAT) to take place at a central location in the factory system. The primary interface of the CAT system is with the common data base of manufacturing information.

A computer-based factory information system (FIS) can also be used to monitor the manufacturing process directly. This ties together the various elements of the factory through computer integration and provides the basis for total automation. Such a system tends to be complex, since it must coordinate many activities in order to maximize the productive efforts.

The benefits of automatic data exchange include the following:

1. The capability to check a product or process design before it is actually placed in production.
2. The critical evaluations necessary to optimize the manufacturing or processing system.
3. More complete test documentation.
4. Automation of the test interface, which allows a quicker reaction to changes.
5. Automation of operations for process planning, which can be used to assign the units to the various manufacturing or processing stations.
6. The use of application software which allows the manufacturing data base to be automatically converted into codes for such functions as design drafting, numerical-control (NC) fabrication, and functional testing.

All of this requires automatic high-speed data acquisition and data transmission to make the test and inspection system fully integrated into the computer-controlled manufacturing system. The requirements for a factory network include the computer language, as well as the methods of data exchange. There are advantages in using those methods and procedures that have been standardized or are likely to become standards in the future.

Many early factory automation systems used a proprietary language and network. The language in some cases may have been an extension of a language standardized by ANSI. Some of these extensions were similar to those due to be covered for inclusion in the next revision of the standard. Often the systems based on nonstandard language extensions were not very different from those that used a proprietary language.

Problems occurred because the native language of each computer was different, and each needed its own version of the high-level languages to translate programs into machine code. In the absence of standards, a program written in BASIC might run only on a machine compatible with the one for which it was designed. Since that time, the high-level languages have become more and more machine independent.

Considerable effort is required to implement a factory network and the requirements for future peripheral equipment may be difficult to forecast, so it is cost effective for the software to be fully portable. It should be usable on the widest possible range of equipment.

It is also desirable to use equipment supported by at least two or more suppliers. A disadvantage of the one-vendor approach is that the options are limited by the selection of systems and equipment available from the chosen vendor.

Some vendors address only one aspect of the factory environment. However, they deal with that aspect so well that many customers choose to integrate their products into an overall factory system.

In order for the software to be used in a variety of applications, the following characteristics should be considered:

1. Standardization of functions.
2. Adjustable addressing in the network.
3. A standardized instruction set.
4. A standardized protocol.

Many manufacturers use special versions of standardized languages and protocols. These versions may consist of the standard version and the manufacturer's additions, or they may be a subset of the standardized version. A special version of a language may not be able to use a standardized compiler.

Languages such as BASIC, FORTRAN, and Pascal may use extensions or enhancements that may be standardized. Only the extensions to BASIC have been used for most instrumentation. Many popular extensions may never be standardized.

An extension of IEEE-802 could be used for a network of test systems. Ethernet and IEEE-802 can interchange data at a rate of 10 Mbits/second, which is 500 times faster than that of RS-232 or RS-449, even at the higher data rate of 9600 baud. The later techniques would be used only at the lowest protocol levels.

In the case of a protocol and network for instrumentation, the standards most frequently used are IEEE-488 and IEEE-802. Other standards for instrumentation include Proway, which is a version of IEEE-802, and Ethernet. An automation system using one of these languages and protocols is bound to have some portability. This allows the hardware and software to be utilized for a variety of purposes in the factory. If another instrument is required to replace a unit being repaired or to balance changes in the production line, another unit from elsewhere in the factory can easily be used. The word size should be at least 16 bits, but processing can be in 32- or 64-bit blocks, depending on the efficiency needed.

A wide variation in machine architectures can make the sharing of information difficult, requiring the development of hardware and software compatibility at the network level. In order to integrate the various functions into a manufacturing environment, it is necessary to consider a number of factors. In extremely high electrical noise environments, low-noise or fiberoptic cable should be used.

The access method may use a form of carrier sense multiple access with collision detection (CSMA/CD) or token passing. The method used depends on the efficiency required and the expected channel traffic. The cost of implementing these techniques rises as the number of nodes in the network is increased.

The protocol should be compatible with the seven-layer OSI structure. Hierarchical data link control (HDLC) can be used to support the data link layer. IEEE-802 or Proway-based techniques can be used to support the lower layers and define the header, data, and trailer fields of the message frame.

If connected systems use a language of the same form as the one that has been standardized, the application software can be compatible with a similar system from a different source. The hardware and software can use standard or proprietary interfaces, algorithms, and protocols. An important factor in selecting a language for application software is the languages used in the other factory networks. But for data to be exchanged between any two networks, not only the language must be the same, but also

1. The protocol, or the rules for data exchange.
2. The word size, or the number of bits in a word.
3. The characterset, or the decimal number assigned to each alphanumeric character and graphic symbol.

The three most popular character sets are ASCII, EBCDIC for alphanumeric characters, and IGES for graphic characters.

The networks between which data exchange is desired are connected by an I/O interface. This section may be passive until specifically addressed by a network. It can then buffer the data to be sent to the other network and determine if the data have been correctly received using a redundancy check.

The interface function becomes more complicated when there are differences in any of the characteristics of the networks being connected. The interface must then harmonize, or eliminate, the differences.

One technique used to do this, especially for simple differences, is to use programmable read-only memory (PROM). A table look-up can be performed, using the incoming bit stream as the index. The bit stream required by the other network for the identical meaning is returned from the table. The message is then reconstructed with the header and the trailer of the data stream which were stripped from the incoming message and placed around the new bit stream. This technique is used in principle in some gateways.

NETWORK PROTOCOLS

In the network environment, *protocol* has come to mean both the format and the set of procedures which allow communications. The layering of protocols is possible because of the stratification of functions among the various parts of the system. Layering allows devices from different sources to communicate as long as they follow the protocol. A standard model for a layered protocol for computer networks has been released by the United Nations Consultive Committee for Telephone and Telegraph (CCITT) through ISO. This is the *Open Systems Interconnection*, also known as the *Open Systems Architecture*. It has been adopted by ANSI, and defines the function and layering of a set of protocols for the interoperation of pieces of equipment that are built independently. This model consists of seven layers or shells. Starting from the highest or seventh layer, the layers are shown in Figure 11-5.

Physical communication between the networks occurs at the physical layer managed by the data link layer, which, in turn, serves the other upper layers.

A standard such as IEEE-488 can be used to implement the first layer. The advantage of using a standard interface such as the 488 bus is the elimination of the need to design an I/O interface for each device. One type of hardware driver, the IEEE-488 controller, provides the handshake, timing, and protocol logic for most instruments.

The use of standards such as this can dramatically lower the cost of factory inspection and test stations. In addition to IEEE-488, the other major transmission media and interconnections most often considered for computer-controlled

LEVEL	TITLE	FUNCTION
7	APPLICATION LAYER	Deals with the specific application, such as editing or communications
6	PRESENTATION LAYER	Provides the techniques for mapping between the various user terminals
5	SESSION LAYER	Allows the controlled access mode to host services such as user LOG IN operations
4	TRANSPORT LAYER	Provides services as flow control, error control, sequencing, and multiplexing
3	NETWORK LAYER	Contains the set of procedures that allows a host to send or receive data on the network. Includes addressing and network control messages
2	DATA LINK LAYER	Provides the line-control procedure which allows the transporting of frames from one computer to another
1	PHYSICAL/ELECTRICAL LAYER	Defines the physical or electrical standard, such as the type of connector used, the voltage levels, and the function of each connector pin

Figure 11-5 OSI layers.

inspection and testing include RS-232C/RS-449, IEEE-802/Proway, and Ethernet.

The RS-232C and RS-449 standards are designed for communications between two specific nodes. They do not contain the mechanisms required for transferring data from one node to another. These standards have been widely accepted for terminal devices and are used in a variety of equipment.

12

Local Network Protocols for Factory Applications

In this chapter we will discuss the current standards for LANs. The protocols used will be referenced to the Open Systems Interconnection (OSI) reference model, introduced in Chapter 11. The growth of LAN standards began with EtherNet and evolved to the current IEEE-802 standards, which offer several different types of LAN configurations.

While RS-232 and IEEE-488 have been applied to a broad range of applications, they generally do not have enough of the transmission control features required for many factory control applications. They are useful for instruments at the basic cell and station levels of the automation system. The interface I/O between the test center level and the factory level uses IEEE-802/PROWAY or another Ethernet-based standard.

Protocols and standards for LANs involve more than those internal to the local network. Standards like RS-232 and IEEE-488 for interfacing the terminals and computers to the LAN are also required. Then there are the protocols for network management and control operations. In the OSI reference model, the protocols internal to a local network occupy the two lower layers: the physical layer and the data link layer.

Physical layer protocols are concerned with the mechanical, electrical, functional, and procedural functions. For example, the RS-232 standard defines a standard 25-pin connector, specific electrical signal levels, the assignment of functions to the pins, and the procedures for initializing communications and for sending and receiving the data.

The data link layer protocol provides functions, such as error control and access control, on a logical data link. Many local network protocols resemble high-level data link control (HDLC) at the data link layer. HDLC uses a cyclic redundancy checksum (CRC) for error detection and polling to control which station has access to a multipoint link.

In networks, where many stations may share a common communications line, one of several access-control schemes may be used. Token-passing schemes related to polling or a contention scheme known as *carrier sense multiple access with collision detection (CSMA/CD)* can be used.

The first Ethernet installation appeared in 1976, and in 1980 the Ethernet specification was released by Digital Equipment Corporation (DEC), Intel, and Xerox. The Ethernet specification was treated as the industry standard for local networks by several independent vendors, and components began appearing for Ethernet networks as a result. The concept of an open system which allows the connection of different vendors' equipment was an important reason for this popularity.

An access method is always required in an Ethernet type of network, since any node may transmit a message when the network is available. If two or more nodes were allowed to transmit at the same time, the message would become garbled beyond recognition.

A contention system often used involves employing a Carrier Sense Multiple Access (CSMA) algorithm. One particular type of CSMA is collisions detection (CSMA/CD). The basic technique is discussed below. The node is monitored before transmission is allowed. The network must be in a clear state for the transmission to be allowed. If the monitoring of two nodes results in a contention problem, the timing procedure is repeated using a random timing algorithm. Since the timing is random, the contention is unlikely to be repeated.

At the link control layer, the CSMA/CD technique is employed at the line control which defines a frame structure for messages similar to that used in HDLC.

Ethernet-based standards use coax cable which is usually less than a few kilometers in length. The network can have a tree-like or dendritic network configuration, which is typical of factory networks. Propagation delays in the cable are short, so transceivers may be used to detect collisions by comparing the outgoing and incoming signals. The minimal configuration consists of a single coax segment up to 500 meters long.

The user devices, workstations, or host computers are connected to the main cable using an access or transceiver cable that may be up to 50 meters in length. Most of the Ethernet protocol logic is implemented in the workstation node.

The access cable is connected to the main cable using a transceiver, which performs signal transmission and reception. The transceiver detects collisions by comparing the outgoing and incoming signals, and transmission is halted when collisions are detected.

A network of more than 500 meters will use multiple segments connected with repeaters. The repeater retransmits each message with a small delay designed to cancel out the distance limitation of a single segment with the baseband signaling employed.

In a large building installation, one segment may be vertical, up a utility shaft, with segments spread out over each floor. A single segment can contain up to 100 stations.

A point-to-point link is used to connect segments that may be separated by some distance. The link may be up to 1000 meters in length and acts as a repeater divided into two sections. The interconnected segments appear to be in one logical space to the CSMA/CD protocol. A maximum of two repeaters can be in the path between any two stations.

IEEE-802

The Ethernet standard is now one of the options within the IEEE-802 local network standards. There have been a few modifications from the 1980 specification. The final form of the Ethernet specification will be discussed as part of the IEEE-802 standard.

The IEEE started project 802 for local network standards in 1980. This project covered the two lower layers of the OSI reference model.

The data link layer is divided into two sublayers: a logical link control sublayer similar to HDLC and a medium access control (MAC) sublayer that uses the CSMA/CD protocol.

An access unit interface is used in place of the transceiver cable in Ethernet. This interface is defined between the component containing both the link layer and physical signaling and the physical medium attachment to the transmission medium.

The IEEE-802 standard is actually a family of standards. The IEEE 802.2 physical layer options include the IEEE 802.3 CSMA/CD bus, the 802.4 token bus, and the 802.5 token ring.

These options allow either CSMA/CD or token passing. The two techniques are not compatible and cannot be used interchangeably. In token or baton passing, only the node that has a special bit, the token pattern, is allowed to transmit. The node is allowed to hold the token for only a certain period of time; then it must be passed on to another node. The nodes can also be assigned a priority so that a more important node has a greater chance to transmit than others.

The three main options involving the MAC sublayer and the physical layer allow a CSMA/CD access protocol on a bus topology and a token access protocol on either a bus or a ring topology.

Along with these three major options, there are a number of suboptions for different transmission media, signaling methods, and data rates. The following list summarizes these options:

1. CSMA/CD

Baseband 50-ohm coax	*Broadband 75-ohm coax*
Manchester coding	
10 Mb/second	2, 10 Mb/second

2. Token Bus

Single-channel, 75-ohm coax	*Broadband 75-ohm coax*
Continuous and Coherent FSK*	Multilevel, Duobinary AM/PSK†
1, 5, 10 Mb/second	1, 5, 10 Mb/second

3. Token Ring

Baseband, 150-ohm, twisted pair	*Baseband 75-ohm coax*
Differential Manchester	
1, 4 Mb/second	4, 20, 40 Mb/second

Although Ethernet and IEEE-802 have common origins, there are a number of important differences between these standards. Ethernet systems use a common set of specifications so that every Ethernet system is compatible with every other. IEEE-802 allows options as discussed above, so there may not always be complete compatibility between all nodes of a network unless each node uses exactly the same options. Both network standards require transceivers at the nodes, but the cables connecting the transceiver I/O to the node are different. There are also differences in addressing, encoding, control, synchronizing, and access methods.

The logical link control sublayer for IEEE-802 includes

1. An interface service specification to the network layer above.
2. The logical link control procedures.
3. An interface service specification to the medium access control sublayer.

These relationships are shown in Figure 12-1.

THE MAP STANDARD

The goal of making factory equipment compatible with OSI has been difficult. One of the main reasons has been the different protocols used by the various vendors. Computer equipment from one vendor could not communicate with the computers or PLCs of another vendor.

The National Bureau of Standards (NBS) has the Transport Control Protocol

*Frequency shift keying.

†Phase shift keying.

Logical link control	Interface service specification
	Logical link control procedures
Medium access control	Link control/MAC interface specification
Physical signaling	
Physical media attachment	Access unit interface
Medium	

Figure 12-1 IEEE-802 logical link control sublayer.

(TCP) and the Internet Protocol (IP). Several other protocols have evolved as de facto standards for telecommunication applications. These include:

1. File Transfer Protocol (FTP)
2. Virtual Terminal Protocol (VTP)
3. Mail Transfer Protocol (MTP)

In most cases, the user can add to the header or trailer fields of a frame the additional protocol information necessary to ensure proper application of the terminal equipment and proper processing of the message or data field of the frame. Exactly what must be added depends on the different pieces of equipment needed for a particular system implementation.

The needs of many users who have to connect diverse devices together has resulted in the development of related OSI networks, such as the Manufacturing Automation Protocol (MAP) and the Technical and Office Protocol (TOP). General Motors and Boeing have been involved in the development of these two networks since 1980.

MAP is a broadband token-bus network designed for the factory environment. Over 500 companies in the United States and Europe, as well as Japan, Canada, and Australia, have been involved in the developments linking the various islands of information of the factory network called with MAP.

MAP's advantage lies in the fact that its specifications have been developed by users rather than vendors. General Motors plans to install MAP in all of its plants.

In a typical factory system (Figure 12-2) a mainframe computer may be running materials requirement planning (MRP), microcomputers may be running computer-aided design with real-time mini- or microcomputers doing quality control, and PLCs scheduling and dispatching parts on the manufacturing line. Other mini- or microcomputers may be controlling robots for spot welding and painting or NC machines for drilling holes and milling parts.

The initial version of MAP has undergone major changes. At the application layer, the protocol called the *Manufacturing Message Format Standard (MMFS)* was superseded by the Electronics Industries Association standard EIA 13 93,

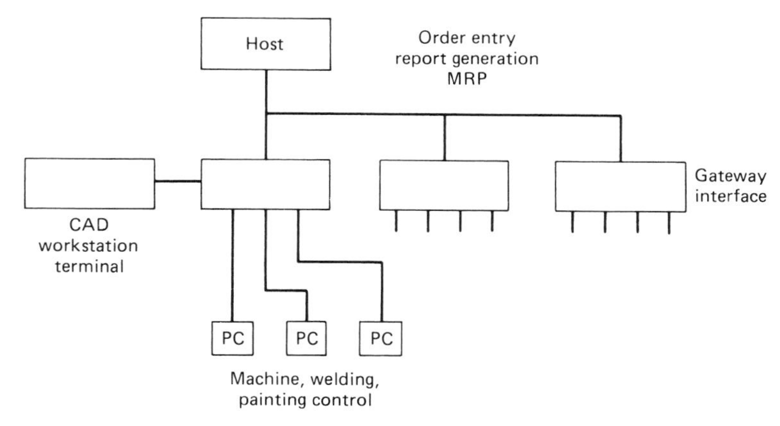

Figure 12-2 Typical network application.

also known as *RS-511*. As a result, most equipment for the implementation of MAP will bypass MMFS as much as possible.

Adjusting to other changes to MAP in a way that is cost effective requires that the changes affect only the software. Several configurations may be used. One is to implement layers 3–7 in software, with layers 1 and 2 and an HDLC interface on a Token Interface Module (TIM) which communicates under X.25. Another configuration is to implement layers 5–7 in software with a MAP interface board to handle layers 1–4.

MAP APPLICATION

For manufacturing and process control, applications standards like MAP and the Proway-LAN standard for industrial data promise to solve many present network bottlenecks.

A typical MAP application may involve the machining, assembly, and testing of components. The factory LAN integrates the computer equipment with automated test systems and PLCs, which in turn may be linked to conveyors, robots, machining equipment, and inspection gauges. Network users can access data collected over the MAP network by a supervisor processor by logging in with ASCII terminals.

Software resident on the supervisor allows users to access factory data over the MAP network for applications such as statistical process control, production reports, monitoring production equipment, and tracking machine tool usage in order to plan for tool maintenance. A central data base maintains information about the operation of manufacturing equipment tied to the network. This information can be used to create reports on machine performances.

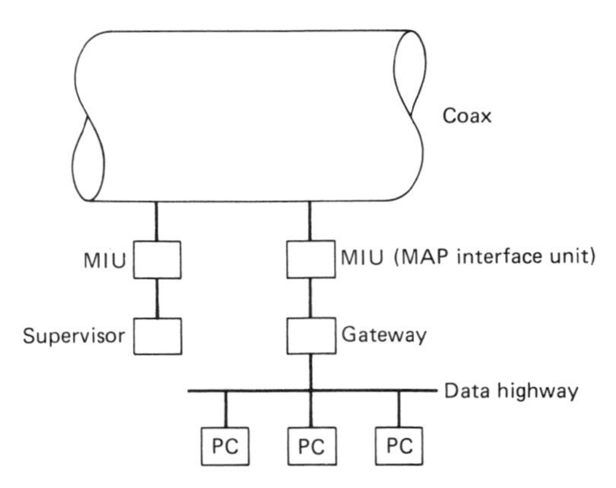

Figure 12-3 Typical MAP network connections.

The PLCs which step control the operation of shop floor equipment are hardwired to robots, numerical controllers, and conveyor belt machinery.

A gateway is provided for PLCs which transmit data to the MAP network (Figure 12-3). The PLCs are connected to a data highway which operates at 57K bits/second.

Many factors must be studied for the implementation of a network on a company's plant floor. Network access methods, difficulty of network implementation, cost of components, maximum area of coverage, and transmission media flexibility are just a few of the more important areas to be considered.

The following is a summary of typical factory network implementations:

1. Baseband
   Carrier sense multiple access with collision
   detection (CSMA/CD)
   Coax cable
   Bus topology (IEEE-802.3)

This network scheme is based on Ethernet. It supports a maximum transmission speed of 10 Mbps and has a maximum length of 1500 meters. The maximum length and other restrictions of this scheme make it difficult for the network to cover large areas, such as an entire factory floor for a large plant. The CSMA/CD access method limits the maximum attainable throughput to about 40% of the transmission speed, or 4 Mbps. The advantage is that these networks are easily installed.

2. Broadband
   CSMA/CD
   Coax cable
   Bus topology (Broadband IEEE-802.3)

The major advantage of this LAN over a baseband bus LAN is that it can operate with longer distances between components. Broadband signaling provides the ability to operate several networks simultaneously over different radio frequency channels on a single coax cable. The disadvantages of a broadband network using CSMA/CD include the added complexity and cost of the cable.

3. Baseband
   Token passing
   Coax cable
   Bus topology (IEEE-802.4)

This network provides the advantages of the token-passing network access technique and has the ability to cover long distances on the factory floor. Its main disadvantage is that the token-passing algorithm is more difficult to implement than contention-based access methods such as CSMA/CD.

4. Broadband
   Token passing
   Coax cable
   Bus topology (IEEE-802.4)

This network was adopted as the first of the seven layers of General Motors' MAP. It is possible to calculate the maximum amount of time a single network node would wait to gain access to a token-passing network. This cannot be done with networks that implement the CSMA/CD network access method. Implementing a broadband token-passing network is more expensive and requires more engineering than implementing a baseband token-passing network. This is a disadvantage of broadband.

5. Baseband
   Token passing
   Twisted-pair
   Ring topology (IEEE-802.5)

This is the scheme used by IBM in its cabling system. It allows users to create a network with existing twisted-pair wiring. Like other token-passing networks, this network provides a higher throughput than contention-based networks using CSMA/CD. The wiring plan and the network access algorithms should take into account the reliability problems of ring technology.

6. Baseband
   Token-passing
   Fiber-optic cable
   Ring topology (ANSI X3T9.5)

LOGICAL LINK CONTROL	MAC service specification
MEDIUM ACCESS CONTROL	MAC PS service specification
PHYSICAL SIGNALING	PS/AU interface specification Access unit interface
PHYSICAL MEDIA ATTACHMENT	Baseband MAU or broadband MAU
MEDIUM	

Figure 12-4 IEEE-802 ISO layers—CSMA/CD access protocol.

This network is different from others in that it uses fiberoptic cable rather than coax cable or twisted pairs. In addition to the advantages of token passing, this type of LAN is insensitive to electromagnetic interference. The optical cable networks have a throughput of 100 Mbps. The major disadvantage is the cost of the fiberoptic cable system.

The CSMA/CD access protocol is one of the options in IEEE-802. This option relates to the ISO layers, as shown in Figure 12-4. This option contains:

1. The MAC service specification at the logical link control ISO layer.
2. The MAC sublayer with CSMA/CD.
3. A physical signaling layer service specification.
4. The interface specification between the physical signaling layer and the access unit (AU).
5. The physical medium attachment unit (MAU) with baseband and broadband options.

MEDIA ACCESS CONTROL SUBLAYER FUNCTIONS

Services provided by the MAC sublayer permit the local logical link control sublayer to exchange data with other sublayers. The MAC sublayer defines the procedures of the CSMA/CD protocol, as well as the MAC frame structure.

The functions provided by the MAC sublayer are similar to those provided in a data link layer protocol. Data encapsulation takes place with framing, addressing, and error detection. Medium access management is provided, with allocation performed using collision avoidance and contention resolution for collisions.

COLLISION DETECTION

The CSMA/CD protocol takes place during the following sequence of events:

1. The MAC sublayer constructs the frame from the data supplied by the logical link control.
2. The carrier-sense signal at the physical layer is monitored for traffic.
3. If the response is positive, the transmission is placed in a hold state.
4. If the response is negative or clear, a frame transmission is initiated after a fixed delay of 9.6 microseconds for a 10-Mbps system to provide recovery time for other stations.
5. The medium is monitored by the physical layer for collisions.
6. A collision-detect signal is sent to the MAC sublayer when a collision occurs.
7. A bit sequence known as a *jam* (32 bits for a 10-Mbps system) is transmitted to ensure that the duration of the collision is sufficient to be noticed by the other transmitting stations.
8. The transmission is terminated and a retransmission is attempted after a random delay.
9. If collisions occur again, the transmission is repeated, and the random delay is doubled with each attempt.*
10. The retransmission is abandoned after 16 tries, and an error alarm occurs.

Collisions occur when several stations attempt to transmit at the same time and their signals interfere with each other. A collision window is used during the first part of the transmission. This allows the transmitted signal to propagate to all of the stations and a return signal to propagate back.

The time needed to acquire the medium is based on the round-trip propagation time of the physical layer. The retransmission delay is a function of the slot time, which should exceed the round-trip propagation time and the jam time. In a 10-Mbps baseband system, the slot time is equal to 512 bit times. The backoff delay is always an integral number of slot times. At the end of the collision window, the station should have control of the medium and subsequent collisions should not occur.

The MAC frame format is similar to the original Ethernet specification, except for two differences. Ethernet addresses are always 6 octets, and a type field is used instead of the length field.

*This is called *binary exponential backoff*. The delay interval is doubled for each of the first 10 retransmissions and then stays the same up to the 16th retransmission.

PHYSICAL SIGNALING WITH THE ATTACHMENT UNIT INTERFACE

The MAC sublayer communicates with the physical signaling sublayer using primitives or rules that are defined in the service specification. Data terminal equipment (DTE) containing the data link and higher-level protocols may be physically separate from the MAU that connects to the transmission medium. The DTE and MAU are connected by a cable with specifications defined in the attachment unit interface (AUI). The interface allows the DTE to be used with the specified types of transmission media. It also allows the DTE to test the interface, the interface cable, the MAU, and the medium.

Manchester coding is used for the transmission of data through the AUI. This type of coding combines the data and clock into a single bit stream. A transmission to the inverse state always occurs in the middle of each bit period. The signal state during the first half of the bit period indicates the data value.

The same framing and preamble are used for the transmission of data across the AUI as used in the MAC sublayer. The AUI will support data rates of 1, 5, 10, or 20 Mbps with a cable of up to 50 meters in length.

OPERATION OF THE MEDIUM ATTACHMENT UNIT

The MAU for a baseband coaxial system originated in the original Ethernet specification. The signaling rate is 10 Mbps, and up to 500 meters of coaxial cable is allowed in a single segment. The total number of MAUs on a segment cannot exceed 100. Multiple segments are connected with repeaters using a bus topology similar to that of Ethernet. There can be up to two repeater units in the path between any two MAUs.

The 50-ohm coaxial cable is normally marked with color rings every 2.5 meters. The MAUs are then attached at these points to keep the signal reflections to a minimum. The connections are made with a piercing tap connector or by cutting the cable. The shield of the trunk coax cable should be grounded at only one point of the cable.

Collisions are detected on the coax cable when the signal level on the cable exceeds the value allowed by two transmitters. The maximum cable connection path for a point-to-point link is 2500 meters, and the data propagation delay for a repeater must be less than 7.5 bit times.

BROADBAND SYSTEMS

Since broadband networks use the same types of coaxial cable and components used in CATV systems, the signal capability of these systems can exceed 300

MHz. Broadband systems use a tree topology and employ splitters. The root of the tree is known as the *head end*. The signals are applied to the cable at directional taps and travel in a single direction.

Broadband systems use 75-ohm coax cable. The trunk sections are normally sheathed in aluminum.

The signal levels are calculated during the design based on the attenuation of the cable lengths and the number of taps and splitters. Then amplifiers are inserted at the proper points to maintain the signal levels. Due to the use of these amplifiers, the system can have a maximum length of several kilometers from the head end.

Either a dual-cable or a single-cable system can be used. The dual-cable system is based on two parallel cables in a tree configuration which are looped together at the head end. One MAU will transmit on the inbound cable to the head end, where the signal is looped to the outbound cable and broadcast to the other MAUs. The single-cable systems uses different frequencies for each direction of transmission.

If the frequency range is halved, the lower half of the frequency range is used for inbound transmissions and the upper half for outbound transmissions. A modulator at the head end shifts the inbound signals up in frequency and retransmits them on the outbound cable.

The transmitter and receiver use different frequencies in the single-cable system. Dual-cable systems do not normally use a head-end modulator with separate transmitter and receiver frequencies.

Collision detection cannot be based on the signal levels because of the variation in signals along the different points of the cable. Collision detection is usually done with a bit comparison.

A copy of the transmitted frame is kept by the MAU, and this is compared with the received frame. A match is required within a certain time window in order to acquire the cable. If other MAUs are transmitting, the match within the window will not be reached.

Data rates range from 1 to 10 Mbps, and frequency allocation is performed in units of 6 MHz, as done for television channel bandwidths.

TOKEN BUS OPERATION

The token access protocol with a bus topology is similar to the CSMA/CD distinct MAU, which is physically separate from the DTE is not used. This is shown in Figure 12-5.

The right to transmit the token is passed among all the stations in a logical circle or ring. Each station knows the address of the station it got the token from (the predecessor) and the address of the station that should get the token next (the successor). The predecessor to successor addresses must be determined and updated. The token is passed from station to station in numerically

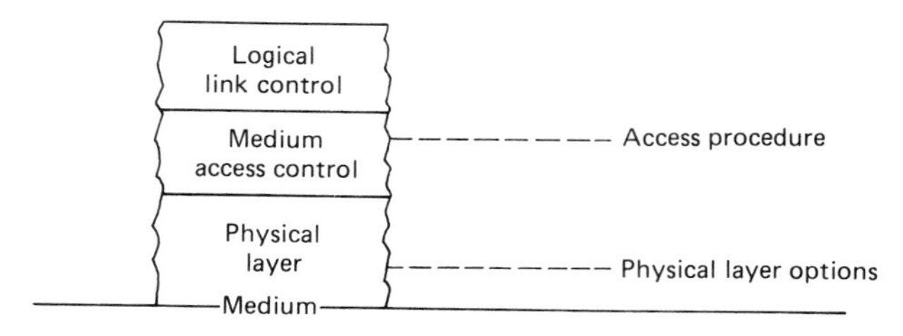

Figure 12-5 Token bus layers.

descending order, but the numerical address of a station does not necessarily relate to its physical position on the bus.

As a station completes the data transaction and the required maintenance functions, it passes the token to its successor. After it transmits the token frame, the station checks the bus. If it detects a valid frame following the token, it assumes that its successor has the token and is transmitting. If a valid frame is not detected, it attempts to recover the token.

When a valid response is not detected after sending the token the first time, successive fallback steps are used to recover the token. The token passing is tried again, and if there is still no response, the sending station assumes that the successor is inoperative.

A special bit frame called a WHO FOLLOWS frame is then sent to the failed successor's address. The network stations then compare the value of this frame with the address of the predecessor.

The station that achieves a match then sends its address in a SET SUCCESSOR frame. The station holding the token uses this address, and the failed station is bypassed on the logical ring.

When there is no response to the WHO FOLLOWS frame, the frame is sent again. If there is still no response, the station sends a SOLICIT SUCCESSOR 2 frame, naming itself as the destination. If any operational stations respond, they will be added to the logical ring using a controlled-contention process that employs response windows.

The response window defines a time interval during which a station must hear the beginning of a response from another station. The window is a controlled interval of time following the transmission of a MAC control frame in which the station sending the frame pauses to detect a response. If the station detects a transmission start during this window period, it will continue to listen to the transmission, even after the response window time expires, until the transmission is complete.

Every station in the ring periodically sends SOLICIT SUCCESSOR frames specifying a range of station addresses between the frame source and the destination address. Stations with an address within this range can enter the ring by responding to the frame.

The sender of a SOLICIT SUCCESSOR frame transmits the frame and waits for a response in the response window after the frame. The responding stations send requests to become the next station in the logical ring.

When the frame sender detects a valid request, it allows the new station to enter the ring by changing the address of its successor to the new station and passing the token to the new successor. Initialization is handled as a special case of adding new stations using a timer.

If multiple stations respond to a SOLICIT SUCCESSOR frame, an arbitration method is required to identify a single responder. A RESOLVE CONTENTION frame is used in which the responding stations must choose a 2-bit value and then listen for either zero, one, two, or three slot times. The process is repeated until only a single responder is left.

A priority system is used to prevent any single station from monopolizing the network. Data frames are assigned a service class which represents one of four priorities. The priority system allocates the network bandwidth to the higher-priority frames and sends the lower-priority frames only when there is sufficient bandwidth. The bandwidth is allocated by timing the rotation of the token around the logical ring. Each class is assigned a target token rotation time.

The station measures the time it takes the token to circulate around the logical ring. If the token returns to a station in less than the target rotation time, the station is allowed to send frames of that particular class until the target rotation time expires. If the token returns after the target rotation time has expired, the station is not allowed to send frames of that priority. A token hold time is also used to limit the number of high-priority frames that can be transmitted before passing the token.

PHYSICAL LAYER AND MEDIUM

Three types of physical layer and medium can be used with the token bus:

1. Single-channel, phase-continuous FSK.
2. Single-channel, phase-coherent FSK.
3. Broadband bus.

Phase-continuous FSK is a type of frequency shift keying in which the translation between the signaling frequencies is continuous, rather than a shift from one frequency to another. The medium in this case consists of a long unbranched trunk cable which uses tee connectors and short drop cables.

The extension to a branched trunk requires repeaters which are used to span the branches. The tee connectors are nondirectional, and the signal from a station propagates in both directions along the trunk cable.

The trunk cable is 75-ohm, CATV-type coaxial cable. The drop cable is 35- to 50-ohm cable and must be less than 35 cm in length.

For phase-continuous FSK, the data-signaling rate is 1 Mbps with differential Manchester encoding. The modulation carrier uses a center frequency of 5 MHz and varies smoothly between the signaling frequencies of 3.75 and 6.25 MHz.

Phase-coherent FSK is also a type of frequency shift keying. Here the two signaling frequencies are integrally related to the data rate. The transitions between the two signaling frequencies are made at zero crossings of the carrier waveform. The medium is 75-ohm cable similar to that used in CATV. Branching is accomplished with splitters, and the trunk cable is connected to the drop cable with nondirectional taps. Extensions use repeaters connected in the trunk cable. Data-signaling rates are 5 and 10 Mbps. The 5-Mbps data rate uses signaling frequencies of 5 and 10 MHz, while the 10-MHz data rate uses signaling frequencies of 10 and 20 MHz.

In the broadband bus option, the medium is also 75-ohm coaxial cable. Either a single-cable split configuration or a dual cable may be used.

Signals are transmitted toward the head end and then retransmitted on the outbound cable or channel. A multilevel duobinary AM/PSK modulation is used in which the carrier is amplitude modulated and phase-shift keyed.

Duobinary signaling allows the data to be transmitted at reduced frequencies. Multilevel means that several distinct amplitude levels are used to represent the data. Three levels are used in this case, with a signal rate of about 1 bit/Hz.

Data-signaling rates are 1, 5, and 10 Mbps, and the required channel bandwidth are 1, 5, 6 and 12 MHz. There is a difference of 192.25 MHz between the inbound and outbound channels.

TOKEN RING

The token ring uses a token MAC with stations connected in a ring configuration instead of the straight bus configuration. The token ring standard is illustrated in Figure 12-6. A trunk coupling unit is inserted in the trunk cable.

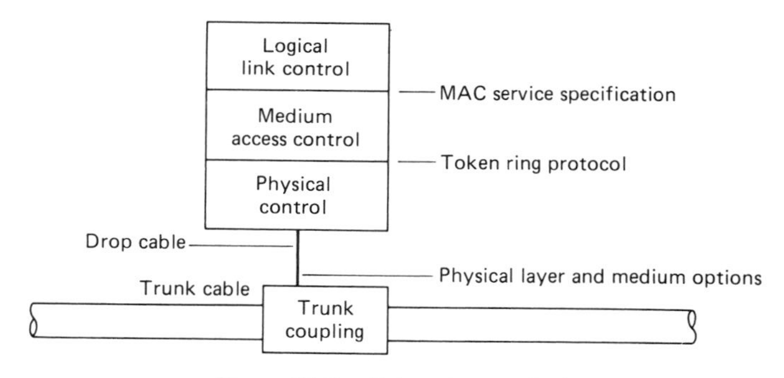

Figure 12-6 Token ring standard.

The frames traverse a complete circle of the ring, returning to the source station, which removes the frame. A station passes the token to its physical successor

1. If it has no more data to transmit.
2. If the token-holding timer has expired.

When the destination address matches the address of a receiving station, the station sends the frame to its logical link control layer. In this way, the physical connections of the medium establish a logical connection of active stations on the ring. The token is used by a station to transmit frames. Each station on the ring will repeat the data it receives.

TOKEN RING OPTIONS

Two physical layer and medium options can be used for the token ring. For lower-cost applications, 150-ohm shielded twisted pairs with data rates of 1 or 4 Mbps can be used. Baseband signaling is used with differential Manchester coding. For higher data rates, coax cable is used, which allows data rates of 4, 20, and 40 Mbps. Baseband signaling with differential Manchester coding is also used for this option.

Acronyms

A	open loop gain
ACK	Acknowledge
ADP	Ammonium Dihydrogen Phosphate
AEN	Address ENable
ALE	Address Latch Enable
AM/PSK	Phase Shift Keying
ANSI	American-National Standards Institute
ASCII	American Standard Code for Information Exchange
ASG	Automatic Speech Generation
ASR	Automatic Speech Recognition
AU	Access Unit
AUI	Attachment Unit Interface
BCD	Binary Coded Decimal
BOB	Bipolar Offset Binary
BTC	Bipolar Two's Complement
CAD	Computer-Aided Drafting or Design
CAT	Computer-Aided Testing
CATV	Community Antenna Television
CCD	Charged-Coupled Device
CCITT	United Nations Consultive Committee for Telephone and Telegraph
CD	Compact Disk
CJC	Cold-Junction Compensation
CMOS	Complementary Metal Oxide Semiconductor
CMRR	Common Mode Rejection Ratio
CPU	Central Processing Unit

CRC	Cyclic Redundancy Check or Checksum
CS	Chip Select pin
CSMA/CD	Carrier Sense Multiple Access with Collision Detection
CSMA	Carrier Sense Multiple Access
CSR	Continuous Speech Recognition
CUSUM	Cumulative SUM
DAC	Digital-to-Analog Converter
DAS	Data Acquisition System
DAT	Digital Audio Tape
DAV	Data Valid on Lines
DCS	Distributed Control System
DDC	Direct Digital Control
DEC	Digital Equipment Corporation
DMA	Direct-Memory Access
DPMI	DOS Protected Mode Interface
DTE	Data Terminal Equipment
DTL	Diode-Transistor Logic
DVM	Digital Voltmeter
EBCDIC	Extended Binary-Coded-Decimal Interchange Code
EIA	Electronics Industry Association
EMI	ElectroMagnetic Interface
EMS	Expanded Memory Specification
EMS	Extended Memory Specification
FET	Field Effect Transistor
FFT	Fast-Fourier Transform
FIS	Factory Information System
FSK	Frequency Shift Keying
FTP	File Transfer Protocol
GM	General Motors
GPIB	General-Purpose Interface Bus
HDLC	High-Level Data Link Control
I/O	Input-Output
IBF	Input Buffer Full
IC	Integrated Circuit
IEEE	Institute of Electrical and Electronics Engineers
INTR	Interrupt Request
IOR	I/O Read
IOW	I/O Write
IP	Internet Protocol
ISA	Instrument Society of America
ISO	International Standardizing Organization
ISO	International Standards Organization

ISO	International Standards Organization
IWR	Isolated Word Recognition
LAN	Local Area Network
LCD	Liquid Crystal Display
LED	Light-Emitting Diode
LPC	Linear Predictive Coding
LSB	Least Significant Bit
LVDT	Linear Variable Differential Transformer
MAC	Media Access Control
MAC	Medium Access Control
MAP	Manufacturing Automation Protocol
MAU	Medium Attachment Unit
MEMR	MEMory Read
MEMW	MEMory Write
MFM	Modified Frequency Modulation
MMFS	Manufacturing Message Format Standard
MOSFET	Metal Oxide Semiconductor Field Effect Transistor
MRP	Materials Requirement Planning
MSB	Most Significant Bit
MTBF	Mean Time Between Failures
MTP	Mail Transfer Protocol
NBS	National Bureau of Standards
NC	Numerical Control
NDAC	Not-Data-Accepted
NEMA	National Electrical Manufacturers Association
NMOS	n-channel Metal Oxide Semiconductor
NRFD	Not-Ready-For-Data
OBF	Output Buffer Full
OCR	Optical Character Recognition
OCR-A	Optical Character Recognition
OSI	Open Systems Interconnection
P&ID	Process and Instrument Design
PBX	Private Branch Exchange
PFK	Programmed Function Keyboard
PID	Proportional-Integral-Derivative
PLC	Programmable Logic Controller
POS	Programmable Option Select
RD	Read
RFI	Radio Frequency Interference
RLL	Run-Length Limited
ROM	Read Only Memory
RTD	Resistance Temperature Detector

RTU	Remote Terminal Unit
RW	Read/Write
SAMA	Society of American Manufacturer's Association
SCSI	Small Computer System Interface
SDLC	Synchronous Data-Link Control
SIL	Single Inline
SIMM	Single Inline Memory Module
SLC	Single Loop Controller
SNA	Systems Network Architecture
SPC	Statistical Process Control
SQC	Statistical Quality Control
TCP	Transport Control Protocol
TIM	Token Interface Module
TOP	Technical and Office Protocol
TTL	Transistor-Transistor Logic
UART	Universal Asynchronous Receiver Transmitter
UPS	Uninterruptible Power Supply
USB	Unipolar Straight Binary
VCPI	Virtual Control Program Interface
VLSI	Very Large Scale Integradation
VOM	Volt-Ohm-Milliammeter
VTP	Virtual Terminal Protocol
WORM	Write-Once/Read-Many
ZWS	Zero Wait State

Bibliography

Aarons, Richard. BASIC, *PC Magazine*. October 1985, 117–119.

Abbott, K.W. Plan and Control Projects with a Personal Computer. *Hydrocarbon Processing*. December 1983, 63–68.

Adams, L. *High Performance Interactive Graphics*. Blue Ridge Summit, PA: TAB, 1988.

Adams, L.F. *Engineering Measurements and Instrumentation*. London: English Universities Press, 1975.

Altman, L., ed. *Microprocessors*. New York: Electronic Magazine Book Series, 1975.

Andrews, M. *Principles of Firmware Engineering in Microprogram Control*. London: Computer Science Press, 1980.

Asaoka, S., Shino, K., Norimura, K., and Yoshioka, M. *Operation of a 1.3 GeV Electron Synchrotron with a Personal Computer*. Report INS-TH-149. Tokyo: Tokyo Uni Institute for Nuclear Studies, July 2, 1982.

Asohima, N. Personal Computer Based Signal Compression Method Applied to the Measurement of the Sound Field in a Pipe. *Journal of the Acoustical Society of Japan*. March 1984, 146–151.

Athans, M., and Falb, P. *Optimal Control: An Introduction to the Theory and Its Applications*. New York: McGraw-Hill, 1966.

Bade, P., Enggebretson, A.M., Heidbreder, A.F., and Niemoeller, A.F. Use of a Personal Computer to Model the Electroacoustics of Hearing Aids. *Journal of the Acoustical Society of America*. February 1984, 617–620.

Baily, S.J. Personal Computers Frugal Path to Specialized Control Systems. *Control Engineering*. July 1984,

Bannister, B., and Whitehead, D.G. *Instrumentation: Transducers and Interfacing.*, 2nd ed. New York: Van Nostrand Reinhold, 1990.

Bartogiak, G. Guide to Thermocouples. *Instruments and Control Systems*. November 1978,

Battista, Fred F. A Toolkit Approach to Operator Interface Systems. *Proceedings of the*

North Coast Conference, Research Triangle Park, NC: Instrument Society of America. May 1986,

Bell, C., and Newell, A. *Computer Structures*. New York: McGraw-Hill, 1970.

Bell, L.H., and Stout, J.L. Future of the Personal Computer as an Interactive Geological Graphics Workstation. *American Association of Petroleum Geologists Bulletin*. April 1984,

Benedict, R.P. *Fundamentals of Temperature, Pressure and Flow Measurements*. New York: Wiley, 1969.

Berger, J., and Tannhauser, D.S. Personal Computer as an Inexpensive Lock-In Analyzer Operating at Very Low Frequencies. *Review of Scientific Instruments*. December 1983, 1781–1783.

Beveridge, G.S.G., and Schechter, R.S. *Optimization: Theory and Practice*. Series in Chemical Engineering. New York: McGraw-Hill, 1970.

Beychok, M.R. Selecting a Personal Computer. *Chemical Engineering*. October 3, 1983, 103–104.

Blasso, L. Flow Measurement Under any Conditions. *Instruments and Control Systems*. February 1975,

Bone, T., and Stoffer, J.O. Use of the Apple II Personal Computer for GPC Data Collection and Analysis. *Journal of Water Borne Coatings*. August 1983, 24–31.

Breuer, M.A., and Griedman, A.D. *Diagnosis and Reliable Design of Digital Systems*. Woodland Hills, CA: Computer Science Press, 1976.

Brooks, F.P. *The Mythical Man-month, Essays on Software Engineering*. Reading, MA: Addison-Wesley, 1975.

_______. An Overview of Microcomputer Architecture and Software. Micro Architecture. *EUROMICRO 1976 Proceedings*, Europe 1976,

Buckley, P.S. *Techniques of Process Control*. New York: Wiley, 1964.

Burton, D.P. Handle Microcomputer I/O Efficiently. *Electronic Design*. June 21, 1978,

Burzio, G. Operating Systems Enhance uCs. *Electronic Design*. June 21, 1978,

Caine, K.E. Personal Computer as an Engineering Tool. *Iron and Steelmaker*. April 1984, 12–16.

Caldwell, W.I., Coon, G.A., and Zoss, L.M. *Frequency Response for Process Control*. New York: McGraw-Hill, 1959.

Camenzind, H.R. *Electronic Integrated System Design*. New York: Van Nostrand Reinhold, 1972.

Caro, Richard H. *Programming for Batch Process Control, Advances in Instrumentation*, Vol 40. Research Triangle Park, NC: Instrument Society of America, 1985, 883–889.

Cecil, D.R. *Debugging BASIC Programs*. Blue Ridge Summit. PA: TAB, 1976.

Chandy, K.M., and Reiser, M., eds. *Computer Performance*. Amsterdam: North-Holland, 1977.

Chang, C.Y., Ksu, W.C., Uang, C.M., Fang, Y.K., Liu, W.C., and Wu, B.S. Personal Computer-Based Automatic Measurement System Applicable to Deep-Level Transient Spectroscopy. *Review of Scientific Instruments*. April 1984, 637–639.

Chintappali, P.S., and Ahluwalia, M.S. The Use of Personal Computers for Process Control. *Energy Progress*. June 1984,

Chintappalli, P.S., and Steely, R.K. Microcomputers as Productivity Tools in Plant Operations. Presented at the AIChE National Meeting, Anaheim, CA, May 1984.

Chornik, B. Application of a Personal Computer for Control and Data Acquisition in an Auger Electron Spectrometer. *Review of Scientific Instruments*. January 1983, 80-84.

Chu, Y. *Computer Organization and Microprogramming*. Englewood Cliffs, NJ: Prentice-Hall, 1972.

Coffee, M.B. Common-made Rejection Techniques for Low-level Data Acquisition. *Instrumentation Technology*. July 1977,

Collier, D. Personal Computers in Industrial Control. *Microelectronics and Reliability*, 1981, 461–465.

Combs, C.F., ed. *Basic Electronic Instrument Handbook*. New York: McGraw-Hill, 1972.

Considine, D.M. *Process Instruments and Controls Handbook*. New York: McGraw-Hill, 1957.

Corwin, T.K., Clarke, P., and Frederick, E.R. Field Application of a Personal Computer for Data Collection and Reduction. *Proceedings of the Specialty Conference on Continuous Emission Monitoring Design, Operation, and Experience, Air Pollution Control Association*. Denver, CO: 1982, 293–302.

Craig, J.C. *True BASIC*. Blue Ridge Summit, PA: TAB, 1986.

Crick, A. Scheduling and Controlling I/O Operations. *Data Processing*, May–June 1974,

Cronin, D.P. The Personal Computer as an Energy Management System. *Energy Technology*. June 1983, 388–396.

Dannenberg, R.B. Resource Sharing in a Network of Personal Computers. Ph.D. thesis CMU-CS-82-152. Work performed at Carnegie-Mellon University, Pittsburgh, December 1982.

Dao, L.V. *Mastering the 8088 Microprocessor*. Blue Ridge Summit, PA: TAB, 1985.

Davis, A., and Molinari, F. Personal Computer Additions Can Speed Your R&D Projects. *Industrial Research & Development*. February 1983, 178–180.

Davis, D.B. Personal Computer Networks Go On Line. *High Technology*. March 1984, 62–68.

Deschoolmeester, D. The Personal Computer: A Necessary Instrument or a Management Toy. *Informatie Netherlands*. January 1984, 37–43.

Diefenderfer, A.J. *Principles of Electronic Instrumentation*. Philadelphia: W.B. Saunders, 1972.

Dijkstra, E.W. *A Discipline of Programming*. Englewood Cliffs, NJ: Prentice-Hall, 1976.

Dodds, D.E. Home Energy Simulation Using Personal Computers. *Proceedings of ENERGEX '83*. Winnipeg, Manitoba: Solar Energy Society of Canada, 1982, 1020–1025.

Doebelin, E.O. *Measurement System—Application and Design*. New York: McGraw-Hill, 1975.

Donovan, J.J. *Systems Programming*. New York: McGraw-Hill, 1972.

Dow, J. One File-Transfer Protocol Serves All Personal Computers. *Electronics*. August 11, 1983, 114–115.

Downey, R.M. *Communicating Between the IBM Personal Computer and the Wang Word-Processing System*. Report UCID-19889. Livermore CA: Lawrence Livermore Laboratory, September 7, 1983.

Eckhouse, R.H., Jr. *Minicomputer Systems*. Englewood Cliffs, NJ: Prentice-Hall, 1975.

Eckman, D.P. *Automatic Process Control*. New York: Wiley, 1958.
E.E.U.A., *Installation of Instrumentation and Process Control Systems*. Handbook No. 34. London: Constable, 1973.
Eisenberg, J., and Hill, J. Using Natural-Language Systems on Personal Computers. *Byte*. January 1984,
Elliott, T.C. Temperature, Pressure, Level, Flow-key Measurements in Power and Process. *Power*. September 1975,
Engel, S., and Granda, R. *Guidelines for Man/Display Interfaces*. Technical Report TR 00.2720, Poughkeepsie, NY: IBM, 1975.
Englemann, B., and Abraham, M. Personal Computer Signal Processing. *Byte*. April 1984, 94–110.
Evans, F.L., Jr. *Equipment Design Handbook for Refineries and Chemical Plants*. Houston, TX: Gulf Publishing, 1971.
Fabrycky, W.J., Ghare, P.M., and Torgersen, P.E. *Industrial Operations Research*. Englewood Cliffs, NJ: Prentice-Hall, 1972.
Farnbach, W.A. Bring Up Your uP Bit-by-Bit. *Electronic Design*. July 19, 1976,
Feiner, S., Nagy, S., and van Dam, A. An Integrated System for Creating and Presenting Complex Computer-based Documents. *Proceedings of the 1981 SIGGRAPH Conference*. Published in *Computer Graphics*, August 1981,
Foley, C., and Lamb, J. Use a Personal Computer and DFT to Extract Data from Noisy Signals. *EDN*. April 5, 1984,
Foley, J.D. Evaluation of Small Computers and Display Controls for Computer Graphics. *Computer Group News*. January-February 1970,
Foskett, R. Torque Measuring Transducers. *Instruments and Control Systems*. November 1968,
Foster, C.C. *Computer Architecture*. New York: Van Nostrand Reinhold, 1970.
Freedman, M.D. *Principles of Digital Computer Operation*. New York: Wiley, 1972.
Friedman, A.D., and Memon, P.R. *Fault Detection in Digital Circuits*. Englewood Cliffs, NJ: Prentice-Hall, 1971.
Fung, K.T., and Toong, H.D. On the Analysis of Memory Conflicts and Bus Contentions in a Multiple-Microprocessor System. *IEEE Transactions in Computers*, C-27, 1. January 1979,
Gammill, R.C. *Personal Computers for Science in the 1980s*. Report P5954. Santa Monica CA: RAND Corp., February 1978.
Garland, H. *Introduction to Microprocessor System Design*. New York: McGraw-Hill, 1979.
Gear, C.W. *Computer Organization and Programming*. New York: McGraw-Hill, 1974.
Grant, E., and Leavenworth, R.S. *Statistical Quality Control*. New York: McGraw-Hill, 1972.
Gregory, B.A. *An Introduction to Electrical Instrumentation*. New York: Macmillan, 1973.
Grimsdale, R.L., and Johnson, D.M. A Modular Executive for Multiprocessor Systems. *Trends in On-Line Computer Control Systems*. Sheffield, England, April 1972,
Groff, G.K., and Muth, I.F. *Operations Management: Analysis for Decisions*. Homewood, IL: Irwin, 1972.
Guedj, R., et al. *Methodology of Interaction*. Amsterdam: North-Holland, 1980.

Gunashingham, H., Ang, K.P., Mok, J.L., and Thiak, P.C. Design of a pH Titrator as a Component Part of a Personal Computer. *Microprocessors and Microsystems*. July-August, 1984, 274–279.

Hall, J. Flowmeters—Matching Applications and Devices. *Instruments & Control Systems*. February 1978,

———. Solving Tough Flow Monitoring Problems. *Instruments & Control Systems*. February 1980,

Hamann, J.C., and Jacquot, R.G. Analog Simulation on the Personal Digital Computer. *CoED Journal of the Computers in the Education Division of ASEE*. April-June 1985, 12–15.

Hamilton, M., and Zeldin, S. Higher Order Software—A Methodology for Defining Software. *IEEE Transactions in Software Engineering* SE-2, 1. March 1976,

Hannemyr, G. High Quality Document Production on a Personal Computer. *Microprocessing and Microprogramming*, the Netherlands. March-April 1983, 233–242.

Harriott, P. *Process Control*. New York: McGraw-Hill, 1964.

Hayes, P., Ball, E., and Reddy, R. Breaking the Man-machine Communication Barrier. *Computer*. March 1981,

Hearing on National Centers for Personal Computers in Education. Washington DC: House of Representatives. Committee on Education and Labor. April 21, 1983.

Heaton, J.E. *The Personal Computer as a Controller*. Ann Arbor, MI: Oracle.

Helmers, C.T., ed. *Robotics Age, in the Beginning*. Rochelle Park, NJ: Hayden, 1983.

Herrick, C.N. *Instrumentation and Measurement for Electronics*. New York: McGraw-Hill, 1972.

Hewlett Packard Personal and Portable Computers, 1979–1986. Citations from the INSPEC Data Base. Report PB86-857042/XAB. Springfield, VA: National Technical Information Service, January 1986.

Hill, F.J., and Peterson, O.R. *Digital Systems: Hardware Organization and Design*. New York: Wiley, 1973.

Hnatek, E.R. *A User's Handbook of Semiconductor Memories*. New York: Wiley, 1977.

———. Current Semiconductor Memories. *Computer Design*. April 1978,

Hodges, D.A. *Semiconductor Memories*. New York: IEEE Press, 1977.

Holton, M.J. Programmable Input/Output System Prevents Communication Bottlenecks. *Advances in Instrumentation*, Vol 39. Research Triangle Park, NC: Instrument Society of America, 1984, 1001–1010.

Hordeski, M.F. Digital Control of Microprocessors. *Electronic Design*. December 6, 1975,

———. Digital Sensors Simplify Digital Measurements. *Measurements and Data*. May-June 1976, 90–91 .

———. When Should You Use Pneumatics, When Electronics? *Instruments & Control Systems*. November 1976, 51–55.

———. Guide to Digital Instrumentation for Temperature, Pressure Instruments. *Oil, Gas and Petrochemical Equipment*. November 1976,

———. Digital Instrumentation for Pressure, Temperature/Pressure, Readout Instruments. *Oil, Gas and Petrochemical Equipment*. December 1976,

———. Innovative Design: Microprocessors. *Digital Design*. December 1976,

———. Passive Sensors for Temperature Measurement. *Instrumentation Technology*. February 1977,

_______. Adapting Electric Actuators to Digital Control. *Instrumentation Technology*. March 1977, 53–57.

_______. Fundamentals of Digital Control Loops and Factors in Choosing Pneumatic or Electronic Instruments, presented at the SCMA Instrumenation Short Course, Los Angeles, CA, April 6, 1977.

_______. Balancing Microprocessor-Interface Tradeoffs. *Digital Design*. April 1977, 66–70.

_______. Digital Position Encoders for Linear Applications. *Measurements and Control*. July-August 1977. 58–59.

_______. Future Microprocessor Software. *Digital Design*. August 1977,

_______. Radiation and Stored Data. *Digital Design*. September 1977,

_______. Microprocessor Chips. *Instrumentation Technology*. September 1977,

_______. Process Controls Are Evolving Fast. *Electronic Design*. November 22, 1977, 142–145.

_______. Fundamentals of Digital Control Loops. *Measurements & Control*. February 1978,

_______. Using Microprocessors. *Measurements & Control*. June 1978,

_______. *Illustrated Dictionary of Micro Computer Terminology*. Blue Ridge Summit, PA: TAB, 1978.

_______. *Microprocessor Cookbook*. Blue Ridge Summit, PA: TAB, 1979.

_______. Selecting Test Strategies for Microprocessor Systems. *ATE Seminar Proceedings, Pasadena, CA*. New York: Morgan-Grampian, January 1982.

_______. Selection of a Test Strategy for MPU Systems. *Electronics Test*. February 1982,

_______. *Trends in Displacement Sensors. Sensors and Systems Conference Proceedings*, Pasadena, CA. Campbell, CA: Network Exhibitions, Exhibitions, May 1982.

_______. *The Impact of 16-bit Microprocessors. Las Vegas: Instrumentation Symposium Proceedings, May 1982*. Research Triangle Park, NC: Instrument Society of America, 1982.

_______. *Diagnostic Strategies for Microprocessor Systems. ATE Seminar Proceedings, Anaheim, CA, January 1983*. New York: Morgan-Grampian, 1983.

_______. *Microprocessors in Industry*. New York: Van Nostrand Reinhold, 1984.

_______. *The Design of Microprocessor Sensor and Control Systems*. Reston, VA: Reston, 1984.

_______. *CAD/CAM Equipment Reliability*. San Francisco: Western Design Engineering Show and ASME Conference, December 5, 1984.

_______. *Specifying and Selecting CAD/CAM Equipment. CADCON West, Anaheim, CA January 14–17, 1985*. New York: Morgan-Grampian, 1985.

_______. *A Tutorial on CIM/Factory Automation*. Anaheim, CA: Western Design Engineering Show and ASME Conference, December 12, 1985.

_______. *CAD/CAM Techniques*. Reston, VA: Reston, 1986.

_______. *Microcomputer Design*. Reston, VA: Reston, 1986.

_______. *Transducers for Automation*. New York: Van Nostrand Reinhold, 1987.

Hornbuckle, G.D. The Computer Graphics/User Interface. *IEEE Transactions* HFE-8 (1). March 1967,

Hougen, J.O. *Measurements and Control Applications*. Research Triangle Park, NC: Instrument Society of America, 1979.

Hughes, J.S. Personal Computers: If You Don't Have One Now, You Soon Will, *InTech*. February 1985, 45–47.

———. *Personal Computers in Process Control. Advances in Instrumentation*, Vol 40. Research Triangle Park, NC: Instrument Society of America, 1985.

IBM PC Multiuser and Networking Systems, 1983–1986. Citations from the INSPEC Data Base. Report PB8 6-856853/XAB. Springfield, VA: National Technical Information Service, January 1986.

IBM Personal Computers and Compatible Equipment, 1975–1983. Citations from the INSPEC Data Base, Report PB84-855956. Springfield, VA: National Technical Information Service, December 1983.

IBM Personal Computers and Compatible Equipment, January-September 1984. Citations from the INSPEC Data Base, Report PB84-875145. Springfield, VA: National Technical Information Service, September 1984.

IBM Personal Computers and Compatible Equipment, April 1985–March 1986. Citations from the INSPEC Data Base, Report PB86-859022/XAB. Springfield, VA: National Technical Information Service, March 1986.

Intel Corporation. *8086 User's Guide*. Santa Clara, CA: Intel, 1976.

Isaacson, P. Ask Portia . . . *Computer Retail News*. June 25, 1984, 23.

Ivanov, V.V., Morenkov, A.D., and Oleinkkov, A.Y. Personal Computers and Measurement-Computation Complexes for Automation of Large-Scale Laboratory Experiments. Instruments and Experimental Techniques English Translation of *Pribory I Teknika Eksperimenta*. September-October 1984, 1106–1109.

Iwaik, T., Imai, Y., Ono, M., Okamoto, T., Kobayashi, M., Matsui, M., Takeo, Y., and Kawamori, Y. The Personal Computer for Measuring Controller. *Anritsu Technical Bulletin*, Japan. May 1979, 82–94.

Jackson, M.A. *Principles of Program Design*. New York: Academic, 1975.

James, D.R., and Griterson, J.B. *Evaluating Personal Computers for Use in Process Control. Advances in Instrumentation*, Vol. 38. Research Triangle Park, NC: Instrument Society of America 1983, 707.

———. Evaluating Personal Computers for Process Control. *InTech*. January 1984, 49–51.

Johnson, S., and Lesk, M. Language Development Tools., *The Bell System Technical Journal*. July-August 1978,

Jones, B.E. *Instrumentation, Measurement and Feedback*. New York: McGraw-Hill, 1977.

Jones, J.C. *Design Methods*. New York: Wiley-Interscience, 1970.

Jutila, J.M. Temperature Instrumentation. *Instrumentation Technology*. February 1980,

Kay, Alan C. Microelectronics and the Personal Computer, *Scientific American*. September 1977,

Keene, B. Solving Mechanical Problems with Personal Computers. *Machine Design*. February 10, 1983, 121–125.

Klingman, E.E. *Microprocessor Systems Design*. Englewood Cliffs, NJ: Prentice-Hall, 1977.

Klipec, B. How to Avoid Noise Pickup on Wire and Cables. *Instruments & Control Systems*. December 1977,

Klopf, P., van Rhee, L.A., and Stuber, W. An Autonomous CAMAC Crate Controlled

by a Personal Computer. *Nuclear Instrumentation and Methods of Physical Research*, August 15, 1981, 435–441.
Knuth, D.E. *The Art of Computer Programming*. Vol. 1: *Fundamental Algorithms*. Reading, MA: Addison-Wesley, 1973.
Kobayashi, K., Watanabe, K., Ichikawa, R., and Kato, A. Personal Computer. (In *Computing and Communications*). *Proceedings of the IEEE*. March 1983, 352–362.
Kofler, G. Personal Instrumentation. A Personal Computer as Part of a Measuring System, *Elekronkschau*, Austria. June 1983, 46–47.
Kohonen, T. *Digital Circuits and Devices*. Englewood Cliffs, NJ: Prentice-Hall, 1972.
Kotelly, G. Personal Computer Networks. *EDN*. March 3, 1983,
Krakowsky, A.M. *Cost Savings Resulting from Standardization and Support of Personal Computers*. Report UCID-19969. Livermore, CA: Lawrence Livermore Laboratory, December 14, 1983.
Kramper, B., and MacKinnon, B. *EPICS Personal Computer Evaluation*. Report No. FERMILAB/TM-1255. Batavia, IL: Fermi National Accelerator Lab, February 1984.
Krigman, A. Selecting Peripherals for Process I/O. *InTech*. April 1984, 49.
Kuck, D.J. *The Structure of Computers and Computations*, Vol. 1. New York: Wiley, 1978.
Kuenning, M.K. *Programmable Controllers: Configuration and Programming*. Greenville, SC: Automated Manufacturing, March 19–22, 1984.
Kwakernaak, H., and Swan, R. *Linear Optimal Control Systems*. New York: Wiley, 1972.
Lawrence, S., and Marcus, L.S. Designing PC Boards with a Centralized Database. *Computer Graphics World*. March 1984, 20–24.
Lee, H. Use of the Personal Computer to Design Processing Conditions for Improving the Accuracy of RTV Silicone/Epoxy Resin Replicas. *SAMPE Journal*. September-October 1985, 22–24.
Levenspiel, O. *Chemical Reaction Engineering*. New York: Wiley, 1962.
Leventhal, L.V. *Microprocessors: Software, Hardware, Programming*. Englewood Cliffs, NJ: Prentice-Hall, 1978.
Lightman, S. Personal Computer Instruments and Personal Preferences. *Test & Measurement World*. March 1985, 93–102.
Liptak, B.G. *Instrument Engineers' Handbook*. Radnor, PA: Chilton. Vol. I, 1969; Vol. II, 1970; Supplement, 1972.
______. *Environmental Engineers' Handbook*, Vols. I–III, Radnor, PA: Chilton, 1974.
______. *Instrumentation in the Processing Industries*. Radnor, PA: Chilton, 1973.
______. Ultrasonic Instruments. *Instrumentation Technology*. September 1974,
______, ed. *Instrument Engineers' Handbook on Process Measurement*. Radnor, PA: Chilton, 1980.
Lorin, H. *Parallelism in Hardware and Software*. Englewood Cliffs, NJ: Prentice-Hall, 1972.
Lupfer, D.E., and Johnson, M.L. *Automatic Control of Distillation Columns to Achieve Optimum Operation*. Pittsburgh: Instrument Society of America, 1974.
Madnick, S.F., and Donovan, J.L. *Operating Systems*. New York: McGraw-Hill, 1974.
Manoff, M. Control Software Comes to Personal Computers. *Control Engineering*. March 1984, 66–68.

Martin, D.P. *Microcomputer Design*. Chicago: Martin Research Ltd., 1975.

Martin, J. *Design of Man-Computer Dialogues*. Englewood Cliffs, NJ: Prentice-Hall, 1973.

Marubayashi, K., Matsumoto, Y., and Tawara, H. Low Cost Personal Computer System for Controlling an X-Ray Crystal Spectrometer in Use for Collision Experiments. *Nuclear Instrumentation and Methods of Physical Research*, the Netherlands. December 1, 1982, 571–576.

Mazur, T. Microprocessor Basics: Part 4: The Motorola 6800. *Electronic Design*. July 19, 1976,

McArthur, L. Automate Your Instrument Maintenance with a Personal Computer. *InTech*. November 1983, 41–42.

McDermott, J. Personal Computer Add-ons and Add-ins. *EDN*. January 20, 1983, 62–82.

McGirt, F. *REMOTE Modem Communicator Program for the IBM Personal Computer*. Report LA-10143-MS. Los Alamos, NM: Los Alamos National Laboratories, June 1984.

McGlynn, D.R. *Microprocessors*. New York: Wiley, 1976.

McKenezie, K., and Nichols, A.J. Build a Compact Microcomputer. *Electronic Design*. May 10, 1976,

McMahon, W.J. Development Tool Helps Write Integrated Software for Personal Computers. *Electronic Design*. August 9, 1984, 171–180.

Meditch, J.S. *Stochastic Optimal Linear Estimation and Control*. New York: McGraw-Hill, 1969.

Merritt, K., and Persun, T. Personal Computers Move into Process Control. *Instruments & Control Systems*. June 1983,

Metcalfe, B. Controller/Transceiver Board Drives Ethernet into PC Domain. *Mini-Micro Systems*. January 1983,

Meyer, J., and Jayaraman, R. Simulating Robotic Applications on a Personal Computer. *Computers in Mechanical Engineering*. July 1983, 15–18.

Mills, J. Use Your Personal Computer for Measurement and Control. *Analog Dialog*, Vol. 16, No. 2. Norwood, MA: Analog Devices, 1982.

Mills, M. Memory Cards: A New concept in Personal Computing. *Byte*. January 1984,

Milne, B. Personal Computers; Instruments. *Electronic Design*. September 29, 1983,

Moss, C.E. *Sophisticated Gamma-Ray Data Acquisition System Based on an IBM PC/XT Computer*. Report LA-UR-85-3755. Proceedings of the IEEE Nuclear Science Symposium, October 23, 1985.

Moss, D. Multiprocessing Adds Muscle to uPs. *Electronic Design*. May 24, 1978,

Motorola Semiconductor. *M6800 Microprocessor Applications Manual*. Phoenix, AZ: Motorola, 1975.

———. *MC68000 Microprocessor User's Manual*. Austin, TX: Motorola, 1979.

Mullins, M. Instrument Controllers—Evaluating Cost and Function, *Test & Measurement World*, April 1984,

Murrill, P.W. *Automatic Control of Processes*. Scranton, PA: International Textbook, 1967.

Myers, G.J. *Reliable Software Through Composite Design*. New York: Petrocelli Charter, 1975.

Nick, J.R. Using Schottky 3-State Outputs in Bus-Organized Systems. *Electronic Design News*. December 5, 1974,

Niles, J.M., Carlson, F.R., Gray, P, Hayes, J.P., and Holmen, M.G. *Technical Assessment of Personal Computers*, Vols. I–III. Report NSF/PRA-7805647(1-3). Los Angeles, CA: UCLA Office of Interdisciplinary Programs, September 1980.

Norton, H.N. *Handbook of Transducers for Electronic Measuring Systems*. Englewood Cliffs, NJ: Prentice-Hall, 1969.

Offereins, R.P., and Meerman, J.W. *Simulation Program (BASIM) for Personal Computers. Proceedings of the 3rd IFAC/IFIP Symposium on Software for Computer Control*. Oxford: Pergamon Press, 1983, 409–413.

Ohkubo, T., and Nakamura, T. *Automatic Data Acquisition System of Environmental Radiation Monitor with a Personal Computer*. Report INS-TS-24. Tokyo: Tokyo University (Institute for Nuclear Study), May 1984.

Oliver, B.M., and Cage, J.M. *Electronic Measurements and Instrumentation*. New York: McGraw-Hill, 1971.

Olmstead, K. The Future Factory—a First Report. *Test & Measurement World*. December 1983,

Ottinger, L. *Using Robots in Flexible Manufacturing Cells/Facilities*. Greenville, SC: Automated Manufacturing, March 19–22, 1984.

Parisi, V.M. *Development of a Computer-Aided Design Package for Control System Design and Analysis for Use on a Personal Computer*. Report AFIT/GE/EE83D-53. Wright Patterson AFB, OH: Air Force Institute of Technology, December 1983.

Park, R.M. Applying the Systems Concept to Thermocouples. *Instrumentation Technology*. August 1973,

Patterson, D.A., and Seguin, C.H. Design Considerations for Single-Chip Computers of the Future. *IEEE Transactions in Computers*, 0–29, February 1980,

Personal Computers in Japan: The Most Up-to-Date Information on Japanese Computer Industries. Report PB84-176510. Work performed at PB Co., Ltd., Tokyo. Springfield, VA: National Technical Information Service, December 1983.

Peters, L.J., and Tripp, L.L. Is Software Wicked? *Datamation*. June 1976,

Peuto, B.L., and Shustek, L.J. Current Issues in the Architecture of Microprocessors. *Computer*. February 1977,

Pinto, J.J. Evolution of the Industrial Process Measurement and Control Computer. *Advances in Instrumentation*, Vol. 38. Research Triangle Park, NC: Instrument Society of America, 1983.

_______. Software for Industrial Microcomputer Applications. *Advances in Instrumentation*, Vol. 40. Research Triangle Park, NC: Instrument Society of America, 1985.

Plumb, H.H. *Temperature: Its Measurement and Control in Science Industry*. Research Triangle Park, NC: Instrument Society of America, 1972.

Pritty, D.W. The Potential of Personal Computers in Laboratory Control Applications. *Journal of Microcomputer Applications*, UK, January 1983, 47–57.

Process Control by Personal Computers. *Wireless World*, UK. September 1983, 54–59.

Riblet, G.P. A Personal Computer Based System for the Rapid Display of Smith Chart Curves Using 6 -Ports. *Proceedings of the Colloquium on Advances in S-Parameter Measurement at Micro-Wavelengths*. London: IEEE, 1983.

Riley, J. *Process Control for a PWB Facility*. Greenville, SC: Automated Manufacturing, March 19–22, 1984.

Rompelman, O., Snijders, J.B.I.M., and Van Spronsen, C.J. The Measurement of Heart Rate Variability Spectra with the Help of the Personal Computer. *IEEE Transactions on Biomedical Engineering*. July 1982, 503–510.

Ross, C.A. *Automation Policies, Practices and Procedures*. Greenville, SC: Automated Manufacturing. March 19–22, 1984.

Roth, G. Integrated Test Functions Aid in Process Control. *Test & Measurement World*. December 1983,

Sandberg, U. Personal Computers in Control and Regulatory Systems: Dream or Nightmare? *Industrielle Dataknik*, Sweden. June 1984, 47.

Sandberg, U., and Faxer, M. Personal Computers for Real-Time Control of Power System Distribution Networks. *IEEE Transactions in Power Apparatus and Systems*. July 1984, 1720–1724.

Saxena, P., and Gupta, L.K. PCs Provide Low Cost Process Monitoring. *Proceedings of the North Coast Conference*. Research Triangle Park, NC: Instrument Society of America, 1986.

Schaeffer, E.J., and Williams, T.J. *An Analysis of Fault Detection Correction and Prevention in Industrial Computer Systems*. West Lafayette, IN: Purdue Laboratory for Applied Industrial Control, Purdue University, October 1977.

Schgor, G. Personal Computers in the Control of Industrial Plants. *Automazione y Strumentazone*, November 1983, 93–96.

Schwartz, P. When Personal Computers Become Measuring Instruments. *Electronique Industrielle*, France, May 1983, 49–53.

Schwiderski, G. Personal Computers—A New Computer Technology for the Steel Industry. *Stahl und Eisen*. West Germany, February 23, 1985, 39–46.

Sheingold, D.H. *Analog-Digital Conversion Handbook*. Norwood, MA: Analog Devices, 1972.

Shinskey, F.G. *Process Control Systems*. New York: McGraw-Hill, 1979.

Shneiderman, B. Human Factors Experiments in Designing Interactive Systems. *Computer*. December 1979,

Sigma Instruments. *Stepping Motor Handbook*. Braintree, MA: Sigma Instruments, 1972.

Singer, A., and Rony, P. Controlling Robots with Personal Computers. *Machine Design*. September 23, 1982, 78–82.

Sinha, N.K. Control System Design with Personal Computers. *Proceedings of the 1983 International Electrical and Electronics Conference*. New York: IEEE Press, 1983, 420–423.

Skrokov, M.R., ed. *Mini and Microcomputer Control in Industrial Processes*. New York: Van Nostrand Reinhold, 1980.

Slomiana, M. Selecting Differential Pressure Instruments. *Instrumentation Technology*. August 1979,

Smith, W.D. An EEG Monitoring System on a Personal Computer. *Proceedings of the 1983 AAMSI Congress*. Bethseda, MD: American Association of Medical Systems and Informatics, 1983, 315–317.

Soisson, H.E. *Instrumentation in Industry*. New York: Wiley, 1975.

Soriano, S. Simulation by Means of a Personal Computer of a PID Regulator Controlled Process. *Automaziane y Strumentazione*, Italy. June 1984, 159–165.

Soucek, B. *Microprocessors and Microcomputers*. New York: Wiley, 1976.

Sroczynski, C. *3D Modeling in Process and Power Plant Design*. CADCON East 84, June 12, Boston, MA, New York: Morgan-Grampian, 1984.

Stone, H.S. *Introduction to Computer Architecture*. New York: McGraw-Hill, 1975.

Sylvan, J. Industrial Monitoring with Personal Computers. *Machine Design*. October 6, 1983, 91-95.

_______. Personal Computers in Distributed Measurement and Control. *Advances on Instrumentation*, Vol. 40. Research Triangle Park, NC: Instrument Society of America, 1985.

Takanishi, I., Tomokiyo, O., and Yokouchi, N. Personal Computer for Measurement and Control. *Hitachi Zosen Technical Review*, Japan. March 1983, 56-63.

Take Measurements with Your Personal Computer. *Measures*, September 13, 1983, 27-29.

Taylor, A.P. Getting a Handle on Factory Automation. *Computer-Aided Engineering*. May-June 1983, 74, 80.

Technical Data Book—Petroleum Refining. Washington, DC: American Petroleum Institute, 1970.

ter Haar Romney, B.M., Nuijen, W.C., and Magielse, A.D.L. Single Fibre Electromyography with a Personal Computer. *Medical Biology and Engineering Computation*. May 1984, 240-244.

Teschler, L. Engineering Software for Personal Computers. *Machine Design*. January 26, 1984, 64-69.

Texas Instruments. *The TTL Data Book for Design Engineers*. Dallas, TX: Texas Instruments, 1973.

_______. *The Microprocessor Handbook*. Houston, TX: Texas Instruments, 1975.

_______. *TMS 9900 Microprocessor Data Manual*. Dallas, TX: Texas Instruments, 1978.

The Personal Computer as a Measuring Instrument. *Regulacion y Mendo Autommatico*, Spain. December 1983, 55-56.

Thomas, T.B., and Arbuckle, W.L. Multiprocessor Software: Two Approaches. *Conference on the Use of Digital Computers in Process Control, Baton Rouge, LA*. February 1971.

Tippie, J.W., and Kulaga, J.E. Design Considerations for a Multiprocessor-Based Data Acquisition System. *IEEE Transactions on Nuclear Science*, August 1979,

Toong, H.D., and Gupta, A. An Architectural Comparison of Contemporary 16-Bit Microprocessors. *IEEE* Micro, May 1981,

_______. New Direction in Personal Computer Software. *Proceedings of the IEEE*, March 1983, 377-388.

Torrero, E.A. Focus on Microprocessors. *Electronic Design*. September 1, 1974,

Torrero, E.A., ed. *Microprocessors: New Directions for Designers*. Rochelle Park, NJ: Hayden, 1975.

Totaro, L. Personal Computer and a Quality Control. *Electronica*, Oggi, Italy. March 1983, 167-192.

Useda, K., and Kinoshita, K. Data Processing System for the Measurement of Thermal Desorption Spectra Using a Personal Computer. *Journal of the Vacuum Society of Japan*. Vol 26 (No 10) 1983, 784-788.

Vacroux, A.G. Explore Microcomputer I/O Capabilities. *Electronic Design*. May 10, 1975,

Van Winkle, M. *Distillation*. New York: McGraw-Hill, 1967.

Voelcker, H.B., and Requicha, A.G. Geometric Modelling of Mechanical Parts and Processes. *Computer*. December 1977,

Wallace, V.L. The Semantics of Graphic Input Devices. *Proceedings of the SIGGRAPH/SIGPLAN Conference on Graphics Languages*. Published in *Computer Graphics*, April 1976,

Warner, J.R. Device-Independent Tool Systems. *Computer Graphics World*. February 1984, 72–73, 99–102.

Wegner, W., ed. *Research Directions in Software Technology*. Cambridge, MA: MIT Press, 1978.

Weir, J.D., and Weir, C.J. Personal Computers for Machine and Process Control: Fast, Inexpensive, Easy Automation. *Elastromerics*. August 1983, 17–18.

Weisberg, D.E. Performance and Productivity in CAD. *Computer Graphics World*. June 1983,

Weiss, B. Evaluating Graph and Chart Output. *Computer Graphics World*. February 1984, 87–90.

Westerhoff, T. Software in the Future Factory. *Test & Measurements World*. December 1983,

Wightman, E.J. *Instrumentation in Process Control*. Woburn, MA: Butterworth, 1972.

Wolf, S. *Guide to Electronic Measurements and Laboratory Practice*. Englewood Cliffs, NJ: Prentice-Hall, 1973.

Yourdon, E., and Constantine, L.L. *Structured Design*. New York: Yourdon, 1975.

Zaks, R. *Microprocessors*. Berkeley, CA: Sybex, 1979.

Zilog Corporation. *Z8000 User's Guide*. Cupertino, CA: Zilog Corp., 1980.

Index